GLOBAL PERSPECTIVES ON THE ECOLOGY

OF HUMAN-MACHINE SYSTEMS

RESOURCES FOR ECOLOGICAL PSYCHOLOGY

A Series of Volumes Edited by:
Robert E. Shaw, William M. Mace, and Michael T. Turvey

Reed/Jones • *Reasons for Realism: Selected Essays
of James J. Gibson*

Warren/Shaw • *Persistence and Change*

Kugler/Turvey • *Information, Natural Law,
and the Self-Assembly of Rhythmic Movement*

McCabe/Balzano • *Event Cognition: An Ecological Perspective*

Lombardo • *The Reciprocity of Perceiver and Environment:
The Evolution of James J. Gibson's Ecological Psychology*

Alley • *Social and Applied Aspects of Perceiving Faces*

Warren/Wertheim • *Perception & Control of Self-Motion*

Thinès/Costall/Butterworth • *Michotte's Experimental
Phenomenology of Perception*

Jansson/Bergström/Epstein • *Perceiving Events
and Objects*

Flach/Hancock/Caird/Vincente • *Global Perspectives
on the Ecology of Human-Machine Systems (Volume 1)*

Hancock/Flach/Caird/Vincente • *Local Applications
of the Ecological Approach to Human-Machine Systems (Volume II)*

GLOBAL PERSPECTIVES ON THE ECOLOGY

OF HUMAN-MACHINE SYSTEMS

edited by

John Flach
Wright State University

Peter Hancock
University of Minnesota

Jeff Caird
University of Calgary

Kim Vicente
University of Toronto

CRC Press
Taylor & Francis Group
Boca Raton   London   New York

CRC Press is an imprint of the
Taylor & Francis Group, an **informa** business

First Published by
Lawrence Erlbaum Associates, Inc., Publishers
10 Industrial Avenue
Mahwah, New Jersey 07430

Transferred to Digital Printing 2009 by CRC Press
6000 Broken Sound Parkway, NW Suite 300, Boca Raton, FL 33487
270 Madison Ave, New York NY 10016
2 Park Square, Milton Park, Abingdon, Oxon, OX14 4RN, UK

Library of Congress Cataloging-in-Publication Data

Global perspectives on the ecology of human-machine systems / edited
by John Flach...[et al.].
     p.   cm.
  Includes bibliographical references and index.
  ISBN 0-8058-1381-0 (acid-free paper). -- ISBN 0-8058-1382-9 (pbk.)
  1. Man-machine systems--Environmental aspects. 2. Human
engineering. 3. Environmental psychology. I. Flach, John.
TA167.G56 1995
620.8'2--dc20   95-6690
                                                       CIP

Publisher's Note
The publisher has gone to great lengths to ensure the quality of this reprint
but points out that some imperfections in the original may be apparent.

To Jimmy

Contents

Contributors

Geoffrey P. Bingham
Indiana University

Mark Chignell
Department of Industrial Engineering
University of Toronto

Marvin J. Dainoff
Miami University

John Flach
Department of Psychology
Wright State University

Oded M. Flascher
Intentional Dynamics Laboratory
Center for Ecologcal Study of Perception and Action
The University of Connecticut

Peter Hancock
Human Factors Research Laboratory
University of Minnesota

Endre E. Kadar
Intentional Dynamics Laboratory
Center for Ecologcal Study of Perception and Action
The University of Connecticut

Alex Kirlik
Center for Human-Machine Systems Research
School of Industrial & Machine Systems Engineering
Georgia Institute of Technology

Leonard S. Mark
Miami University

Michael M. Muchisky
Indiana University

Annelise Mark Pejtersen
Risø National Laboratory

Jens Rasmussen
Risø National Laboratory

Robert E. Shaw
Intentional Dynamics Laboratory
Center for Ecologcal Study of Perception and Action
The University of Connecticut

Kim Vicente
Department of Industrial Engineering
University of Toronto

William H. Warren
Brown University

Rik Warren
Armstrong Laboratory
Wright-Patterson Air Force Base

David D. Woods
Cognitive Systems Engineering Laboratory
Ohio State University

Brian S. Zaff
Logicon Technical Services

Preface

"The great mystery, after all, is not the answers that scientists contrive, but the questions they are driven to pose. Why? Why this question rather than another? Why this search, hope, despair, rather than another? Why this ill-lit, nil understood, hobo path? And why the outrageous confidence, born of no evidence, to tred it?"
- Kauffman, S.A. (1993). *The origins of order.* (pp. vii-viii), Oxford University Press.

In the developed world, our ecology is technology. There are few individuals on this planet whose everyday lives are not substantively affected by the action of technical systems albeit at differing levels of sophistication. At the confluence of technology and ecology we see two components of a singular opportunity. The first is the chance to test the principles of ecological psychology against human factors concerns regarding the design and operation of human-machine systems. The second is the chance to pose meaningful questions to the ecological theorist about just which "real" world they choose to focus their efforts on.

The first effort is particularly relevant at present because the information-processing approach, which sets the foundation for much of what we now know as human factors, has broached problems that expose some of the weaknesses of its theoretical basis. Lest some believe us over ardent ecologists, let us state for the record that we still see much that is vital in the information-processing approach, where the nature of the "information" to be processed can be specified with more precision. Further, the wise man does not throw away hard won knowledge and basic understanding in whatever paradigm such understanding is couched, and we would loath to be thought foolish. Indeed, those in human factors are frequently marked by an eclectic pragmatism, especially when the practitioner is "required" to produce an immediate, ready-made answer.

In short, we see the ecological approach as one that offers an alternative view. A view that has provided us with additional insights into how people work with machines. Its value is not as much in the answers it offers, but in the questions it raises.

An eventual integration between information processing and the ecological approach is not one we would rule out and is briefly discussed by several of the authors. However, we believe that examining the human-environment linkage as the basic "unit of analysis" is a critical approach for human factors. Indeed, with its emphasis on action in technical systems, it is central to that enterprise to consider the nature of the context of behavior in which action is embedded. Examining the main effects of human capability is important, but interactions supersede main effects, and situational demands can modify behavior to such an extent that our original knowledge of the isolated human ability might serve to mislead us in richer and more complicated environmental settings.

We are neither priests nor uncritical disciples of the ecological movement. The ecological approach is a strange and curious attractor. Some of us are more attracted than others. Our individual positions range from evangelical to cautiously optimistic, to curious, to argumentatively skeptical. We still have raging arguments among ourselves about some of the basic concepts of the ecological approach. For example, the concept of *affordances* and its theoretical and practical value remains a major bone of contention. We are not above throwing stones at the ecological theorists, some of whom explicitly seem to ignore the world in which humans live and are happier with insects and their intentionality. We have news for those individuals concerning the "real" world.

However, like others in human factors, we have been faced with sometimes critical questions of human behavior in technical systems and, having gone to our cache of theoretical weapons, have found the cupboard uncomfortably bare. Having taken knowledge where we can, we offer the following texts to our companions in the hope that they too might find something useful for their own thoughts and work. This book is directed to behavioral scientists and engineers struggling to design future environments. If you do not already feel as though you are standing on the "edge of chaos," read on. Our goal is to agitate, to stimulate, to challenge, to press the edge of the envelope, to question assumptions that are at the heart of the human enterprise. Insofar as useful information is found or insofar as you are disturbed or unsettled, we in small part, and the authors, in large part, have succeeded. The need and urgency for good human factors applications in a world immersed in technology is obvious. If the following volumes can help to guide and shape the human use of technology to some small degree, we are content that our efforts have been worthwhile.

This work has been produced in two volumes. Volume 1 takes a more global theoretical perspective on human factors and ecological psychology. Volume 2 looks more at local applications of an ecological approach to particular design problems. Across the chapters there is a diverse and sometimes conflicting and contradictory set of perspectives. Ecological theory and its application to human-machine systems is a new and vital enterprise still struggling to adapt and define its niche.

Volume 1 is divided into two sections. The first section provides a selection of theoretical perspectives. Flach opens with a brief historical perspective on the development of an ecological approach to human-machine systems; he attempts to put his own stamp on Gibson's concepts of information and affordance. Hancock and Chignell lay out a set of fundamental principles for the discipline of human factors as an enterprise at the heart of science and technology. Vicente assesses some of the implications for an ecological approach to human factors. This chapter originally appeared in the *Human Factors Bulletin* and received the award as best paper of 1990 from the Human Factors Society. Kirlik's chapter takes inspiration from Brunswik's perspective on ecologically valid research design. He challenges psychologists to a higher level of metacognitive awareness of the intuitions that guide their choice of experimental tasks and stimuli. Rasmussen and Pejtersen integrate a wealth of field experience in domains of nuclear power plant control, information systems, and hospitals into a taxonomic framework to guide generalizations across domains and from the laboratory to the field. Woods, who also has spent much of his career in the wilderness of design and system development, provides a theoretical perspective for the design of representations for supporting communication within the joint cognitive system of human, machine, and work domain. Finally, Flach and Warren present a framework for addressing the emergent properties arising from the interactions of perception and action.

The second section of Volume 1 addresses the central but highly controversial issue of *affordances*. Warren's chapter is a revision of his paper originally presented at the conference on event perception and action. This paper is arguably the first to explicitly consider the concept of affordance and its implication for understanding the "fit" between a human and the built environment. He introduces the concept of *intrinsic measurement* as fundamental to what we mean by "fit." Zaff also considers the implications of affordances for design. He brings a perspective that examines the ability to perceive the affordances of others (a fundamental challenge for designers). Dainoff and Mark examine affordances in the office. They link the concept of *affordances* with Rasmussen's abstraction hierarchy in an attempt to provide a context for asking relevant questions about the design of workstations in the office. Shaw, Flascher, and Kadar extend the concept of *intrinsic measurement* and consider the role of the resulting "pi" numbers for understanding the workspace of dynamic, intentional systems. Finally, Bingham and Muchinsky illustrate how the concept of affordances can inform research on the deceptively simple problem of the grasp and control of objects.

The concept of affordance surfaces repeatedly throughout both volumes. Inconsistencies are apparent. However, the hope is that the variance will provide a context against which the invariance (if indeed there is any stability to this concept of affordance) will be revealed. Again, this is a major source of contention and debate among the editors.

In Volume 1, the discussion is often directed at the metaphysical assumptions underlying a science of behavior that can guide the design and development of technology. Arguments are often presented at an abstract general level. These arguments are directed to the academic exercise of defining who we are and why we do what we do. Although this is a necessary and important exercise, it can often be perceived as a "cacophony of sound and fury signifying nothing." Seeing how these arguments translate into the practical concerns of design require no small leap of induction. Volume 2 takes a more practical approach. There we consider specific applications where an ecological approach is stimulating research. Applications range from vehicular control (tractors, automobiles, and aircraft); to navigation and orientation; to display design (process control and virtual realities); to crew coordination; to problem solving and decision making in naturalistic settings.

As with any production of this scope, contributions are made by many who do not get credit as authors or editors. Our sincere thanks to Rob Stephens, Jonathon Sweet, Steve Scallen, and Shannon Skistad for their assistance in production. Thanks to the Ecological Psychology Seminar class at Wright State University for editorial comments on early versions of many of the chapters. Particular thanks to Bart Brickman, Rob Hutton, and Charlie Garness for their editorial comments.

     John Flach   *Peter Hancock*   *Jeff Caird*   *Kim Vicente*

RESOURCES
for ECOLOGICAL PSYCHOLOGY

Edited by
Robert E. Shaw, William M. Mace, and Michael Turvey

This series of volumes is dedicated to furthering the development of psychology as a branch of ecological science. In its broadest sense, ecology is a multidisciplinary approach to the study of living systems, their environments, and the reciprocity that has evolved between the two. Traditionally, ecological science emphasizes the study of the biological bases of *energy* transactions between animals and their physical environments across cellular, organismic, and population scales. Ecological psychology complements this traditional focus by emphasizing the study of *information* transactions between living systems and their environments, especially as they pertain to perceiving situations of significance to planning and execution of purposes activated in an environment.

The late James J. Gibson used the term *ecological psychology* to emphasize this animal-environment mutuality for the study of problems of perception. He believed that analyzing the environment to be perceived was just as much a part of the psychologist's task as analyzing animals themselves, and hence that the "physical" concepts applied to the environment and the "biological" and "psychological" concepts applied to organisms would have to be tailored to one another in a larger system of mutual constraint. His early interest in the applied problems of landing airplanes and driving automobiles led him to pioneer the study of the perceptual guidance of action.

The work of Nicolai Bernstein in biomechanics and physiology presents a complementary approach to problems of the coordination and control of movement. His work suggests that action, too, cannot be studied without reference to the environment, and that physical and biological concepts must be developed together. The coupling of Gibson's ideas with those of Bernstein forms a natural basis for looking at the traditional psychological topics of perceiving, acting, and knowing as activities of ecosystems rather than isolated animals.

The purpose of this series is to form a useful collection, a resource, for people who wish to learn about ecological psychology and for those who wish to contribute to its development. The series will include original research, collected papers, reports of conferences and symposia, theoretical monographs, technical handbooks, and works from the many disciplines relevant to ecological psychology.

Series Dedication

To James J. Gibson, whose pioneering work in ecological psychology has opened new vistas in psychology and related sciences, we respectfully dedicate this series.

Chapter 1

The Ecology of Human-Machine Systems: A Personal History

John M. Flach
Wright State University
Armstrong Laboratory,
Wright-Patterson Air Force Base

I first became interested in the ecological approach to psychology when, as a graduate student at Ohio State, I heard Rik Warren describe the properties of flow fields and how they might be specific to properties of locomotion such as heading, altitude, and speed. It occurred to me that these descriptions of optical structure may have far more relevance to understanding a skill, such as landing a plane, than the changes in slope or intercept of a reaction time function that were, at that time, central to the chronometric analyses of mind that dominated much of my graduate training (even though these reaction times may have been measured while the operator was simultaneously flying a simulator). As I learned more about the ecological approach it seemed obvious to me that challenges such as understanding automobile driving and flight were important to the evolution of Gibson's theories about behavior. It was surprising to me that those interested in human factors and those interested in ecological approaches to behavior were not actively embracing each other's theories and problems. However, I have always tended to underestimate the inertia in systems. Although the merging of ecological theories and human factors challenges has not happened as quickly as I expected, I think there is a gradually accelerating movement toward communion. This book is perhaps evidence of this movement and will hopefully be a stimulus to encourage a continuing movement toward communion.

To set the stage for this book, I would like to briefly present my personal perspective on the events that have made such a book inevitable. To begin, as I mentioned earlier, the obvious roots for a union of human factors challenges with ecological perspectives are Gibson's early works on automobiles (Gibson & Crooks, 1938) and his work in aviation (Gibson, 1944,1947/1958; Gibson, Olum, & Rosenblatt, 1955). However, Gibson was not alone in his insights. Langewiesche (1944), in his analysis of the information for landing and approach, anticipated many of the ideas about optical sources of information for controlling flight. Also, the need for an ecological approach, that is, an analysis whose scope is the human environment as a system, was voiced by Taylor (1957) as he described the importance of research on human-machine systems for the future of basic psychology:

> "It drew attention to the fact that in many circumstances the behavior of the man was inseparably confounded with that of the mechanical portions of his environment. This meant that psychologists often could not study human behavior apart from that of the physical and inanimate world - that all along they had been studying the behavior of man-machine systems and not that of the men alone. The inseparability of the behavior of living organisms from that of the physical environment with which they are in dynamic interaction certainly argues against maintaining separate sciences and construct languages: one for the environment, the other for that which is environed." (Taylor, 1957, pp. 257-258)

One of the earliest papers, other than Gibson's, to analyze the information in optic flow relative to vehicular control was David Lee's (1976) paper, "A Theory of Visual Control of Braking Based on Information About Time-to-Collision." In this paper, Lee introduced the higher order optical variable tau. Lee wrote:

> "A mathematical analysis of the changing optic array at the driver's eye indicates that the simplest type of visual information, which would be sufficient for controlling braking and would also be likely to be easily picked up by the driver, is information about time-to-collision, rather than information about distance, speed, or acceleration/deceleration. It is shown how the driver could, in principle, use visual information about time-to-collision in registering when he is on a collision course, in judging when to start braking, and in controlling his ongoing braking." (p. 437)

Perhaps, the first laboratory dedicated specifically to an ecological approach to problems of human-vehicular systems was not established until the late 1970s and early 1980s. During this time, the Aviation Psychology Laboratory at Ohio State was under the direction of Dean Owen. Dean assembled the components for a visual flight simulation system to evaluate the optical flow field as a source of information for flight control. During the 1980s, a number of theses and dissertations evaluated sources of information for judgments about altitude and speed. Much of this research is reviewed in a chapter written by Owen and Warren (1987) in *Ergonomics and Human Factors: Recent Research* (Mark, Warm, & Huston, 1987). Many of Owen's students have gone on to continue active research careers in Aviation Human Factors. I was lucky to be on the fringes of this group during my graduate training at Ohio State in the early 1980s. This association had a major impact on my future research agenda.

The analysis of optical information specific to vehicular control continues to be one of the areas of active interchange between ecological theory and human factors problems. In 1986, Rik Warren and Alex Wertheim organized an international workshop on the perception and control of self-motion held at the Institute for Perception TNO, Soesterberg, The Netherlands. This workshop brought together a diverse group of researchers, and although many among the group would not consider themselves "Gibsonian," the influence of Gibson's work was clearly evident. This workshop led to the publication of a book, *Perception and Control of Self-Motion* (Warren & Wertheim, 1990). In Spring of 1989, Walt Johnson and Mary Kaiser hosted a workshop on Visually Guided Control of Movement at the NASA Ames Research Center, Moffett Field (Johnson & Kaiser, 1990). Many important issues with regard to the nature of information in optic flow and the importance of the coupling between perception and action were discussed at this workshop (see also Flach, 1990a).

Outside the area of vehicular control, Warren's paper, "Environmental Design as the Design of Affordances," presented at the Third International Conference on Event Perception and Action, Uppsala, Sweden (1985), was a landmark in the merging of human factors with ecological theory. Warren writes:

> *"Analyzing an affordance requires a task-specific description of an ecosystem that considers the relevant organism and environmental variables and the biomechanics of the task. The fit between organism and environment must be measured relationally, using methods of*

intrinsic measurement, and can be characterized in terms of optimal points at which performance is most efficient or comfortable, and critical points, at which performance breaks down." (p. 2)

The role of intrinsic measurement as a basis for scaling the affordances in artifactual environments has become a cornerstone for building ecological theories of the workplace (see also Mark, 1987; Mark & Voegele, 1987, Warren, 1984, 1987; Warren & Whang, 1987). In 1988, *innovation*, which is the journal of the Industrial Designers Society of America, featured three articles that address the concept of intrinsic measurement and its importance for design (Mark & Dainoff, 1988; Rutter & Newell, 1988; Rutter & Wilcox, 1988).

Another important area of research that has seen the impact of an ecological approach is the problem of interface design for process control and decision support. Dave Woods was one of the first to bring ecological theory to bear on this problem. At a NATO Advanced Study Institute on Intelligent Decision Aids in Process Environments, Woods (1986) wrote:

"The important point for the development of effective decision support systems is the critical distinction between the available data and the meaning of the information that a person extracts from that data (e.g., S. Smith 1963). The available data are raw materials that the observer uses to answer questions (questions that can be vague or well formed, general or specific). The degree to which the data help answer those questions determines the informativeness or inferential value of the data. Thus, the meaning associated with a given datum depends on its relationship to the context or field surrounding the data including its relationship to the objects or units of description of the domain (what object and state of the object is referred to), to the set of possible actions and to perceived task goals (after Gibson, 1979, what that object state affords the observer). The process is analogous to figure-ground relations in perception and shows that information is not a thing-in-itself but is rather a relation between the data, the world the data refers to, and the observer's expectations, intentions, and interests. As a result, informativeness is not a property of the data field alone, but is a relation between the observer and the data field." (p. 163)

Rasmussen (1986) also references Gibsonian theory in his book, *Information Processing and Human-Machine Interaction: An Approach to Cognitive Engineering.* Rasmussen wrote:

"As I understand Gibson's concept of direct perception, the "dynamic world model" is in the present context very similar to the mechanisms needed for the "attunement of the whole retino-neuro-muscular system to invariant information" (Gibson, 1966, p. 262), which leads to the situation where "the centers of the nervous system, including the brain resonate to information." This selective resonance relies on the existence of a generic dynamic model of the environment. The implications of Gibson's view of perception, as based on information pickup instead of sensation input, are in many ways compatible with [my own model]. To Gibson, perception is not based on processing of information contained in an array of sense data. Instead the perceiver, being attuned to invariant information in space and time in the environment, samples this invariant information directly by means of all senses. That is, arrays of sense data are not stored or remembered. They have never been received; instead the nerve system "resonates." In my terms, the world model, activated by the needs and goals of the individual, is updated and aligned by generic patterns in the sensed information, but the idea of an organism "tuning in" on generic time-space properties is basically similar and leads to the view of humans as selective and active seekers of information at a high level of invariance in the environmental context. The subconscious dynamic world model or the attunement of the neural system leads to the situation where primitive sense data are not processed or integrated by symbolic information processes as Minsky suggests, but the generic patterns in the array of data in the environment are sampled directly by high-level questions controlling the exploratory interaction involving all senses."
(pp. 90-91)

The Fourth International Conference on Event Perception and Action at Trieste in 1987 included a symposium chaired by Sebastiano Bagnara entitled "Errors in Human-Machine Interaction." Presenters included Mancini, Rasmussen, Reason, and Vicente. However, almost no one except for the presenters and myself attended this session. However, this insulation from the rest of the conference helped to cement a close bond between myself, Vicente, and Rasmussen which has greatly influenced my thinking about how to attack the problems of interfaces in complex systems. The ideas presented by Rasmussen and Vicente led to a paper published in the *International Journal of Man-Machine Studies* (1989) — "Coping with Human Errors Through System Design: Implications for Ecological Interface Design."

It is notable that Norman in his popular book *The Psychology of Everyday Things* (1988) adopted Gibson's term affordance. He wrote:

"There already exists the start of a psychology of materials and of things, the study of affordances of objects. When used in this sense, the term affordance refers to the perceived and actual properties of the thing, primarily those fundamental properties that determine just how the thing could possibly be used." (p. 9)

A careful reading of Norman's presentation shows that it is somewhat at odds with Gibson's view of affordance (in a footnote, Norman clearly acknowledges the conflict between his view and Gibson's view of affordance). Norman confuses the affordances of an object with the information that specifies the affordances. For example, he writes that "affordances provide strong clues to the operations of things" (p. 9). Clearly, Norman is moving toward an ecological approach, but there is still a strong influence of the more traditional information processing approach, in which meaning must be constructed from the clues available. This is somewhat reminiscent of Neisser's transition from *Cognitive Psychology* (1967) to *Cognition and Reality* (1976). Just as with Neisser, this revision in thinking will likely disturb traditional cognitive psychologists, but will not be satisfactory to many already entrained in the ecological approach. This probably applies, as well, to Woods and Rasmussen's work. However, it is encouraging to see a number of significant contributors to the development of cognitive engineering moving somewhat in the direction of an ecological approach.

Although many will strongly object, I think that this movement toward a middle (perhaps higher) ground is healthy for our science. One of the benefits of the challenge in applying psychology to problems of human-machine systems (as opposed to conducting research exclusively within narrowly defined experimental paradigms or toy laboratory worlds) is that it provides an acid test for dogmatism of any sort. More than anything these challenges teach humility and open-mindedness.

In summer 1989, a symposium was organized for the Fifth International Conference on Event Perception and Action at Miami University entitled "The Ecology of Human-Machine System." Presenters included Stappers and Smets, Woods, Moray, Kugler, and Vicente and Rasmussen. This session was very well attended and stimulated much interesting discussion. I think that most notable were some challenges from Stappers and Smets about how to go beyond purely kinematic models of optic flow. This is particularly significant to my thesis here, as these important theoretical issues were being raised from within a department of industrial design engineering. It is important to see that applied problems are beginning to reflect back in a way that challenges our theories and advances our basic science of behavior. This is a theme that Taylor (1957) emphasized as one of the promises of engineering psychology. Several publications resulted from this symposium (Flach, 1989; 1990b; Vicente & Rasmussen, 1990).

It is important for us to realize that hypothesis testing is not the only way to validate our theories. Design of products and the success or failure of those products is another way of validating the implicit and explicit theories that guided the design. Whereas hypothesis testing emphasizes internal validity, design emphasizes external validity. Good science demands a balance between these two forms of validity. Kirlik's chapter (this volume) discusses design as experiment. Also, I think the work of the Form Theory Group at Delft demonstrates this alternative quite well (see Smets, 1994). The key difference between their work and work of other design groups is the explicit role perceptual theory plays in their designs.

In 1990, the problems of human-machine systems were discussed at both the International Society of Ecological Psychology's Spring meeting in Champaign-Urbana, at which Alex Kirlik's discussion of the affordance properties of a complex helicopter control scenario stimulated much interest, and at the First European Meeting at Marseille which included presentations by myself, John Paulin Hansen from RISO National Laboratory in Denmark, and several presentations from the Industrial Design Group at Delft Technical University. Also, in 1990, Vicente wrote an article entitled "A Few Implications of an Ecological Approach to Human Factors," that was published in the *Human Factors Society Bulletin*. It was awarded the best paper award at the 1991 annual meeting of the Human Factors Society.

A symposium entitled "An ecological approach to human-machine-systems" was held at the 1992 Annual Meeting of the Human Factors Society (Flach & Hancock, 1992). Presentations were given by myself and Peter Hancock and a panel of reactors included Alex Kirlik, Frank Moses, Donald Norman, and Kim Vicente. This symposium was filled to overflowing and elicited a largely enthusiastic response from the audience. A criticism of the presentations was that they were too abstract. People wanted more practical examples of how an ecological theory can affect how they approach design problems. We hope that this book will provide a more balanced presentation with attention to both the theoretical and the practical implications of an ecological approach.

This is a biased and necessarily brief history of the growing attention that ecological theory is getting with respect to applications in human-machine systems. I apologize to those whose contribution I have failed to acknowledge. I consider myself fortunate to have been swept up in this current of ideas. It is our hope that this book captures others who have an interest in taking experimental psychology beyond the laboratory. The ecological approach promises a basic science that does more than play 20 questions with nature (Newell, 1973), which is not mere puzzle solving. It promises a basic science that will be able to inform and guide us as we shape an environment within which we can thrive.

To conclude this chapter, I address the two most critical (and perhaps the most misunderstood) concepts of an ecological approach to psychology; *affordance* and *direct perception*. First, I address the concept of affordance. In order to talk about human-environment systems, a language for describing the environment is needed. Traditionally, psychology has looked to classical physics for that language. The language of classical physics, which describes objects in terms of grams, centimeters, seconds, and so on, was chosen explicitly so that descriptions of the objects were observer independent. This strategy has been very successful, but has a limited scope, even for describing the physical world (e.g., at the level of quantum mechanics this strategy leads to some puzzling contradictions such as the dual particle/wave nature of light). Gibson (1979) opted for a different strategy that he calls ecological physics. This strategy adopts a language that is observer dependent for describing the environment. Thus, the world is described as graspable, walk-on-able, sit-on-able, pass-through-able, step-on-able, climb-over-able, and so on. This kind of description is better suited to the study of *perception*, which is about the relation between the observer and the environment. This type of description has the promise of capturing the "meaningful" or "functional" properties of the environment. Whereas the traditional dimensions from classical physics are afunctional. Meaning requires further processing (e.g., a computation of the ratio between the object's size in centimeters and the size of my own hand in centimeters). Classical physics was designed to describe the environment independent of an observer. Ecological physics is designed to describe the environment relative to a specific observer.

Thus, the underlying assumption is that biological systems perceive the world, not on the basis of extrinsic units (e.g., centimeters) that need to be processed before they are meaningful, but in terms of intrinsic units such as eye height, leg length, or hand size that have explicit meaning for the system. If this assumption is true, then for us to understand perception we must also be able to describe the environment in similar terms. Thus, an ecological physics is required. Note that this is a realist approach. The properties of the environment are not mental constructions. However, the measurement scale we use to describe these properties is a theoretical construction of the scientist. Whether we scale the properties of the environment in terms of inches, centimeters, or hand size does not make the size of the object less real. For classical physicists the observer-independent properties are most useful in their efforts to build a description of the world in terms of observer independent laws. For psychology, however, the goal is to discover the observer dependent laws (i.e., the laws that relate observers to the environment). An ecological approach starts with the assumption that these laws can best be discovered using an intrinsic measurement scale that captures the functional properties of the environment (i.e., affordances).

The second concept is that of direct perception. The issue here is whether the mapping between structure in a perceptual field (e.g., an optic array) and affordances in the environment are specific or not. Direct perception requires that these mappings are specific. To the extent that the mappings are specific, then the observer can directly pick up the affordances in the environment. A theory of direct perception does not require that there are absolutely no ambiguities for an observer (even though the mapping between the perceptual field and the affordances is specific, the observer may not be skillful in picking up the structure). A critical distinction between traditional and ecological approaches, however, is how ambiguities are resolved. A direct perceptual system resolves ambiguities through acting on the environment (looking, touching, manipulating, etc.), rather than through indirect means such as inferring, computing, or assuming.

For building an ecological approach to human-machine systems it is not necessary to uncritically accept the dogma of direct perception. Rather, I think that direct perception poses two fundamental challenges for cognitive engineering. First, in trying to understand skilled behavior (e.g., vehicular control, naturalistic decision making), the concept of direct perception challenges us to evaluate the structure of the perceptual fields to see whether we can find specific mappings to functional properties of the task and second to see if there is a correlation between the structures in the perceptual fields and the actions of a skilled operator. As Neisser (1987) noted:

> "If we do not have a good account of the information that perceivers are actually using, our hypothetical models of their "information processing" are almost sure to be wrong. If we do have such an account, however, such models may turn out to be almost unnecessary." (p. 11; also cited in Vicente, 1990)

The first challenge when modeling skilled performance is to avoid mentalistic constructs whenever possible. The second challenge is in the area of the design of interfaces. Previously, I have characterized the problem of display design as inverse ecological optics (Flach, 1988). Ecological optics is the study of the structures within reflected light (i.e., the optic array) that are specific to properties of environments and observers (e.g., time to contact). Display design is the creation of a perceptual array, the structure of which maps directly to (specifies) the functional properties of the work domain. Thus, the challenge is to build an interface that can support direct perception. The implication is that a poor interface is one in which ambiguities can not be resolved by activity of the observer, where assumptions, computations, and inferences are required. Vicente and Rasmussen (1990); and Rasmussen and Vicente, (1989) called this the problem of ecological interface design. Where traditionally there has been much concern with building interfaces that match the operator's "mental model," ecological interface design focuses on matching the structure in the interface to the natural constraints of the work domain in a way to inform and guide the operator as he navigates within that work domain. With an ecological interface the need for assumptions, computations, and inferences (and perhaps for a mental model) are reduced or eliminated all together. Thus, I recommend that we consider direct perception as a possibility, which can be realized whenever there are specific mappings between structure in a medium (e.g., optic array or electronic display) and the affordances available to a behaving system.

The concepts of affordance (meaning) and direct perception (specificity of information in sensory arrays) will arise repeatedly throughout this volume. By the end of this book, you may have a more informed and richer understanding than my own. And perhaps will be better prepared for the challenges of shaping our future in a technological world.

References

Flach, J. M. (1988, June). *The ecology of human-machine systems: Problems, prayers, and promises.* Paper presented at the spring meeting of the Society for Ecological Psychology. Yellow Springs, OH.

Flach, J. M. (1989). An ecological alternative to egg-sucking. *Human Factors Society Bulletin, 32*, 4–6.

Flach, J. M. (1990a). Control with an eye for perception: Precursors to an active psychophysics. *Ecological Psychology, 2*, 83–111.

Flach, J. M. (1990b). The ecology of human machine systems I: Introduction. *Ecological Psychology, 2*, 191–205.

Flach, J. M. & Hancock, P. A. (1992). An ecological approach to human-machine systems. *Proceedings of the Human Factors Society 36th Annual Meeting* (pp. 1056–1058). Santa Monica, CA: The Human Factors Society.

Gibson, J. J. (1944). History organization and research activities of the Psychological Test Film Unit, Army Air Forces. *Psychological Bulletin, 41*, 57–468.

Gibson, J. J. (1958). Motion picture testing and research. In D. Beardslee & M. Wertheimer (Eds.), *Readings in perception* (pp. 181–195). Princeton, NJ: Von Nostrand (Original work published 1947).

Gibson, J. J. (1966). *The senses considered as perceptual systems.* Boston: Houghton Mifflin.

Gibson, J. J. (1979). *The ecological approach to visual perception.* Boston: Houghton Mifflin.

Gibson, J. J., & Crooks, L. E. (1938). A theoretical field analysis of automobile-driving. *American Journal of Psychology, 51*, 453–471. (Reprinted in E. Reed and R. Jones (Eds.), *Reasons for realism: Selected essays of James J. Gibson.* (pp. 119–136). Hillsdale, NJ: Lawrence Erlbaum Associates.)

Gibson, J. J., Olum, P. & Rosenblatt, F. (1955). Parallax and perspective during aircraft landings. *American Journal of Psychology, 68*, 372–385.

Johnson, W. W., & Kaiser, M. K. (1990). *Visually guided control of movement.* (NASA Conference Publication 3118). Moffett Field, CA: NASA Ames Research Center.

Kirlik, A. (1994). Requirements for psychological models to support design: Toward ecological task analysis. In J. M. Flach, P. A. Hancock, J. K. Caird, and K. J. Vicente (Eds.), *Global perspectives on the ecology of human-machine systems.* Hillsdale, NJ: Lawrence Erlbaum Associates.

Langewiesche, W. (1944). *Stick and rudder.* New York: McGraw Hill.

Lee, D. (1976). A theory of the visual control of braking based on information about time-to-collision. *Perception, 5,* 437–459.

Mark, L. S. (1987). Eyeheight-scaled information about affordances: A study of sitting and stair climbing. *Journal of Experimental Psychology: Human Perception and Performance, 13,* 361–370.

Mark, L. S., & Dainoff, M. J. (1988). An ecological framework for ergonomic research. *Innovation, 7,* 8–11.

Mark, L. S., & Vogele, D. (1987). A biodynamic basis for perceived categories of action: A study of sitting and stair climbing. *Journal of Motor Behavior, 19,* 367–384.

Mark, L. S., Warm, J. S. & Huston, R.L. (1987). (Eds.) *Ergonomics and human factors: Recent research.* New York: Springer-Verlag.

Neisser, U. (1967). *Cognitive psychology.* New York: Appleton-Century-Crofts.

Neisser, U. (1976). *Cognition and reality.* San Francisco: Freeman.

Neisser, U. (1987). From direct perception to conceptual structure. In U. Neisser (Ed.), *Concepts and conceptual development: Ecological and intellectual factors in categorization.* (pp. 11–24). Cambridge, MA: Cambridge University Press.

Newell, A. (1973). You can't play twenty questions with nature and win: Projective comments on the papers of this symposium. In W. G. Chase (Ed.), *Visual information processing.* (pp. 283–308). New York: Academic Press.

Norman, D. A. (1988). *The psychology of everyday things.* New York: Basic Books.

Owen, D. H., & Warren, R. (1987). Perception and control of self-motion: Implications for visual simulation of vehicular locomotion. In L. S. Mark, J. S. Warm, and R. L. Huston (Eds.), *Ergonomics and human factors: Recent research.* (pp. 40–70). New York: Springer-Verlag.

Rasmussen, J. (1986). *Information processing and human-machine interaction: An approach to cognitive engineering.* New York: Elsevier Science .

Rasmussen, J., & Vicente, K. J. (1989). Coping with human errors through system design: implications for ecological interface design. *International Journal of Man-Machine Studies, 31,* 517–534.

Rutter, B. G., & Newell A. (1988). Body scaling. *Innovation, 7,* 14–16.

Rutter, B. G., & Wilcox, S. (1988). An ecological view of hand function. *Innovation, 7,* 12–13.

Smets, G. (1994). The theory of direct perception and telepresence. In P. A. Hancock, J. M. Flach, J. K. Caird, and K. J. Vicente (Eds.), *Local applications in the ecology of human-machine systems.* Hillsdale, NJ: Lawrence Erlbaum Associates.

Smith, S. L. (1963). Man-computer information transfer. In J. H. Howard (Ed.), *Electronic information display systems.* (pp. 284-299). Washington D.C.: Spartan Books.

Taylor, F. (1957). Psychology and the design of machines. *American Psychologist, 12,* 249–258.

Vicente, K. J. (1990). A few implications of an ecological approach to human factors. *Human Factors Society Bulletin, 33,* 1–4.

Vicente, K. J., & Rasmussen, J. (1990). The ecology of human machine systems II: mediating "direct perception: in complex work domains. *Ecological Psychology, 2,* 207–250.

Warren, W. H., Jr. (1984). Perceiving affordances: visual guidance of stair climbing. *Journal of Experimental Psychology: Human Perception and Performance, 10,* 683–703.

Warren, W. H., Jr. (1985, June). *Environmental design as the design of affordances.* Paper presented at the Third International Conference on Event Perception and Action, Uppsala, Sweden.

Warren, W. H., Jr. (1987). Visual guidance of walking through apertures: Body-scaled information for affordances. *Journal of Experimental Psychology: Human Perception and Performance, 13,* 371–383.

Warren, W. H., & Whang, S. (1987). Visual guidance of walking through apertures: Body scaled information for affordance. *Journal of Experimental Psychology: Human Perception and Performance, 13,* 371—383.

Warren, R., & Wertheim, A. H. (1990). (Eds.). *Perception and control of self motion.* Hillsdale, NJ: Lawrence Erlbaum Associates.

Woods, D. D. (1986). Joint cognitive system paradigm for intelligent decision support. In E. Hollnagel, G. Mancini, and D. D. Woods (Eds.), *Intelligent decision aids in process environments.* (pp. 153-173). New York: Springer-Verlag.

Chapter 2

On Human Factors

Peter A. Hancock
University of Minnesota

Mark H. Chignell
University of Toronto

The Secret of Machines

We can pull and haul and push and lift and drive,
We can print and plough and weave and heat and light,
We can run and race and swim and fly and drive,
We can see and hear and count and read and write...

But remember please, the Law by which we live,
We are not built to comprehend a lie.
We can neither love nor pity nor forgive -
If you make a slip in handling us, you die.

2.0 Introduction and Overview

This chapter develops a descriptive theoretical structure for human factors. The structure is based on a view of technology as the principal method through which humans expand their bounds of perception and action but also as the medium through which control is arbitrated in systems of increasing complexity and abstraction which explore the new 'territory' revealed. The theory presents a broad rationale for the contemporary impetus in human factors and historical motivations for its growth. It is suggested that human factors is unlike other traditional divisions of knowledge and is more than the mere haphazard interdisciplinary collaboration between the engineering and the behavioral sciences. In identifying the opportunities and constraints intrinsic to emergent dynamic operational spaces derived from the interplay of human, machine, task, and environment, we point to a future for human factors as the essential link in the co-evolutionary development of biological and nonbiological forms of intelligence, the failure of which will see the certain demise of one and the fundamental impoverishment of the other.

'Science above all things is for the uses of life.'

2.0.1 Preamble

Rudyard Kipling's "The Secret of Machines," is as appropriate for the supervisor of contemporary complex systems as it was for the individual worker in the 19th century factory. Slips in handling machines can and do lead to fatal consequences. Yet, we have built a global society whose dependence on such systems grows daily. Human factors is at the heart of this development, seeking on the one hand to maximize the benefit derived from technology, while on the other, exercising a continual vigilance over its darker side. In what follows, we present a framework that views human factors as something more than a convenient fusion of knowledge from disparate disciplines. What emerges is no traditional academic pursuit. Rather, as the above quotation from Francis Bacon intimates, it reveals human factors arguably as the motivation of science and by extension at the very heart of the human enterprise itself.

2.1 Structure and Aims of This Chapter

Human factors is frequently represented as a discipline that makes science and technology more appropriate and palatable for human consumption. However, Bacon's view of science suggests that human factors should not simply ameliorate the adverse impact of technology after the fact. Human factors should be seen to motivate science, engineering, and systematic empirical human exploration in the first place.

In order to provide definitive motivation for science, human factors must present a clear vision of what the "uses of life" are. We need to understand how people use perception, cognition, and action to decide on goals and carry out meaningful and useful tasks in the pursuit of those goals. This necessary understanding includes a rational analysis of tasks, a psychological analysis of human behavior and capability, and an engineering analysis of how humans interact with tools and systems in performing these tasks in differing environments. It should also include purposive and proactive accounts that are consistent with the requirements for explaining goal-oriented reasoning and behavior.

Therefore, we address the question of how humans use technology in the goal-oriented and task-oriented exploration and manipulation of their perceivable environment. Powers (1974, 1978) has indicated that behavior is a goal directed process that is organized through a hierarchy of control systems. Higher-order systems perceive and control an environment composed of lower-order systems, with only the first-order systems interacting directly with the external world. Human behavior has been characterized as an inner loop of skilled manual control and perceptual processing which is embedded within an outer loop of control that includes knowledge-based problem solving. Moray (1986), for instance, gives the example of a nested series of goals working from an extreme outer loop (influencing society and raising children) to inner loop processes such as controlling the position of a car's steering wheel in order to drive that car to a destination.

In this chapter, we explore human capabilities for perception and action and the role of technology in redefining the bounds and the nature of those capabilities. We also examine the linkage between perception and action and the way in which that relationship is changing as both actions and perceptions are elaborated by evermore sophisticated technologies. Initially, we generate a description of the limits to unaided and aided action and contrast these with bounds to unaided and aided perception. We suggest that the "tension" which results from the disparity between what can be perceived compared with what can be controlled provides the major motivational force for human exploration.

In respect of such exploration, technology generates the dual but opposing effects of increasing the range of action while simultaneously expanding the range of perception. The further these respective bounds are extended, the more complex (Hancock, Chignell, & Kerr, 1988) are the technical systems needed to support exploration and the more reliance, at the present time, is placed on metaphorical representation of the control spaces involved. In elaborating our overall theme we use a relativistic framework to describe the expanded vista of capability that accompanies the sequence from homeothermy, through tool genesis

(Oakley, 1949), and intelligent prosthetics, to the contemporary potential for a 'supercritical' society.

In examining a further duality of technological innovation, we contrast the potential for catastrophic failure that accompanies transition at the edge of chaos, with the knowledge that we cannot 'directly will to be other than we are.' While articulating the constraints on human nature and ability, we are also augmenting our basic abilities as we systematically explore and engineer our environment to 'create' our future selves. Our future selves are bound to the co-evolution of biological and non-biological forms of life. Intrinsic to our whole argument is the centrality of ecological principles and in particular the goal-oriented interaction between humans and the perceivable environment as the basic unit of analysis for human factors.

---

1. The perception-action link may explain *how* humans explore the environment. The perception-action gap may explain *why* they explore the environment.

2.2   PERCEPTION AND ACTION IN SPACE AND TIME

It may be observed that the personal and collective odyssey of humankind is to find and establish our respective place in space and time. While this journey might be considered from one perspective as an artistic endeavor, we focus here on the task-oriented use of technology to provide a degree mastery over the perceivable environment as parsed by the constructed metrics of space and time.

The exploration of space and time is motivated by goals, which can be expressed as desired future states of the environment, and evaluated by the associated perceptual experience that they engender. Strategies are ways of achieving goals, and tasks are the steps by which goals are achieved (see also Shaw & Kinsella-Shaw, 1989). While a goal is a desired future state of a system, a task is a subsidiary component that implies a specific transformation. Goal achievement is composed of the successful and integrated completion of more than one task. Actions which subsume tasks transform the state of the world. From a thermodynamic perspective, actions typically result in a disturbance to local entropy (Swenson & Turvey, 1991) and the expenditure of energy toward a more ordered[1] state of the system. Thus temporal direction is apparently implied in the performance of tasks. Simple temporal progression of a system without alteration cannot be regarded as a task within this definition.

The cost of transformation, or demand of any task may not be specified *in vacuo*. In human performance such costs are typically expressed as a function of the time taken to traverse from initial state to goal state and the accuracy of that transition, in essence space-time synchronization. Transformation cost can also be expressed in terms of effort. Machines, as creators and manipulators of energy, act to increase the number of possible paths by which a specific goal can be achieved. Technology thus serves to broaden the horizon of the possible. While machines serve to open the window of opportunity, environmental constraints frequently frame and limit what is possible.

The environment often presents hurdles and obstacles that the operator anticipates. Indeed one hallmark of expertise is the ability to project expected demands and preempt their more adverse consequences. However, the environment can also present unexpected perturbations which interrupt on-going tasks and can, under certain circumstances, pervert goals altogether by removing them from the range of possible outcomes. With respect to operations then, human-machine systems seek to expand ranges of possible activity while the environment restricts that activity. Unfortunately, this antagonistic aspect of interaction has permeated much design. Thus systems often seek to conquer and control the environment, rather than recognize and harmoniously incorporate intrinsic constraints. (Although conquering nature should be acknowledged as a predominantly Occidental pre-occupation, see McPhee, 1989). Ultimately, it is the ability to recognize and benefit from mutual constraints and limitations to action that characterizes "intelligence" on behalf of systems.

Goal-oriented behavior is defined in part by reacting to

---

[1]A question for the second law of thermodynamics is the intrinsic appointment of a normative arbiter who dictates what connotes order and therefore the distinction between 'more' and 'less' ordered. While this arbiter is conjured by the physicist the problem of ordering is apparently obviated. However, if the arbiter is invoked by the psychologist, the perceived nature of order becomes more equivocal as does entropy and by extension, time. The same question can be extended to the first law where we might ask whether burning a piece of ink-splattered paper is the same as setting fire to the U.S. Constitution, a document which embodies both informational and societal significance. While the physical effect may be identical, for human observers they are certainly not perceived as equivalent events. (see also Gibson, 1975)

Time

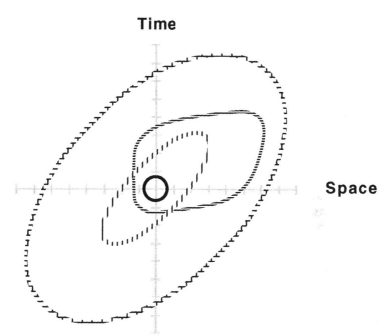

Space

Figure 2.1. *Human-range (the inner circle), perceptual-range (the horizontal line envelope), prosthetic-range (the vertical line envelope), and universal-range (the outer envelope) expressed as functions of space and time[2]. Note that these regions are approximations and are not drawn to represent definitive boundaries.*

environmental constraints that limit the range and effectiveness of perception and action. Despite the arbitrary nature and the relativity of space and time, the limitations of perception and action can be considered initially within a framework that views space and time, artificially, as orthogonal. The environment may be scaled from the small to the large and the brief to the prolonged. Within these continua there are relative ranges of space and time with respect to our own size and our own perception of duration. A representation in terms of orders of magnitude is given in Figure 2.1. We place ourselves at the origin and it can be observed that our collective recorded history is an account of

[2]Scales of spatial representation are clearly illustrated in the text *Powers of Ten* (Morrison, Morrison, Eames, & Eames, 1982). It is of course interesting to note that these authors achieved their illustration by fixing the orientation of one axis and subsequently adjusting the bounds of the remaining axes. Hence, *three-dimensionality* is strongly, if intrinsically emphasized. It is also clear that there is no equivalent temporal Powers of Ten. Indeed, it is an instructive exercise in imagination to attempt its construction. The reader is invited to do so.

Scale (sec)	Time Units	System	Level of Interaction
10^7	months		
10^6	weeks		Society
10^5	days		
10^4	hours	Task	
10^3	10 min	Task	Work Group
10^2	minutes	Task	
10^1	10 sec	Unit Task	
10^0	1 sec	Operations	Cognition
10^{-1}	100 ms	Acts	
10^{-2}	10 ms	Neural Circuit	
10^{-3}	1 ms	Neuron	Neurophysiology
10^{-4}	100 microsec	Organelle	

Figure 2.2. *A time scale of human actions (After Newell, 1990).*

our physical, but not spiritual, displacement from this central location. (see also Gooddy, 1988).

Human response in spatial dimensions has been studied extensively, particularly in psychophysics. Fine spatial discrimination can be measured by Vernier acuity in vision, two-point threshold for touch, and auditory spatial localization and discrimination in hearing.

Human interest in the dimension of time has a particularly long history, going back to the use of astronomical tables in early religion and agriculture (Fraser, 1966). Due to the need to coordinate and synchronize the actions of a large number of people and things, the social importance of temporal scaling is now reflected in the omnipresence of timekeepers. Figure 2.2 (after Newell, 1990) shows part of the time axis of Figure 2.1, interpreted in terms of human action (see also Iberall, 1992, p. 45).

2.2.1 *Humanrange*: Boundaries To Unaided Action

One of the initial lessons in comparing the range of human perception and action with the entire range of space-time is that direct human experience is a relatively small subset of the entire space-time continuum. In defining the boundary of *humanrange* we can begin by indicating the limits of unaided action on the spatial axis. Unaided is here meant to signify without the assistance of other entities including natural or manufactured tools or machines. It is clear, given the physical constraints of our musculo-skeletal system, that unaided we cannot directly manipulate objects less than some fractions of an inch in size. Also, in respect of an upper boundary, we might be able to throw a stone some hundred yards or so but without some form of assistance we could not exceed this distance to any great degree. We should note immediately that the specification of this spatial boundary include intrinsic temporal assumptions. That is, throwing the stone implies a force exerted over a short duration. As becomes immediately apparent spatial constraints cannot be specified independent of time and vice-versa and this mutuality is as important for the behavioral sciences as it is for the physical sciences (Locke, 1690; and see also Hancock & Newell, 1985). With respect to the boundaries of time, the lower threshold can be viewed as the duration which divides the performance of two separate acts. The upper temporal boundary is, putatively, the length of an individual's lifetime. However, this latter definition, like each of the others, may not go unchallenged. It is a defendable assertion that humans leave partial representations of themselves through communication, procreation, or *re*creation.

Even over a lifetime, unaided by any tools or prosthetics, a human may achieve a considerable manipulation of the environment. However, history informs us that few existing archaeological monuments were not constructed without the use of the then existing highest state of technology. Indeed it might be argued that no totally unaided human manipulations of the environment have survived prolonged periods. It is clear that in the overall picture, the spatial and temporal dimensions are interdependent and the collective range over which an unaided human may exert action (*humanrange*) is highly restricted in comparison to the limits of unaided perception to which we now turn.

2.2.2   *Perceptualrange*: Boundaries to Unaided Perception

If the boundaries to unaided action are relatively restricting, the same cannot be said of the boundaries of unaided perception or *perceptualrange*. We will deal first with perception at the lower bounds. The lower temporal boundary is usually represented by events that are separated by fifty to one hundred milliseconds in duration (Stroud, 1955, see also Poppel, 1988, but see Vroon, 1974). This period is projected to represent the perceptual moment (but see Gibson, 1975) or in the terms given by Clay (cited by James, 1890) and subsequently Minkowski (1908), the 'specious present.' Depending upon what it is we wish to observe, various limits to spatial perception might be suggested. Unaided, the human observer can see objects down to quite small sizes

Figure 2.3. *William Blake represents the eternal reaching of human nature in the illustration "I want, I want."*

and from their actions infer the presence of even smaller particles. However, without aid, empirical microbiology might be somewhat limited as illustrated in one of Gary Larson's wonderful cartoons. But it is not the lower bounds of space and time, nor even the upper temporal boundary that represent such a vast contrast with action limits. Rather it is the bounds to unaided spatial perception. As can be seen from the superimposed envelope of *perceptualrange* in Figure 2.1, it is the vast regions of space which we may perceive unaided, but over which we cannot act that represents the major disparity. It is therefore no coincidence that astronomical observation provided the major early impetus for what we now recognize as the scientific enterprise (Koestler, 1959).

It is, we suggest, the 'tension' created by this dissociation between perception and action that provides a basic motivation for exploration. The contemporary vehicle for such exploration is applied science in the form of technology. 'Reaching' as the metaphor for exploration and knowledge is not new and nowhere is this urge more clearly represented than in the illustration by William Blake, reproduced in Figure 2.3. In this picture he expresses the essence of the human desire to reach beyond frustrating restrictions on action. Here again we see that Blake's example is taken from the large scale of space, a reaching toward the nearest celestial body (the moon). Our manifest inability to exercise influence over far distant objects has been clear for many millennia. The plethora of non-holonomic overtones in Blake's illustration have been explored by others (see Bronowski, 1958), however, there yet remains more irony and pathos to be distilled.

2.2.3   *Prostheticrange*: Boundaries to Aided Action

With respect to the process of exploration and manipulation of the environment with the aid of external implements, technology has always served two antagonistic purposes. As we observed earlier, tools have increased the ranges of space and time over which an individual may act. However, their use also results in the furtherance of the boundaries of the regions of space and time which can be observed. With respect to aided action, the envelope is expanded some orders of magnitude over the meager range of unaided exploration. Contemporary boundary markers to this *prostheticrange* are represented by elementary particle manipulation in the lower spatial range to the Voyager Spacecraft and its physical presence beyond the edge of the solar system at the upper spatial range. It might be argued that humankind has exercised influence over a much larger range when we consider the information intrinsic to radiowaves that have left this planet within the last century. The choice of the precise nature of which physical manipulation is used as a criterion is one that may be challenged. However, as this simply extends the envelope by some multiple it is not a question to which we direct particular concern here.

On the temporal scale, we have become familiar with picosecond-based measures (Rifkin, 1988) at the lower boundary, while storage and dynamic knowledge representation of expert systems promises the use of technology to preserve at least a small portion of individual knowledge or expertise beyond our traditional lifespan (Moravec, 1988). It may further be argued that procreation, *r*ecreation, and information communication through traditional media also perpetuate some portion of many individuals. According to allometric scaling, humans should live on average to 27 years of age (see Schroots & Birren, 1990; and Yates, 1988). Already, our use of technology staves off death more than three to four-fold our expected life-span. Also, there is a trend with improvements in nutrition, personal fitness, and medical faculties for individuals to live even longer. However, it is one of the basic human characteristics to countenance one's own demise. In spite of limits on individual perception of spans of time, the same scale for upper temporal boundaries cannot be applied to the things we create with technology. We have direct evidence that the constructions of our forebears have lasted several thousand years and we project that our own manipulations might last into the hundreds of thousands of years (e.g., nuclear waste). However, it is important to distinguish, at a number of levels, between mere persistence of effect versus creative and generative actions.

2.2.4   *Universalrange*: Boundaries to Aided Perception

Outside prosthetic scale, we have located *universalrange*. This represents the boundaries of what we may perceive when aided by technology. In essence, it represents the known universe and like other envelopes is still at present expanding (that is our knowledge of it is still expanding, whether there is and/or will be continued physical expansion is a question upon which cosmologists seem unable to agree). An individual looking out into space is looking back in time. The interdependence of space and time has been recognized by physicists for over three centuries, while this combination (space-time) has also been explored with respect to human behavior (Hancock & Newell, 1985). We use this approach to pursue our argument below. With the aid of contemporary technology our range of perception is vastly increased. The resolution of the Hubble telescope promised to expand *universalrange* and improve our knowledge of entities between ourselves and that threshold. Like other forms of expensive and complex technical systems it proved unfortunately vulnerable to failure (see Perrow, 1984; Reason, 1990). At the aided lower end of the spatial scale, where observation fades through metaphor to concept, we have begun to recognize the interaction of the conceivable, the perceivable, and the fusion of the potential with the actual. Comparable recognition at the other extreme boundary of space-time would represent a significant step forward. Nor is it happenstance that the very large co-varies with the prolonged and the very small co-varies with the exceptionally brief and the emergent long axis need not necessarily be space, time, or even space-time.

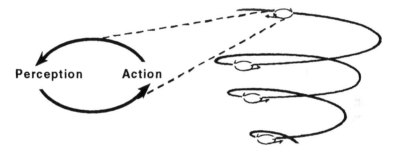

Figure 2.4. *Perception-Action Loops expand and interpolate into individual and collective Perception-Action Spirals*

2.2.5   Synthesizing Scales

We have suggested that there is a continual tension between these respective regions of perception and action, as humankind seeks to physically control that which they can perceive. If the ecological approach can be characterized as a looping of the perception-action cycle, what we have presented here represents an extension to this concept, seeing exploration more as a perception-action spiral. In this conception there is a continual expansion of the ranges over which the perception-action loop occurs, (see Figure 2.4). It is indeed the specific purpose of technology to synthesize the inequities in the envelopes of perception and action. Therefore, we can recognize a companion view of technology as the vehicle which brings *universalrange* to *humanrange* by representing entities at our level. We have not explored this form of mutuality here directly but recognize its validity.

However, in addition to the tension created by the dissonance between regions of perception and action, there is also a growing dissonance between actions and experience. If a human operator (or supervisory controller) interacts with a system via aided perceptions and aided actions, then the directness of everyday experience is, at present, replaced by a more abstract and indirect relationship between the person and the environment. This is particularly true for extensions beyond regions in which light magnification simply rescales the display. Such a difference is represented by the respective disparities in the Envelopes of Figure 2.1. This leads to our second observation:

> 2. The further the envelopes expand away from the relatively fixed region of human-range, the further divorced are actions from experience.

The corollary of this growing indirectness is that as we progress in our efforts to perceive and influence the very large and the very small, we have begun to rely progressively on metaphorical representations of these entities with which, by constraint, we have had no direct experience, although virtual reality promises a potential resolution for such dissociation as we discuss below. The advisability of this strategy and some potential remedies are the topics of other chapters in the present text. Thus technology must seek not only to expand *universalrange* but also to provide a representation of its content in a manner coherent with *humanrange*. This represents a major challenge to future development of technology in general, and to the discipline of human factors in particular.

As we explore ranges of our 'universe' that are further from our personal experience, we have traditionally dealt with progressively more interconnected and interactive systems (Perrow, 1984). Indeed, it is the emergent properties of these interactions which frequently provide the challenge, the uncertainty, and the novelty which is sought alongside the expansiveness of exploration. However, the problem mounts as we move further from *humanrange* and as we use progressively indeterminant prosthetics to do so. It is, of course, a step of imagination to understand that scale, and therefore complexity are *only* relative to the entity under examination, whether it be human, machine or the human-machine dyad.

2.3   Perception-Action in Space-Time

While the diagram in Figure 2.1 provides a useful first representation of the envelopes of perception and action, it is an essentially 'static' representation of the dimensions of space and time. However, we are interested in a task-oriented approach where the human operator navigates through space and time in a goal-oriented fashion. One framework for understanding such navigation is the space-time diagram provided by Minkowski, (see Figure 2.5). For our present purpose we

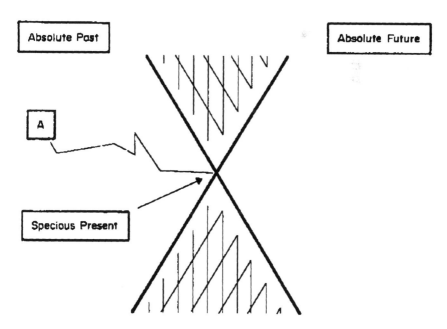

Figure 2.5. *Minkowski Space-Time Diagram.*

can capture the essence of this representation in the often reproduced

quotation from Minkowski (1908). He said that:

> *"space by itself and time by itself are doomed to fade away into mere shadows and only a union of the two would preserve an independent reality."*

He went on to ask:

> *"who has been at a place except at a time, and who has experienced time except at a place."*

Minkowski's space-time is a four dimensional manifold, usually represented for the purposes of simplicity as a two-dimensional diagram as given in Figure 2.5, where the multiple spatial dimensions are compressed to a single axis. The diagram possesses three areas and a point of intersection. The first quadrant represents the absolute past and contains the sum of previous events, drawn through what Minkowski termed "world lines," illustrated here by the line A. The sum of all world lines represents the world (or the universe, depending upon one's perspective). Each world line leads to an intersection between the absolute past and the absolute future, which intersection is labeled the 'specious' present (see also E.R. Clay in James, 1890). This represents the transition point between past and future and, as will be argued, it implies the presence of a sentient observer whose constitution dictates the nature of that present. Transition through the specious present reveals the absolute future as it passes to become a physically deterministic segment of the absolute past. On either side of the present, lies a symmetric region of absolute elsewhere. The constraint expressed by the hashed region is intentional in that the observer in this framework cannot experience these regions of existence.

It is important to envisage what the observed quadrants are composed of. There are a number of representations and perhaps one of the best ones is line-drawing description by Kugler and Turvey (1987), shown in Figure 2.6. Here, the authors have presented a snapshot of this 'landscape,' although the mutual interaction of environment and entity is not illustrated, and by constraint, the dynamics cannot be shown on the present static illustration.

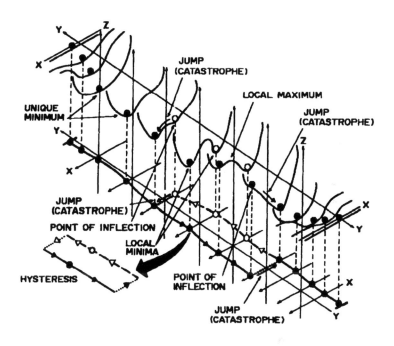

Figure 2.6. *Following the energetically constrained ball. Illustration of an energetic landscape with local catastrophes, discontinuities, and assymetries noted. From Kugler and Turvey (1987), reproduced with permission.*

Since our goal is to understand the task carried out by a goal-oriented human operator, we are interested in representing navigation within space-time. In Figure 2.6 we see a person or system, given by the letter A, weaving a path through space-time. In tasks frequently there are multiple perturbations. Thus the operator or system must adapt to changes and unexpected inputs. In the case of an aircraft, for instance, the position and orientation of the craft must be maintained in spite of changes in windspeed, engine power, and other parameters. Tasks, where a path or goal must be pursued in spite of perturbations are control tasks. Such tasks may be more or less difficult depending on the tools available and the degree of dynamism and uncertainty in the task environment.

Indeed, within such tasks, the degree of 'intelligence' on behalf of the operator is also problematic until the constraints of the environment are laid bare (Kirlik, 1994, Simon, 1981) The origin of intention however remains the *magnum opus* (Iberall, 1992). A dramatic representation of a dynamic control problem is given in the following quotation from "Treasure Island" by Robert Louis Stevenson (1883).

> *"I found each wave, instead of the big, smooth glossy mountain it looks from shore, or from a vessel's deck, was for all the world like any range of hills on the dry land, full of peaks and smooth places and valleys. The coracle, left to herself, turning from side to side, threaded, so to speak, her way through these lower parts, and avoided the steep slopes and higher, toppling summits of the wave. 'Well now,' I thought to myself, 'It is plain I must remain where I am, and not disturb the balance; but it is plain, also, that I can put the paddle over the side, and from time to time, in smooth places, give her a shove or two towards land'."*

Although Stephenson (as Jim Hawkins) was controlling his coracle, many of the principles that apply in steering a craft through a rough sea also apply in navigating through space-time in other task situations. (Indeed, the relationship goes beyond analogy) The organism or system needs to establish some stable platform of operation (by which we include here the self-similar sequence from cellular integrity through homeothermy, which at a meta-level can be viewed as the self-similar sequence phylogeny, ontogeny and technology, see Hancock, 1993b). But, all the while the organism, or the human-machine system, faces the vagaries of an uncertain environment which acts to modify what level of stability can be established and adaptability attained.

Dynamic navigation (in the same way that Jim Hawkins tries to navigate above) is the key adaptive capability of humans' and is aided (hopefully) by the technology we create (Hancock & Chignell, 1987). However, the technological aiding that supports dynamic navigation can be a two edged sword, as indicated in the following principle.

---

3. **Dynamic aiding through automation currently assists the operator in maintaining control of the system at the cost of distancing the operator from the physical cause and effect and moment-to-moment status of the system.**

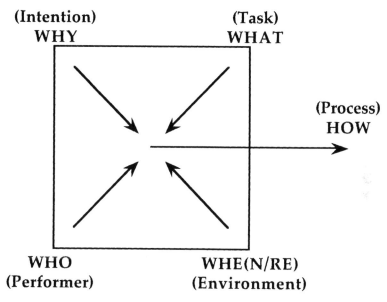

(Intention)
WHY

(Task)
WHAT

(Process)
HOW

WHO
(Performer)

WHE(N/RE)
(Environment)

Figure 2.7. *Structure of and constraints on a general decision point. The outcome process is dictated by the confluence of constraints from the four cited elements. The location of intention (why a task is performed) and the operator (who performed it), as precursors to action at the left of the diagram is intentional as the time course of emergence ripples across the left-right arrangement.*

Thus when dynamic aiding breaks down it is more difficult for the operator to intervene successfully. One particularly clear example of failure in dynamic aiding occurred in an aircraft incident where a thrust imbalance built up between the aircraft engines. The autopilot corrected for the imbalance as long as it could after which the aircraft went into a prolonged dive. In this case there were subtle cues as to change in the status of the aircraft and the way the autopilot was adapting to them, but these went unnoticed and unobserved. This form of failure may be anticipated as the distancing of operator from system under automated operation increases (Parasuraman, Molloy, & Singh, 1993).

2.3.1   The Unaided Individual in Space-Time

Task performance consists of navigation in space-time. Analysis of human-machine tasks is complicated by the interaction between the automated and human components of the system. Thus it will help to consider how the unaided individual navigates in space-time in order to understand the general principles and problems of such navigation.

If we view the range of human exploration as an expansion of the potential paths of progress in the quadrant of the absolute future, where this exploration is fueled by the disparity between scales of perception and scales of action, then we can now ask what strategies are used in the exploratory process. It is, we think, insufficient simply to state that humans do explore, or simply to affirm that exploration is always going on, all the time, for all individuals. Rather, we view exploration as a result of goal-oriented decision-making and subsequent task performance in striving to achieve these goals.

This form of exploratory strategy can be viewed as a decision point (cf., Newell, 1986) followed by a sequence of actions, which themselves trigger subsequent decisions or represent achievement of the goal itself. In Figure 2.7, we have illustrated such a point as an expansion upon Newell's task-environment-organism triad to include the intention for action (the why), and the aiding that technology renders, as included in the process of achievement (the how).

The point of departure in task-oriented navigation is the present location, always at the specious present. The goal is then defined as some *desired location in the absolute future.* Implicit in this formulation is the notion of planning. Planning implies the recognition of achievable goals and the elimination of unachievable goals. It implies the integration of what is known, into a strategy to achieve what is desired. In Holland's (1992) terms it requires the synthesis of feedback with the potential for prediction. In ecological terms it requires a blurring of the specious present to include an elaboration of perception-action linkages beyond the specious present. In our terms it recognizes that dynamic navigation is not merely a reactive response to the instantaneously presented conditions but is a proactive stance that helps mold an individual, and the instantiation of their future selves (predicated on their own actions). Given that a sequence of actions can be performed based on the decision point illustrated in Figure 2.7, we expect that there will be significant periodicities in behavior that act rather like waves propagating along a channel (see Iberall's 1992 excellent article).

A goal is most simply represented as a single location. However, humans and systems frequently seek to achieve multiple goals (Rasmussen, 1986). A specific sequence of actions may even be directed towards the most desirable compromise that exists between two or more competing goals. In the special case of a singular goal state, the intention of the operator is then to traverse the region between the present state and the goal state. Constrained by the factors noted in Figure 2.7, there are a finite number of possible paths which permits this transition (Shaw & Kinsella-Shaw, 1989). *It is the choice between these paths which represents the strategies of exploratory behavior.* Thus attention is a facet of behavior, enabled by change, to provide a priority for one course of action over its competitors. Once the decision has been made, an individual strategy is constructed as *the linked sequence of planned actions directed to attain the prescribed goal.* One of the problems in designing modern human-machine systems is that they may unintentionally violate the following principle.

> **4. Human goal-directed strategies that work well for unaided navigation in a non-technical environment may sometimes be inappropriate in a technologically advanced context.**

Consider the role of energy in the past, and throughout the whole of human evolution, the need for individual *energy minimization* as an endogenous strategy has dominated (and see Swenson & Turvey, 1991). However, the advent of technology frequently fractures this energetic 'cost' constraint. Hence, we have an organism (the human operator) with a vestigial strategic imperative (energy minimization), placed in a largely 'manufactured' environment in which such a strategy is frequently no longer appropriate. Design then becomes a process of recognition of these vestigial imperatives and amending perception-action-machine linkages accordingly. Such design by extension creates the future conditions from which new strategies and new operators emerge.

2.4   Metaphor, Systems, Control, and the Use of Ecological Principles

As we have noted, the further that a system operates from *humanrange* in space and time, the more abstract the representation of the control spaces. *Humanrange* is meant to imply some direct empathy in terms of object size and event duration. However, it does not exclude empathy with control spaces that emerge from human-machine interaction. For example, in the origin and development of the industrial revolution, we find frequent reference to skilled workers who operate through an intrinsic 'feel' for the process under their control. To such individuals, augmented information as to the status of a single variable in the process often meant little more than distraction as it was the confluence and emergent properties of multiple interactive factors upon which their skilled, intuitive or empathic grasp was based. In many cases, measures of absolute level were of limited use compared to relative values. With the evolution of technology, the ability to intuitively grasp interactive states of ever more complex systems, became a progressively more precarious operational strategy. This was especially true for new systems, which themselves did not possess sufficient history such that a skilled cadre of 'masters' could be assembled.

One natural reaction to the reduction in "process empathy" was the attempt to provide ever greater amounts of information on the principle that embedded within this matrix must be the right informational "answer" to the sequence of dynamic questions posed to the operator. However, system complexity soon defeated skilled intuition, a defeat exacerbated by the proliferation of physically confusing and ambiguous analog displays. Such proliferation eventually resulted in computer mediation, removing the operator one more step physically, and further distancing them representationally, from an empathic grasp of the process itself.

The solution to the problem of decreased process empathy did not, and does not, lie in first aid approaches to poorly conceived display designs. Rather, it is founded on a fundamental re-evaluation of the theoretical basis for displaying system functions. Although metaphor and ecology do not seem to sit well in the same sentence, we must recognize that a profound change in display strategy must be grounded upon knowledge of how the perceiver views the world as a display in the first place.

Preliminary attempts at this strategy have generated a class of displays, labeled direct displays. They are founded upon the critical notion of affordance, (although there is *still* no consensus as to what an affordance is, see Turvey, 1992)[3]. An affordance is purportedly a theoretical construct that addresses the perception of meaning in displays and environments. Gibson (1979, p. 127) explains that:

> *"The* affordances *of the environment are what it* offers *the animal, what it* provides *or* furnishes, *either for good or ill. The verb to* afford *is found in the dictionary, but the noun* affordance *is not. I have made it up. I mean by it something that refers to both the environment and the animal in a way that no existing term does. It implies the complimentarity of the animal and the environment."*

He concluded that:

> *"The possibilities of the environment and the way of life of the animal go together inseparably. The environment constrains what the animal can do, and the concept of a niche in ecology reflects this fact. Within limits, the human animal can alter the affordances of the environment but is still the creature of his or her situation. There is information in stimulation for the physical properties of things, and presumably there is information for the environmental properties. ... Affordances are properties taken with reference to the observer. They are neither physical nor phenomenal."* (Gibson, 1979, pp. 143)

Therefore, an affordance is apparently *a functional relationship*

---

[3]There are perhaps few concepts as important and contentious as that of affordances. Reed (1988, p. 231) notes that:

> *"Affordances are the functional properties of objects as, for example, the affordance of a heavy stick or rock for pounding. Any particular object will probably have many affordances. An apple may be eaten, thrown, juiced, or baked to name but a few of its affordances. Yet a given object will also lack many affordances. An apple is no use as a brick or as kindling."*

And yet it is. The problem being, it doesn't serve this function very well. It is clear that for an ecological approach to human-machine systems to work, the concept and functional utility of affordances are critical (see Hancock, 1993c). It is equally clear that further elucidation is still needed.

between objects or properties within the environment and the capabilities of a sentient organism. The advantage of such a conception is the obviation of translation and representation on behalf of the observer. Given an affordance, ambiguity is, again purportedly, eliminated and thus action is uniquely specified. Degrees of degeneration from this 'best of all possible worlds' occur as we introduce individual perceivers, who may or may not assimilate said affordance, and for whom the action specified may vary according to the goal or intention. This degree of degeneration resurrects the concept of valence, in which actions are not uniquely specified but probabilistically specified, with respect to the individual actor.

The generation of "direct" displays (Bennett & Flach, 1992; Flach & Bennett, 1992; Vicente & Rasmussen, 1990) is an attempt to disambiguate conditions and, by implication, to make affordances visible. The search for a framework for seeking, validating and generalizing affordances across multiple conditions is still ongoing (cf., Pittenger, 1991). In designing systems and their displays, we have the advantage of not only benefiting from naturally occurring affordances, but seeking the potential for created affordances, or of exploiting culturally defined expectations. Naturally, there is considerable debate over what exactly constitutes a direct display (as there is on affordances themselves), and the extent to which such displays should rely on metaphor in converting displays into a more easily assimilated and understood form. One promising avenue for enhancing directness of displays is the use of virtual reality. However, in spite of problems in defining and implementing directness in displays, it is clear that the evidence against the past approach, where operators are provided with a "data dump" of undigested information about the system, is considerable and growing. It is *only* through application of ecological psychology and its principles to this realm of human factors that substantive *near term* gains can be expected in process empathy, and in awareness (Smith & Hancock, 1993) of critical system parameters and states.

What then is the role of human factors in enhancing process empathy? Clearly it is necessary to implement more direct displays and to implement appropriate metaphors and models where needed. However, there is also a need for a broader view of technological application. Technology is driven by forces which require that the boundaries we have identified are progressively enlarged. Systems are required to be faster, greater in physical size or at least density per unit, and when connected with other units they grow in complexity. Often such progress can appear relatively "mindless." That is, driven by external forces beyond the control of any single individual with few concerns whether such progress is 'beneficial' in a specific sense. (The latter must nearly always be the case since the modern democratic society specifies no explicit goals other than global assertions such as 'freedom from want', or everyone has the right to 'life, liberty, and the pursuit of happiness'). Thus:

> **5. There is an important need for a rigorous model of human factors that assumes a leadership role in directing technological innovation. This requires judgments on the best "uses of life."**

While human factors has frequently been characterized as merely facilitating human interaction with machine systems, its most fundamental role is in the active direction of technology. This is why human factors is a socio-political endeavor, whether or not its practitioners conceive of it as such (Hancock, 1993a). In essence, we need to elevate the question 'why?,' and the fundamental issue of intentionality to its appropriate and pre-eminent position (cf., Iberall, 1992, Nickerson, 1992).

2.5   Technology and Nature: Symbiosis Versus Antagonism

In a text on the ecology of human-machine systems, it should be emphasized that the environment in which we live is largely a manufactured one. The adaptations which we now make are to conditions that our forebears created, which were in turn adaptations of earlier conditions. However, there is a tendency to contrast modern human built conditions with more 'natural' earlier environments. There is in fact no obvious metric or scale of technological naturalness. Horse-drawn transportation is not intrinsically more natural than steam driven transportation (Pirsig, 1974). The bridle, the saddle, and the spoked wheel are all technologies that adapt and change the environment to suit human tasks and goals. Thus it is not the adaptation of the environment itself which leads to problems, but rather the indirectness and abstraction that has come to characterize such adaptation in today's complex systems. Thus methods of adaptation in complex systems are needed that allow both direct and empathic awareness of system states and parameters.

The artificial division between 'natural' and unnatural technology finds one of its roots in the origins of science. Indeed, one unique facet of human beings is that they are wielders of science, which is perhaps a stronger distinction than previous divisions based on tool use (Oakley, 1949) or even language. Prior to the 14th Century it was considered both logical and feasible to countenance non-rational explanations of many world characteristics. Indeed, it might be observed that the propensity to invoke paranormal explanations is perpetuated well into the present. However, the growth of science acted to provide a single unified account of all experienced phenomena (cf., Bronowski, 1978). Thus the appeal of science is to the discovery and manipulation of the laws of nature not their suspension or transgression. Hence the origin of Bacon's observation on the paradox that "we cannot command nature except by obeying her."

Embedded, deep in this concept is a notion of harmony or symbiosis, a face of natural philosophy that has been tarnished and degraded by the flash and sparkle of technology. From an initial stance of understanding and cooperating with nature, we have rapidly evolved to an attitude of subjugation and indeed opposition to nature, whether explicitly stated as in the "unsinkable" Titanic or implied as in our confidence in many contemporary processes (and the various processes of oversight and regulation that have been constructed). In reality, technology has generated a veneer that has fostered in us a self-importance in which we wish only to command, and fail at many levels to obey. In consequence, we see in many systems the characteristics of command and control where "pressure" in all its physical and metaphorical forms appears where nature is constrained, blocked, and opposed. Thus:

---

6. **The apparent dichotomy between nature and technology is false, and in reality technology is currently the vital vehicle for human growth and adaptation.**

2.6 Adaptive Systems

We need to develop adaptive systems in which human, machine, and environment can operate together harmoniously. Elsewhere we have considered the nature of such adaptive human-machine systems (Hancock & Chignell, 1987), and how such adaptation may occur through the medium of intelligent interfaces (Hancock & Chignell, 1988, 1989). Here, we wish to extend this argument by considering human-machine symbiosis in a broader context and by projecting such interaction beyond the immediate present. While the future of machine intelligence and co-evolution is uncertain, we can still expect the general principles of human-machine systems to apply.

2.6.1 Humans and Technology as Co-Evolutionary Agents

In a small way now, but in a much more profound way in the near future, intelligent technology will assume a major role as an agent in the process of co-evolution. Such growth is contingent upon the degree of nascent intentionality, which in present technical systems is related to the question of machine intelligence (see Anderson, 1964; Turing, 1950). Since there is no fundamental requirement that an agent which influences evolution in others possess intentionality, or even sentience, we are not as yet in direct competition with our technology. Yet our dependence on technology already influences who we are and who our progeny will be. It is unclear what our relationship will eventually be to intelligent systems, and how non-biological intelligence will develop in the future (see Moravec, 1988). However, the growth of machine "intelligence" promises to be a global phase transition from our present state to one of which as yet we can only glimpse dim shadows.

Indeed, there are many regions of task-space-time that cannot be explored separately by human or machine alone. We have seen earlier how these regions can be envisaged as energetic 'landscapes' of constraint and opportunity. What we have to appreciate is that humans have evolved specifically to deal with one form of 'landscape' and in doing so have accumulated a wealth of context specific expertise and capabilities.

2.6.2   The Edge of Chaos[4]

The hallmark of the expert is the ability to operate at the edge of chaos without falling into chaos nor receding to immutability. At either end of the continuum expressed in Figure 2.8, meaningful exploration is not possible. As can be seen, at one end of the continuum we have an absolute minimum. In this situation, the system or person has reached such a state in which they are unable to move of their own volition. Here, adaptation fails. More importantly, exploration has failed and knowledge of one's past and expectations of one's future are of no value (Holland, 1992) unless some exogenous change is enacted. The other end of the continuum is chaos and eventually random noise. Again adaptation, exploration, feedback and projection are of no value since output is not connected to input. Between these extremes we have

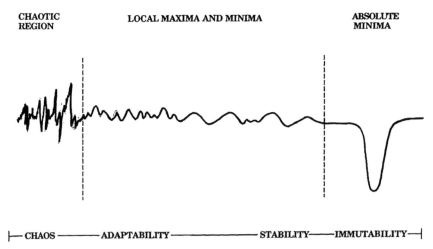

CHAOTIC   LOCAL MAXIMA AND MINIMA   ABSOLUTE
REGION   MINIMA

├── CHAOS ──────── ADAPTABILITY ───────────── STABILITY────IMMUTABILITY──┤

Figure 2.8. *Cross-section of an energetic landscape from pure randomness beyond the chaotic at one extreme to terminal stasis at the other. Adaptation fails at one extreme for lack of stability, it fails at the other extreme for lack of motility.*

---

[4]The phrase, and an intriguing phrase it is, comes from the work of the Santa Fe Institute group doing work on complexity and self-organization. The originators of the phrase N. Packard and C. Langton are those to whom attribution is due (see Kauffman, 1993).

regions of local minima and regions of chaos (Thompson & Stewart, 1986). The contention here is that experts and by extension 'skilled' or advanced human-machine systems, exist and will need to explore the edge of chaos.

Should we cross chaotic thresholds, unintended outcomes are frequently the result of our actions. We then withdraw to more stable regions, if we can, and adopt more conservative strategies immediately following failure. Subsequently, we return to the frontier as we recover the necessary confidence. However, it is this process of hunting at the edge of chaos that allows the most thorough exploration of the possibilities the 'landscape' presents and represents the epitome of adaptability. This is somewhat analogous to the test pilot's task of "pushing the envelope" (Wolfe, 1979) and can be a design principle for "direct" displays which seek to make the thresholds in such landscapes manifest (see Hansen, 1994; and also Vicente, 1995). Some contend that it is only in exploring the boundaries of existing ecological niches that life itself can develop, and by analogy we can expect the edge of chaos to be a breeding ground for the co-evolution of future human-machine systems.

Engineering for success includes: Immediate recovery from error, when error is recognized; explicit recognition of the purposive function, to understand and express the generative and explorative nature of these systems; and a facile interchange of knowledge and information between human and machine elements in languages each understand (Hancock & Chignell, 1989). In light of the adaptive and explorative nature of the enterprise, it is clear that these necessities arise from co-evolving human-machine systems that push the envelope of existing technology and knowledge.

---

7. If we are to safely explore the edges of chaos in emergent 'human-machine' landscapes, we must 'engineer' in the success of the many strategies that originally made humans so successful.

It is critically important to recognize that the 'landscape' or playground in which humans evolved is both *qualitatively* and *quantitatively* different from the 'landscapes' that occur when humans work with machines in complex systems. That is, while we were experts for the vast majority of our development, technology has radically altered the nature of the 'landscapes' in which we operate. So, it is not merely the outward manifestation of change in the physical environment (the dark satanic mills), but it is the energetic nature of the operational workspaces which have been altered.

Some would argue that the 'laws' of nature are immutable and thus the same constraints apply fundamentally to human-machine systems as they do to humans alone. This is a seductive appeal since parsimony and elegance are so prized by the scientific enterprise. Yet such constraints are as much on the mind as on the physical body and emergent technologies offer realms of exploration in which our hard-won and prized laws no longer apply unequivocally, such a technology is virtual reality which we examine below. Whatever we take into 'possible worlds' or advanced human-machine systems, we still take all the residual characteristics of humans, which made them so dominant in the previous conditions, and now transfer them to 'new' conditions. Little wonder that many human capabilities do not transfer well. Further, in altering the energetic 'landscape,' technology introduces a number of pitfalls, *sui generis*. It is not merely that the steamroller of technology rearranges the landscape, BUT that it puts in its own (frequently hidden) catastrophic potential. Both Perrow (1984) and Reason (1990) have articulated forms of system error or 'resident pathogens' which are so easily recognized *post hoc* but frequently so hard to identify *a priori*.

We may be able to gain insight into the potential failure of human-machine systems from some recent work on biological organization. Kauffman (1993), in his recent text, transposed Raup's (1986) data, concerning extinction events in the Phanerozoic, into a ln-ln plot. This illustration is reproduced in Figure 2.9. As can be seen, there are few large extinction events when many families are destroyed and many more minor extinction events when few families are destroyed. Kauffman (1993, p. 268) observed that:

> *"A first general conclusion is the insight that* <u>*coevolutionary avalanches propagate*</u> *through ecosystems, that such avalanches have characteristic frequency-versus-size distributions which change depending on the parameters of the system. In particular, the distribution of avalanche sizes depends on how solid the frozen state is. If we tentatively accept Raup's data as weak evidence, the frozen state is modestly firm. Using Raup's data and improved evidence, we may ultimately be able to build a theory linking both ecosystems structure and extent of external perturbations to the size distribution of coevolutionary avalanches and to such phenomena as the distribution of extinction events.*

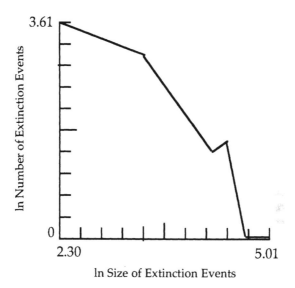

Figure 2.9. *Kauffman's (1993) replot of Raup's (1986) data on the log number of extinction events versus the log size of those events. If a derived power law relationship held for human-machine failures also, it would imply a commonality between biological and technical systems. This is an empirically testable proposition of particular significance.*

A second and critical result is this: <u>Perturbations of the same initial size can unleash avalanches on a large variety of length scales</u>. This conclusion is clear and important. In these simulations, the perturbation in each case is a change in the external world of a single randomly chosen species in the ecosystem. If we may tentatively assume that avalanches can be linked to extinction events, then these results strongly suggest that <u>uniform</u> alterations in the external world during evolution can cause a <u>diversity</u> of sizes in extinction events. This possibility stands in contrast to the generally held hypotheses that small and large extinction events are associated with small and cataclysmic changes in the external world. Since the external environment has almost certainly undergone changes on a variety of scales, I do not wish to assert that high variability in extinction sizes does not in part reflect a heterogeneity in intensity of causes. But these results place part of the responsibility for extinction-size diversity on the dynamics of coupled ecosystems and on the ways in which damage propagates.

But the third conclusion is the most important: Raup's data suggest that ecosystems do coevolve to the edge of chaos. More precisely, the data suggest that ecosystems are slightly within the frozen regime. Thus we ourselves hover on the edge of a new view of coordinated coevolutionary processes among interacting adaptive entities."

If we can consider human-machine systems as interacting adaptive agencies, a suggestion promulgated here, then the insights rendered by Kauffman and Raup have particular pertinence for all of human factors. Given the power law relationship, as suggested by Raup's data, then an empirically testable proposition would relate the frequency of failure in complex technical systems to the size of such failures. Affirmation of this observation would support the notion that technology is bound by the same constraints as biological systems and hence insights about the latter, their adaptation, proliferation and failure may be directly applied. Denial of the power law assertion would question the *qualitative* way in which human-machine assemblies resemble ecosystems and would imply the need for additional principles. Our persuasion is that there is much in common but potential differences make assertions of equivalency hazardous at present.

2.6.3   Virtual Reality

While laws by definition imply ubiquity, imagination is not bound by the constraints on matter. Hence, we can, and frequently have, created worlds in which the suspension of laws is at least as interesting as their application. Virtual reality promises a further step along this road of freedom. However, current use of virtual displays are much directed to the practical and the relationship to direct displays is clear. The first step in using technology to enhance directness is an explicit recognition, intrinsic to the Fitts' list, that we will be at greatest advantage when using the 'best' capabilities of both human and machine. However, unlike earlier attempts at allocation of function, we can no longer talk of discrete division of labor, but must consider more a companion and complementary approach (Jordan, 1963) in which sharing of effort and action is emphasized. One of the pre-eminent human capabilities is navigation in a spatial world. Collision detection, collision avoidance, object manipulation and spatial orientation are some of those functions which, after millennia of refinement, it would be unwise to lose as characteristics of human performance in system control. For this reason we believe that some form of virtual reality is likely to be inherent in the human-machine interface of the near future. While there are many barriers to facile interaction at present (see Kozak, Hancock, Arthur, & Chrysler, 1993), the undoubted advantage of being able to use visuo-spatial abilities is the dominant motivation for future use of virtual reality.

In the past, technology has tended to distance the operator and reduce empathic awareness. However, newer technologies may repair some of the damage done by earlier technological intervention. For instance, the graphical user interface, with its concept of direct manipulation restores a sense of directness when interacting with the file system of a computer. Similarly, the following principle may be applied.

> 8. Virtual reality can create a simulated world where the operator can again experience directness and empathic awareness, even in complex systems where there may in fact be several layers of automation between the operator and the underlying process.

However, we should not use VR only to recreate physical worlds that already exist. VR should be used to create possible worlds that enhance the probability of successful operation of complex systems. VR should also be used to modify displacement in space and time, and return operators to the region of humanrange, even in universalrange systems. We also envisage using VR to represent the energetic 'landscapes' we referred to earlier. One application that we are currently working on is direct expression of a risk-space (see Smith & Hancock, 1993). Thus VR can use all the best facets of human capabilities while allowing navigation and manipulation in combined system performance 'spaces.' These spaces do not have to be 'real', but can represent emergent properties of the system that are most relevant for its operation.

2.6.4   Multi-Operator, Multi-Machine Systems

In our discussion thus far we have talked of human-machine systems in the singular. Our discussion has largely centered upon one human and one-machine, something of the order of a single seat fighter aircraft. However, it is clear that most so-called 'complex' systems are multi-operator, multi-machine systems, in which co-operation across operators is as important as co-operation across machines, see Figure 2.10.

In the "Society of Mind," Minsky (1985) posited that the brain operates rather like a group of individual elements each clamoring to be heard. Similarly, many evolutionary arguments militate for some degree of progressive integration of separate elements. Using such a notion as a basis, we can then project that in so-called 'cyberspace' it is possible to create a 'society of minds' in which multiple individuals are involved. However, in this artificial environment, operators do not have to be singly mapped to actors. That is, we can imagine one-to-many and many-to-one mappings in which the actions of a 'cyberspace' actor are the result of the collaboration of multiple real-world operators. We can envisage 'ghosts' on one's shoulder who are present but exert no influence, we can envisage 'parasites' that 'live' on differing entities but occasionally transfer themselves. Indeed, most of the conceptions of cyberspace activity rapidly adopt biological analogies. In respect to our present endeavor, we expect that collaboration would include machine 'companions.' At that point, we expect that such interaction would initiate the first beginnings of an emergent entity which we have called "The Supercritical Society."

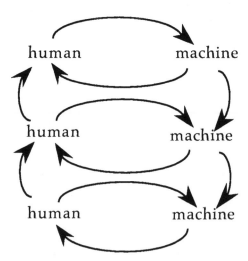

Figure 2.10. *Multiple interaction loops for multiple human-machine agents. Note also this can take the form of an expanding spiral, as in Figure 2.3.*

2.6.5 The Supercritical Society

The present interconnectedness of complex technical systems lays the base for an emergent entity. Many human operators input to systems which then influence the operations of other operators and other systems. It is therefore clear that design of appropriate multi-operator, multi-machine systems will be a major challenge and task for human factors in the twenty-first century. The critical first step will be to specify what such an emergent entity like a supercritical society might be like and might aim to do. We do not, at present, have a strong conception of its nature, function, or supra-emergent properties. Certainly, such an agglomerate will be more global in its consciousness and may focus on questions of collective interest such as environmental balance and meta-knowledge integration. However, regardless of what form the future supercritical society takes it is clear that it will have a fundamental impact on each contributing individual and the way they relate to their world. Indeed, it will become progressively more difficult to distinguish where an individual mind ends and a machine begins (see Anderson, 1964; Moravec, 1988). Yet, in a real sense technology has already created the people we are and we will no doubt adapt to the technologies of the supercritical society as we have done, with varying degrees of success, to all previous technologies.

2.7 Summary

Human factors is the critical bridge between people and technology. However, it supersedes even this role in its function as the directed and conscious way in which we manipulate our environment and by implication our future selves. Gibson (1979) himself had much to say about this technical activity. He noted:

> "In the last few thousand years, as everybody now realizes, the very face of the earth has been modified by man. The layout of surfaces has been changed, by cutting, clearing, leveling, paving, and building. Natural deserts and mountains, swamps and rivers, forests and plains still exist, but they are being encroached upon and reshaped by man-made layouts. Moreover, the _substances_ of the environment have been partly converted from the natural materials of the earth into various kinds of artificial materials such as bronze, iron, concrete, and bread. Even the _medium_ of the environment — the air for us and the water for

fish—is becoming slowly altered despite the restorative cycles that yielded a steady state for millions of years prior to man.

Why has man changed the shapes and substances of his environment? To change what it affords him. He has made more available what benefits him and less pressing what injures him. In making life easier for himself, of course, he has made life harder for most of the other animals. Over the millennia, he has made it easier for himself to get food, easier to keep warm, easier to see at night, easier to get about, and easier to train his offspring.

This is not a new environment—an artificial environment distinct from the natural environment—but the same old environment modified by man. It is a mistake to separate the natural from the artificial as if there were two environments; artifacts have to be manufactured from natural substances. It is also a mistake to separate the cultural environment from the natural environment, as if there were a world of mental products distinct from the world of material products. There is only one world, however diverse, and all animals live in it, although we human animals have altered it to suit ourselves. We have done so wastefully, thoughtlessly, and, if we do not mend our ways, fatally.

The fundamentals of the environment—the substances, the medium, and the surfaces—are the same for all animals. No matter how powerful men become they are not going to alter the fact of earth, air, and water—the lithosphere, the atmosphere, and the hydrosphere, together with the interfaces that separate them. For terrestrial animals like us, the earth and the sky are a basic structure on which all lesser structures depend. We cannot change it. We all fit into the substructures of the environment in our various ways, for we were all, in fact, formed by them. We are created by the world we live in."(Gibson, 1979, pp. 129-130).

We were indeed created by the world we live in, but we need to recognize the reciprocal effect in our creation of that world far more explicitly. In his essay on the origins of human factors, Karl Smith (1987) wrote:

"It is the science of fitting the job to the worker and learner as well as fitting machines to people. It transcends limited engineering formulations of the user-machine relationship and deals preeminently with how people have created themselves by systematically human-factoring their environment."

In a similar vein, Gregory (1981) has argued that:

> "The immense importance of technology in moulding how we think has implications still only dimly appreciated, ... It does, however, imply that any psychology based only on our biological origins is going to be inadequate: what is amazing about man is how far he has escaped his origins. This is through the use of tools, and the effect on us of technology that the tools have created."

While it is important to consider the application of the principles of ecological psychology to human-machine systems, it is facile to consider a science of behavior that fails to recognize the irrevocable effect that technology has had and does have on all individuals. In that way, the present text seeks to contribute to the ecological approach in a non-trivial manner by encouraging further consideration of the manufactured, as well as so called 'natural' environment. As ecological psychologists have been accused of being Aristotelian by nature, it is perhaps apposite to give Aristotle the last word. To paraphrase one of his assertions, he noted that:

> "We cannot directly will to be different than we are. It is therefore only by changing our environment that we change ourselves."

While we can argue over the relative veracity of this statement, we cannot deny that our environment exerts a powerful influence in shaping ourselves. In contemporary society, there are many forces which conspire to push back the boundaries of exploration through the proliferation of technical capability. The individuals who act to reconcile these developments with human abilities are by contrast relatively few. Failures of this reconciliation process are chronicled in the spectacular system failures that regale the daily news media. Yet these acute insults to societal functioning may be relatively trivial compared with the chronic effects of continuing maladaptive relationships between people and the technical, manufactured environments in which they live. If it is true that our effective legacy lives though our progeny and the environment which we sculpt for them, then human factors cannot be considered just an appendage to the applied practice of psychology or engineering. Human factors is much more than a mere appliance science. Rather, it is an endeavor central to the success of the human enterprise itself. And we must treat it as such.

Acknowledgments

We would like to thank John Flach, Vic Riley, Kim Vicente, and Karl Newell for their insightful comments on an earlier version of this work.

2.8 References

Anderson, A. R. (Ed.), (1964). *Minds and machines*. Englewood Cliffs, NJ: Prentice-Hall.

Bennett, K. B., & Flach, J. M. (1992). Graphical displays: Implications for divided attention, focused attention and problem solving. *Human Factors, 34,* 513-533.

Bronowski, J. (1958). *William Blake*. London: Cox & Wyman Ltd.

Bronowski, J. (1978). *Magic, science, and civilization*. New York: Columbia University Press.

Flach, J. M., & Bennett, K. B. (1992). Graphical interfaces to complex systems: Separating the wheat from the chaff. *Proceedings of the Human Factors Society, 36,* 470-474.

Fraser, J. T. (Ed.), (1966). *The voices of time*. New York: Braziller.

Gibson, J. J. (1975). Events are perceivable but time is not. In J. T. Fraser and N. Lawrence (Eds.), *The study of time II.* (pp. 295-301). Berlin: Springer-Verlag.

Gibson, J. J. (1979). *The ecological approach to visual perception.* Boston: Houghton-Mifflin.

Gooddy, W. (1988). *Time and the nervous system.* New York: Praeger

Gregory, R. L (1981). *Mind in science*. London:Penguin.

Hancock, P. A. (1993a). What good do we really do? *Ergonomics in Design, 1,* 6-8.

Hancock, P. A. (1993b). *The natural philosophy of human factors.* Unpublished Manuscript.

Hancock, P. A. (1993c). Evaluating in-vehicle collision avoidance warning systems for IVHS. In E. J. Haug (Ed.), *Concurrent engineering: Tools and techniques for mechanical system design.* (pp. 947-958). Berlin: Springer-Verlag.

Hancock, P. A., & Chignell, M. H. (1987). Adaptive control in human-machine systems. In P. A. Hancock (Ed.), *Human factors psychology.* (pp. 305-345). Amsterdam: North-Holland.

Hancock, P. A., & Chignell, M. H. (1988). Mental workload dynamics in adaptive interface design. *IEEE Transactions on Systems, Man, and Cybernetics, 18,* 647-658.

Hancock, P. A., & Chignell, M. H. (Eds.), (1989). *Intelligent interfaces: Theory, research, and design.* Amsterdam: North Holland.

Hancock, P. A., Chignell, M. H., & Kerr, G. (August, 1988). Defining task complexity and task difficulty. *Proceedings of the XXIV International Congress of Psychology.* Sydney, Australia.

Hancock, P. A., & Newell, K. M. (1985). The movement speed-accuracy relationship in space-time. In H. Heuer, U. Kleinbeck, and K. H. Schmidt (Eds.), *Motor programming and acquisition.* (pp. 153-188). Berlin: Springer.

Hansen, J. P. (1994). Representation of system invariants by optical invariants in configural displays for process control. In P. A. Hancock, J. M. Flach, J. K. Caird and K. J. Vicente (Eds.), *Local applications in the ecology of human-machine systems.* Hillsdale, NJ: Lawrence Erlbaum Associates.

Holland, J. H. (1992). *Adaptation in natural and artificial systems.* Cambridge, MIT Press (First Edition, University of Michigan, 1975).

Iberall, A. S. (1992). Does intention have a characteristic fast time scale? *Ecological Psychology, 4,* 39-61.

James. W. (1890). *Principles of psychology.* New York: Holt.

Jordan, N. (1963). Allocation of functions between man and machines in automated systems. *Journal of Applied Psychology, 47,* 161-165.

Kauffman, S. A. (1993). *The origins of order: Self-organization and selection in evolution.* Oxford: Oxford University Press.

Kirlik, A. (1994). Requirements for psychological models to support design: Towards ecological task analysis. In J. M. Flach, P. A. Hancock, J. K. Caird, and K. J. Vicente (Eds.), *Global perspectives on the ecology of human machine systems.* Hillsdale, NJ: Lawrence Erlbaum Associates.

Koestler, A. (1959). *The sleepwalkers: A history of man's changing vision of the universe.* New York: Hutchinson.

Kozak, J. J., Hancock, P. A., Chrysler, S., & Arthur, E. (1993). Transfer of training from virtual reality. *Ergonomics, 36,* 777-784.

Kugler, P. N., & Turvey, M. T. (1987). *Information, natural law, and the self-assembly of rhythmic movement.* Hillsdale, NJ, Lawrence Erlbaum.

Locke, J. (1690). *An essay concerning human understanding.* New York: Dover.

McPhee, J. (1989). *The control of nature.* New York: Farrar, Straus, Giroux.

Minkowski, H. (1908/1923). Space and time. In H. A. Lorentz, A. Einstein, H. Minkowski., and H. Weyl (Eds.), *The principles of relativity*. London: Dover.

Minsky, M. (1985). *The society of mind*. New York: Simon and Schuster.

Moravec, H. (1988). *Mind children: The future of robot and human intelligence*. Cambridge: Harvard University Press.

Moray, N. (1986) Monitoring behavior and supervisory control. In K. R. Boff, L. Kaufman, and J. P. Thomas (Eds.), *Handbook of perception and human performance*, Vol. II. New York: Wiley.

Morrison, P., Morrison, P., Eames, C., & Eames, R. (1982). *Powers of ten: About the relative size of things in the universe*. New York: Scientific American Library.

Newell, A. (1990). *Unified theories of cognition*. Cambridge, Massachusetts: Harvard University Press.

Newell, K. M. (1986). Constraints on the development of coordination. In M. G. Wade and H. T. A Whiting. (Eds.), *Motor development in children: Aspects of coordination and control*. (pp. 341-360). Dordrecht: Martinus Nijhoff.

Nickerson, R. S. (1992). *Looking ahead: Human factors challenges in a changing world*. Hillsdale, NJ: Lawrence Erlbaum.

Oakley, K. P. (1949). *Man the tool maker*. London: British Museum.

Parasuraman, R., Molloy, R., & Singh, I. (1993). Performance consequences of automation-induced complacency. *International Journal of Aviation Psychology, 3*, 1-23.

Perrow, C. (1984). *Normal accidents: Living with high-risk technologies*. New York: Basic Books.

Pirsig, R. M. (1974). *Zen and the art of motorcycle maintainance: An inquiry into values*. London: Bodley Head.

Pittenger, J. (September. 1991). *Elaborating tau*. Paper presented at the 25th Reunion of the University of Minnesota, Center for Learning, Perception, and Cognition. Minneapolis, MN.

Poppel, E. (1988). *Mindworks: Time and conscious experience*. Boston: Harcourt, Brace, Jovanovich.

Powers, W. T. (1974). *Behavior: The control of perception*. London: Wildwood House.

Powers, W. T. (1978). Quantative analysis of purposive systems: Some spadework at the foundations of scientific psychology. *Psychological Review, 85*, 417-435.

Rasmussen, J. (1986). *Information processing and human-machine interaction: An approach to cognitive engineering*. Amsterdam: North-Holland.

Raup, D. M. (1986). Biological extinction in earth history. *Science, 231,* 1528.

Reason, J. (1990). *Human error.* Cambridge: Cambridge University Press.

Reed, E. S. (1988). *James J. Gibson and the psychology of perception.* New Haven: Yale University Press.

Rifkin, J. (1988). *Time wars.* New York: Bantam

Schroots, J. J. F., & Birren, J. E. (1990). Concepts of time and aging in science. In J. E. Birren and K. W. Schaie (Eds.), *Handbook of the psychology of aging.* (pp. 45-64). San Diego, CA: Academic Press.

Shaw, R., & Kinsella-Shaw, J. (1988). Ecological mechanics: A physical geometry for intentional constraints. *Human Movement Science, 7,* 155-200.

Simon, H. A. (1981). *The science of the artificial, 2nd ed.* Cambridge, MA: MIT press.

Smith, K., & Hancock, P. A. (January, 1993). *Situation awareness is adaptive, externally-directed consciousness.* Proceedings of the Conference on Situational Awareness, Orlando, FL.

Smith, K. U. (1987). Origins of human factors science. *Human Factors Society Bulletin, 30* (4), 1-3.

Stevenson, R. L. (1883). *Treasure island.* London: Puffin, 1946.

Stroud, J. M. (1955). The fine structure of psychological time. In H. Quastler (Ed.), *Information theory in psychology,* Illinois Free Press.

Swenson, R., & Turvey, M. T. (1991). Thermodynamic reasons for perception-action cycles. *Ecological Psychology, 3,* 317-348.

Thompson, J. M., & Stewart, H. B. (1986). *Nonlinear dynamics and chaos.* New York: Wiley.

Turing, A. M. (1950). Computing machinery and intelligence, *Mind, 59,* 433-460.

Turvey, M. T. (1992). Affordances and prospective control: An outline of the ontology. *Ecological Psychology, 4,* 173-187.

Vicente, K .J. (1995). A few implications of an ecological approach to human factors. In J. M. Flach, P. A. Hancock, J. K. Caird and K. J. Vicente (Eds.), *Global perspectives on the ecology of human-machine systems.* Hillsdale, NJ: Lawrence Erlbaum Associates.

Vicente, K. J., & Rasmussen, J. (1990). Ecology of human-machine systems II: Mediating 'direct perception' in complex work domains. *Ecological Psychology, 2,* 207-249.

Vroon, P. A. (1974). Is there a quantum in duration experience. *American Journal of Psychology, 87,* 237-245.

Wolfe, T. (1979). *The right stuff.* New York: Farrar, Straus, Giroux.

Chapter 3

A Few Implications of an Ecological Approach to Human Factors

Kim J. Vicente
University of Toronto

3.0   Introduction

Recently, a relatively new approach to human factors based on an ecological perspective has been receiving an increasing amount of attention (Flach, 1989, 1990; Gaver, 1991; Kirlik, Miller, & Jagacinski, 1993; Vicente, 1991; Vicente & Rasmussen, 1990, 1992). Some have argued that an ecological approach to human factors is necessary if the gap between the worlds of basic research and applied design are to be bridged. A perspective that claims to contribute to the integration of basic and applied concerns deserves serious consideration. The specific purpose of this chapter is to describe in detail the additional benefits of adopting an ecological approach to human factors. First, a general overview of the ecological approach is presented. The remainder of the chapter will be structured into four sections. The first three sections outline the implications of the ecological perspective for human performance modeling, task analysis, and human factors experimentation. The final section provides an example from the area of cognitive engineering illustrating how application of the ecological approach to human factors can lead to fruitful research.

3.1   What Defines the Ecological Approach?

The *ecological* perspective takes its name from an approach to psychology that was advanced first by Brunswik (1956) and subsequently by Gibson (1966, 1979). Although differing in the details

54

of their respective formulations, these researchers shared the view of psychology as the study of the interaction between the human organism and its environment. This approach contrasts with the more traditional view of psychology as the study of the human organism, which was and continues to be the more prevalent perspective (e.g., Banaji & Crowder, 1991). This latter view will be referred to as the *organismic* approach. These two views of psychology — ecological and organismic — lead to two very different conceptions of human factors research.

When applied to human factors, the ecological approach suggests that the fundamental unit of analysis is the human-machine system (Flach, 1989). The human operator and work environment are reciprocally coupled and cannot be studied independently of one another. As a result, an ecological approach to human factors begins by studying the constraints in the environment (i.e., task or work domain) that are relevant to the operator. Not any description of the environment will do, however. To be useful for understanding behavior, such a description must be defined with respect to system goals (see the discussions of Gibson's, 1979, concept of *affordances* in Flach, 1989; and Vicente & Rasmussen, 1990). This emphasis on describing the environment contrasts with the organismic approach which, instead of acknowledging the importance of context on behavior, attempts to minimize or, occasionally even worse, ignore the influence of the environment. Meister (1989) has also drawn a distinction between the ecological and organismic approaches, referring to them as the *systems* and *psychological* perspectives, respectively. In fact, the ecological approach has much in common with systems theory, as Gibson (1979, p. 2) himself pointed out.

What implications does the ecological approach have for the practice of human factors?

3.2   Implications for Human Performance Modeling

One of the areas that the ecological approach can impact is that of human performance modeling. A good way to point out these implications is to begin with Simon's (1981) description of an ant traversing the terrain of a beach: "Viewed as a geometric figure, the ant's path is irregular, complex, hard to describe. But its complexity is really a complexity in the surface of the beach, not a complexity in the ant" (p. 64). The implication of this parable is of fundamental importance: In order to understand behavior , it is necessary to have separate but commensurate descriptions of the functional "landscape" in which behavior takes place and the psychological mechanisms that are generating that behavior.

Within the context of human factors, this suggests that when developing a model of human performance, it is important to have a functional description of the domain in which behavior is taking place (i.e., to have a description of "the beach"). This is essential if one is to be able to correctly partial out and properly attribute the respective influences of the environment and the organism. In the earlier case, the ant's behavior is constrained by the layout of the beach. The ant itself is probably contributing very little. However, if researchers did not have a description of the beach, they could fall into the trap of attributing more information processing capabilities to the ant than it actually possesses. That is, one could incorrectly attribute the complexity in the ant's trajectory to complex psychological mechanisms inside the ant (see Boer & Kugler, 1977, who describe another example of this class of error in the area of biology). Similarly, adopting an organismic approach could lead one to attribute elaborate mental constructs and processes to the human, instead of first finding out how the environment can inform and thereby support skilled behavior. The point has been elegantly made by Neisser (1987): "If we do not have a good account of the information that perceivers are actually using, our hypothetical models of their 'information processing' are almost sure to be wrong. If we *do* have such an account, however, such models may turn out to be almost unnecessary" (p. 11).

Kirlik, Miller, and Jagacinski (1993) provide a good example of how a goal-relevant description of the environment can lead to a parsimonious account of skilled human performance. Their subjects were engaged in a complex, supervisory control task that one might think would require a considerable amount of cognitive processing. However, Kirlik et al. found that it was possible to account for highly skilled behavior in this domain with a model that relied almost exclusively on perception and action. This approach was made possible by first describing the environment as a dynamic set of constraints on productive action and then by identifying the displayed information capable of specifying these constraints. A model with perceptual mechanisms attuned to these sources of information was then constructed. Experiments revealed that the model provided a good fit to experts' behavior. Thus, Kirlik et al. were able to show that expert behavior in such a complex environment could be modeled by assuming that subjects' actions were guided by environmental constraints that were perceptually specified, rather than being selected by an analytical problem solving process, despite the fact that the environment did not uniquely constrain behavior (i.e., there were many different ways to perform the task). This parsimonious modeling approach was presumably made possible because subjects had sufficient practice to allow them to identify perceptual information capable of guiding action. In other systems, a stronger emphasis on problem solving may be required if experts are faced with unfamiliar and unanticipated situations, or if the perceptual information they receive is impoverished. In such cases, psychological constraints will play a larger role in explaining behavior. Even then, however, it is important that the human factors analyst begin by describing the environment so that the sources of constraint can be correctly partitioned and properly attributed.

One implication of an ecological approach, then, is that conducting a careful analysis of the task environment in which behavior takes place can lead to more parsimonious models of human performance. Such an analysis will help avoid the potential trap associated with an organismic approach, namely, developing a theory of the operator which is actually a theory of the task.

3.3   Implications for Task Analysis

Given the emphasis on analyzing the environment, it should come as no surprise that the ecological approach also has important implications for task analysis. Traditional task analysis methods (cf. Meister, 1985) typically result in a description consisting of a single, temporal sequence of overt behaviors. This description represents the normative way in which the task is to be performed. A closer look at Simon's parable suggests that this method of task analysis does not account for the variability in behavior that is observed in complex systems. For example, putting the ant on a different part of the beach will result in a different set of behaviors to reach the same goal state. Furthermore, introducing external disturbances (e.g., wind gusts) will also cause the ant to take a different set of actions to reach the same end point because it must compensate for the negative influence of the disturbance. Finally, if it were the case that ants had different "strategies" (i.e., methods) for navigating on the beach, then one would again expect performance to vary as a function of the strategy adopted. Thus, there are at least three factors that cause the ant's behavior to vary from one trajectory to the next for a given task: changes in initial conditions, unpredictable disturbances, and the use of multiple strategies. A description of behavior in terms of a single sequence of overt actions, as provided by traditional task analysis, could not capture the variability in the ant's behavior.

The generic sources of variability identified in the ant's trajectory can also be found in the behavior of operators of complex human-machine systems. First, because it is often not possible to predict the exact state that the system will be in when a particular task will be performed, it is very difficult to anticipate what actions need to be carried out to achieve the desired goal state. Furthermore, it may not even be possible to anticipate the instigating event requiring action (e.g., unpredictable faults in process control plants, or constantly changing demands in flexible manufacturing systems). If the instigating event cannot be predicted, then one would be very hard pressed to determine what the appropriate corrective actions would be. As a result, in these situations it is virtually impossible to conduct a traditional task analysis based on description of behavior. Second, in open systems that are subject to external disturbances, there will again be a great deal of variability in performance because the operator must counteract the disturbance to satisfy system goals. Because the disturbance is unpredictable, the operator's compensatory actions also cannot be predicted. Third, it has also been established that the same task can be performed in very different ways as a function of the strategies adopted by operators (Pejtersen, 1988; Rasmussen, 1986). This means that behavior will vary as a function of the operators' preferred strategies, not only across individuals, but also within individuals on different occasions as well. The more complex and open-ended a work domain is, the more influential all of these factors will be. For example, in investigating how professional technicians trouble shoot electronic equipment, Rasmussen (1986) found that no two sequences of actions were identical, even through the subjects were performing the same task every time (i.e., finding the faulty component). Clearly, traditional task analysis cannot capture such richness of behavior.

To account for the variability of real-life performance, a task analysis methodology needs to provide separate descriptions of at least three classes of constraints: (a) the functional problem space in which behavior takes place (the "beach" described in the previous section), (b) the generic tasks that are to be accomplished by the operator (the products to be achieved), and (c) the set of strategies that operators can use to carry out those tasks (the processes by which they can be achieved). At any particular time, the specific pattern of behavior exhibited by an operator will emerge from the interaction between these three classes of constraints. In other words, actions will vary as a function of what part of the system is being worked on (including its state), which work activity is currently being pursued, and what strategy has been selected. At the very least, these three factors need to be taken into account if the limitations of traditional task analysis are to be overcome.

It should not be forgotten that the choice of task analysis method has important implications for design. Adopting a traditional task analysis methodology based on a single sequence of behaviors as a guide for design will result in an artifact that will support one way of performing the task. In complex, real-life situations, however, the task will be done in different ways as a function of the current context and the operator's subjective preferences. Therefore, traditional task analysis may result in an artifact that will not support, or perhaps may even impede, alternate ways of achieving system goals. The design of a system that will accommodate and support the rich variability in human performance must be based on a task analysis methodology that incorporates at least the three classes of constraints just listed. An example of such a methodology is briefly described later in the chapter (see also Rasmussen & Pejtersen, 1995).

3.4   Implications for Experimentation

One of the basic tenets of the ecological approach is that the environment has a strong influence on behavior. It follows, therefore, that if different researchers are investigating the same phenomenon, but use different tasks in their experiments, their results may conflict. Unless the influence of the context in which behavior takes place is directly addressed, the conflicting results will remain mysterious.

DeSanctis (1984) provided a revealing case study that illustrates this point. She reviewed a large number of studies investigating the use of computer graphics as decision aids. The basic question guiding most of this research is: Are computer graphics better than more traditional information presentation techniques (e.g., tables of numbers) in supporting effective decision making? DeSanctis's review indicates that this body of literature is filled with contradictory findings. There seems to be no consensus on the effects of computer graphics on decision making performance. Further investigation reveals that different studies, while addressing the same research question, adopted quite different decision problems. DeSanctis (1984) concluded that "the best method of data display may vary as a function of the task to be accomplished by the user" (p. 475).

The conflicting findings described by DeSanctis are a direct result of an organismic approach to experimentation. This is clearly revealed in the question being asked: Which method of data presentation is "better," A or B? This type of experimentation is characteristic of technology driven research (Woods & Roth, 1988). From an ecological perspective, the question has an obvious answer: It depends! This does not mean that experimental results are chaotic and unpredictable. Rather, the regularities in behavior only appear when they explicitly acknowledges the mediating effects of the task environment. Acknowledging the influence of the task on behavior would lead one to ask a different question to address the very same problem. Instead of asking which is "best," one would ask: Under what conditions is method A better, and under what conditions is B preferable?

The work of Benbasat, Dexter, and Todd (1986) provides an exemplary application of this type of approach to the use of computer graphics as decision aids. They conducted three experiments to evaluate the influence of graphical and color-coded information presentation on several decision making factors including decision quality, decision time, and information utilization. The collective results of this research program lead to the following conclusions:

> *"The influence of presentation mode on human performance and the perceived value of information is related to how well it supports the solution approach to a particular task. The benefits of graphics are limited to reducing decision making time but only when the graphical report has been designed to directly assist in solving the task. Multicolor reports aid in decision making, but only in specific circumstances, that is, their benefits are not pervasive. It appears that color is more advantageous when associated with graphical reports, for certain decision maker types, during learning periods, and in time constrained environments."* (Benbasat et al., 1986, p. 1094)

The clarity of the findings derived from this series of studies provides quite a contrast to the inconclusive and conflicting experimental evidence derived from the studies reviewed by DeSanctis (1984).

A natural consequence of adopting an ecological approach to human factors experimentation is that it reveals a new set of research issues that would not be addressed from an organismic perspective. For instance, given the emphasis that the ecological approach places on studying the task environment, it becomes important to determine if there are regularities in the way in which tasks affect behavior. In other words, is there a generic way of describing tasks which will allow one to predict the psychological resources that different classes of tasks will elicit? Clearly, an answer to this question would represent a significant contribution to both basic and applied human factors concerns.

The recent work of Hammond and colleagues represents a very important step toward addressing this problem. For instance, Hammond, Hamm, Grassia, and Pearson (1987) have developed a theoretical account of how task properties tend to induce different types of cognitive activities (intuitive, quasi rational, and analytical), and how performance can vary as a function of the correspondence between task properties and the type of cognitive activity in which subjects engage. A rigorous experiment was conducted using expert subjects to test the theory. The results lend support to their hypotheses, thereby shedding light on the relationship between task properties and different forms of cognitive activity. Hamm (1988) has conducted a complementary study investigating how analytic and intuitive cognitive activity can vary from moment to moment as a function of both task and subject characteristics. Finally, Hammond (1988) has taken up the important problem of developing a theory of tasks. Such a taxonomy is essential if one is to understand the mediating effects of task environments on behavior. The research program represented by these studies is a paradigmatic example of ecologically oriented basic research. (Not surprisingly, Hammond was a student of Brunswik's.)

Adopting an ecological approach also encourages one to address the converse question of whether there are any psychological properties that are invariant over different tasks. This would also represent an important contribution to our knowledge. However, answering this question requires that one partial out the behavioral details associated with the particular task that subjects were performing. Returning to the ant example, if the constraints due to the environment are partialled out, one may find some sort of regularity in the ant's behavior that is invariant across trajectories. A plausible candidate would be a tendency to follow the path of least effort. Such an invariant would never be visible to a human factors researcher taking an organismic approach. From that perspective, the ant's behavior would seem to be unpredictable and random, because its path would be quite different, depending on the landscape of the terrain it was traversing. From an ecological perspective, however, the ant may always be doing the same thing: traversing the path of least effort.

Viewing behavior as the interaction between an adaptive organism and its dynamic environment can not only help to resolve what seem to be inconclusive and conflicting experimental findings, but it can also encourage researchers to pursue questions that could represent significant contributions to our understanding. The important questions of how classes of tasks regularly influence behavior, and of whether there are patterns of behavior that are invariant over tasks, are only likely to be parsimoniously illuminated by adopting an ecological perspective.

3.5   Cognitive Engineering: An Ecological Frontier

So far, this chapter has tried to argue that adopting an ecological approach to human factors can have important implications. It can lead to models, methods, and experiments that differ considerably from those of traditional human factors. One point that has yet to be mentioned, however, is that an ecological approach to human factors already exists in the area which has come to be known as *cognitive engineering* (Rasmussen, 1986; Woods & Roth, 1988). Cognitive engineering deals with the human factors challenges associated with introducing information technology into complex work domains, such as power plants, air traffic control, flexible manufacturing systems, and hospitals. In dealing with these complex systems, researchers have been forced to confront the inadequacies of traditional human factors practices (Rasmussen, 1988) and in the process have independently adopted many of the fundamental tenets of the ecological approach (Flach, 1989, 1990; Vicente, 1991; Vicente & Rasmussen, 1990, 1992; Woods & Roth, 1988).

The work of Rasmussen (1986) is particularly notable since it attempts to address the three major issues described in this chapter (see also Rasmussen & Pejtersen, 1995). With regard to modeling, Rasmussen's abstraction hierarchy is a framework for describing the functional landscape ("the beach") in which behavior takes place in a goal-relevant manner, thereby allowing one to partition environmental constraints from psychological constraints. One can think of the abstraction hierarchy as describing the nested set of affordances in a work environment (cf. Vicente & Rasmussen, 1990). Rasmussen has also developed a comprehensive methodology for cognitive work analysis which overcomes the limitations of traditional task analysis by taking into account the variability of performance in real-life, complex work domains. All three sets of constraints described earlier in the section on task analysis are taken into account by this methodology. Finally, although very few steps have been taken in this direction, experimentation motivated by the conceptual tools developed by Rasmussen is beginning to appear. One example is briefly described in the following section (see Vicente, 1991, for a more detailed account).

3.5.1 An Example: Ecological Interface Design

The research described in this section focused on the design of interfaces for complex human-machine systems. Unfortunately, many questions associated with this applied problem remain to be investigated. A theoretical framework, called Ecological Interface Design (EID), was developed to address some of these topics (Vicente & Rasmussen, 1990 1992). EID is based on Rasmussen's (1986) abstraction hierarchy and skills, rules, and knowledge taxonomy and consists of three prescriptive design principles. The objective of these principles is to exploit the powerful capabilities of perception and action, while at the same time provide the necessary support for more effortful and error-prone problem-solving activities (see Vicente & Rasmussen, 1990, for a discussion of the conceptual ties between EID and ecological psychology). Initial research evaluated how well an interface based on the principles of EID allows operators to cope with problem solving activities associated with unfamiliar and unanticipated events. According to EID, to properly support such knowledge-based behavior, an interface should display the physical and functional properties of the work domain in the form of a multilevel representation based on the abstraction hierarchy. A review of the literature in this area (Vicente, 1991) revealed that no experiment has ever compared a multilevel interface based on an abstraction hierarchy representation with another type of interface. Thus, an experiment was undertaken to address this important research need.

The experiment was conducted within the context of DURESS (Dual Reservoir System Simulation), a thermal hydraulic process control simulation. The performance of two interfaces was compared: a traditional interface based on a physical (P) representation, and an EID interface based on a multilevel physical/functional (P+F) representation. To evaluate how well these two interfaces support knowledge-based problem solving activities, a methodology based on psychological research on the relationship between expertise and memory recall was adopted. Thus, subjects were presented with a dynamic scenario of DURESS behavior and were asked to diagnose the event and to recall the state of the system. There were three types of events: normal, fault, and random. Two groups of subjects were used: theoretical experts in thermal hydraulics and novices. Collectively, the findings are consistent with the following conclusion: An interface based on an abstraction hierarchy can provide more support for knowledge-based behavior than an interface based on physical variables alone. There was also some evidence to indicate that a certain amount of theoretical expertise is required to fully realize this interface advantage. Collectively, the results provide some initial support for the utility of the EID framework as well as indicating that Rasmussen's conceptual framework provides a coherent and productive framework for experimentation.

Although the research just described is limited in many ways (cf. Vicente, 1991), it does show that cognitive engineering concepts which are closely tied to those of ecological psychology can be meaningfully applied to practical human factors problems. Furthermore, the preliminary empirical evidence available suggests that the approach may be a useful one, thereby indicating that this research path is a fruitful one to pursue. Perhaps even more important is the fact that the research that has been conducted under the rubric of EID is beginning to help engineers in industry answer some pressing design problems. More specifically, this research has influenced the design of Toshiba's advanced control room for their next generation of nuclear power plants (Monta et al., 1991). Toshiba has explicitly adopted the EID framework proposed by Vicente and Rasmussen and some of the specific interface features designed by Vicente (1991) in the study just described. The fact that these ideas are convincing enough to have been adopted by industry indicates that cognitive engineering, or more broadly, an ecological approach to human factors, can help to bridge the gap between basic research and significant applied problems. This is noteworthy because traditional organismic approaches to human factors have had limited success in technology transfer to industry (cf. Flach, 1989, 1990).

3.6   Conclusions

The unique demands associated with complex systems have perhaps forced cognitive engineers to embrace the values of an ecological approach earlier than most human factors engineers, but this does not mean that the perspective is limited to this class of problems. The research of Mark and Dainoff (1988) on the ergonomics of chair design is evidence that the ecological approach can be applied to a wide range of human factors issues, not just those of complex human-machine systems (see also Dainoff and Mark's application, 1995, of Rasmussen's abstraction hierarchy to the ergonomic design of workplaces). It is also worthwhile noting that human-computer interaction researchers involved with designing interfaces for office systems are also beginning to realize the advantages of the ecological approach (Gaver, 1991). A glance at the wide range of topics addressed by the various contributors to this volume reinforces the notion that the ecological approach to human factors has a very wide scope of applicability.

Only the future will tell how much of an influence an ecological approach will have on human factors as a discipline. This chapter has tried to outline some of the reasons why the path is worthwhile pursuing. The initial indications are that adopting an ecological approach to human factors can lead to the development of theories, methods, and empirical work which builds on the efforts produced by more traditional approaches and thereby broadens the scope of the science of human factors.

Acknowledgments

I would like to thank John Flach, Peter Hancock, Kelly Harwood, Alex Kirlik, and Jens Rasmussen for many enlightening discussions and comments on the ideas presented in this chapter. This chapter has been adapted with permission from *Human Factors Society Bulletin*, Vol. 33, No. 11, 1990. Copyright @1990 by The Human Factors Society, Inc. All rights reserved.

3.7   References

Banaji, M. R., & Crowder, R. G. (1991). Some everyday thoughts on ecologically valid methods. *American Psychologist, 46*, 78–79.

Benbasat, I., Dexter, A. S., & Todd, P. (1986). An experimental program investigating color-enhanced and graphical information presentation: An integration of the findings. *Communications of the ACM, 29*, 1094–1105

Boer, C., & Kugler, P. (1977). Archetypal psychology is mythical realism. *Spring: An annual of archetypal psychology and Jungian thought* (pp. 131-152). Zurich: Spring Publications.

Brunswik, E. (1956). *Perception and the representative design of experiments* (2nd ed.) Berkeley: University of California Press.

Dainoff, M. J., & Mark, L. S. (1995). Use of a means-end abstraction hierarchy in conceptualizing the ergonomic design of workplaces. In J. M. Flach, P. A. Hancock, J. K. Caird, and K. J. Vicente (Eds.), *Global perspectives on the ecology of human-machine systems*. Hillsdale, NJ: Lawrence Erlbaum Associates.

DeSanctis, G. (1984). Computer graphics as decisions aids: Directions for research. *Decision Sciences, 15*, 463–487.

Flach, J. M. (1989). An ecological alternative to egg-sucking. *Human Factors Society Bulletin, 32 (9)*, 4–6.

Flach, J. M. (1990). The ecology of human-machine systems I: Introduction. *Ecological Psychology, 2*, 191–205.

Gaver, W. W. (1991). Technology affordances. In S. P. Robertson, G. M. Olson, and J. S. Olson (Eds.), *Reaching through technology: CHI '91 conference proceedings*. (pp. 79–84). New York: Association of Computing and Machinery.

Gibson, J. J. (1966). *The senses considered as perceptual systems*. Boston: Houghton-Mifflin.

Gibson, J. J. (1979). *The ecological approach to visual perception*. Boston: Houghton-Mifflin.

Hamm, R. M. (1988). Moment-by-moment variation in experts' analytic and intuitive cognitive activity. *IEEE Transactions on Systems, Man, and Cybernetics, SMC-18*, 757–776.

Hammond, K. R. (1988). Judgement and decision making in dynamic tasks. *Information and Decision Technologies, 14*, 3–14.

Hammond, K. R., Hamm, R. M., Grassia, J., & Pearson, T. (1987). Direct comparison of the efficacy of intuitive and analytical cognition in expert judgement. *IEEE Transactions on Systems, Man, and Cybernetics, SMC-17*, 753–770.

Kirlik, A., Miller, R.A., & Jagacinski, R.J. (1993). Supervisory control in a dynamic and uncertain environment II: A process model of skilled human-environment interaction. *IEEE Transactions on Systems, Man, and Cybernetics. SMC-23*, 929-952.

Mark, L. S., & Dainoff, M. J. (1988). An ecological framework for ergonomic research: The chair as a case in point. *Innovation, 7*, 8–11.

Meister, D. (1985). *Behavioral analysis and measurement methods*. New York: Wiley.

Meister, D. (1989). *Conceptual aspects of human factors*. Baltimore, MD: Johns Hopkins University Press.

Monta, K., Takizawa, Y., Hattori, Y., Hayashi, T., Sato, N., Itoh, J., Sakuma, A., & Yoshikawa, E. (1991). An intelligent man-machine system for BWR nuclear power plants. *Proceedings of AI 91: frontiers innovative computing for the nuclear industry* (pp. 383–392). Jackson, WY: ANS.

Neisser, U. (1987). From direct perception to conceptual structure. In U. Neisser (Ed.), *Concepts and conceptual development: Ecological and intellectual factors in categorization.* (pp. 11-24). Cambridge, MA: Cambridge University Press.

Pejtersen, A. M. (1988). Search strategies and database design for information retrieval in libraries. In L. P. Goodstein, H. B. Andersen, and S. E. Olsen (Eds.), *Tasks, errors and mental models: A festschrift to celebrate the 60th birthday of Professor Jens Rasmussen.* (pp. 171–190). London: Taylor & Francis.

Rasmussen, J. (1986). *Information processing and human-machine interaction: An approach to cognitive engineering.* New York: North-Holland.

Rasmussen, J. (1988). Information technology: A challenge to the Human Factors Society? *Human Factors Society Bulletin, 31(7),* 1–3.

Rasmussen, J., & Pejtersen, A. M. (1992). Virtual ecology of work.. In J. M. Flach, P. A. Hancock, J. K. Caird, and K. J. Vicente (Eds.), *Global perspectives on the ecology of human-machine systems.* Hillsdale, NJ: Lawrence Erlbaum Associates.

Simon, H. A. (1981). *The sciences of the artificial, 2nd ed.* Cambridge, MA: MIT Press.

Vicente, K. J. (1991). Supporting knowledge-based behavior through ecological interface design (Unpublished Dissertation). Urbana, IL: Department of Mechanical and Industrial Engineering, University of Illinois at Urbana-Champaign.

Vicente, K. J., & Rasmussen, J. (1992). Ecological interface design: Theoretical foundations. *IEEE Transactions on Systems, Man, and Cybernetics. SMC-22,* 589-606.

Vicente, K. J., & Rasmussen, J. (1990). The ecology of human-machine systems II: Mediating "direct perception" in complex work domains. *Ecological Psychology, 2,* 207–250.

Woods, D. D., & Roth, E. M. (1989). Cognitive engineering: Human problem solving with tools. *Human Factors, 30,* 415–430.

Chapter 4

Requirements for Psychological Models to Support Design: Toward Ecological Task Analysis

Alex Kirlik

Center for Human-Machine Systems Research
School of Industrial & Systems Engineering
Georgia Institute of Technology

4.0   Introduction

Modern psychology judges its progress and products by a variety of criteria. Reviewing a number of paradigms in current cognitive psychology, Claxton (1988) suggested that the research community gives no less than thirteen answers to the question: "How do you tell a good cognitive theory when you see one?" (pp. 1-31) Each of the 13 criteria he mentions (e.g., experimental, computational, evolutionary) has enough adherents so that research programs are judged successful, even if their products meet perhaps only 1 of these standards of merit. Research activity in current cognitive science thus resembles a massively parallel search, in which most of Claxton's 13 criteria for scientific success are suspended on any one search path so that individual research efforts can proceed unencumbered by a diverse set of otherwise paralyzing constraints. For example, in certain paradigms computational realization is the primary concern, mathematical formalization the major constraint in others, and in still others a necessary condition for a theoretical model may be a demonstration that the proposed cognitive mechanisms and processes could have emerged through human development or evolution. The eventual success of this divide-and-conquer venture, of course, hinges not so much on whether each of the many paradigms meets its own goals, but rather on whether we are somehow able to integrate the resulting array of research products into useful and coherent theory.

It is natural to wonder whether we have decomposed our research efforts in a way that will allow for eventual theoretical unification. One primary concern is whether the many research paradigms that comprise cognitive science are moving along diverging or converging paths. Perhaps this is a question best left for time to decide. I am concerned, however, that although strict and dogmatic adherence to a single scientific criterion may lead to individually successful hill climbing, when considered overall we may find we have all climbed different hills and, if anything, actually increased the difficulty of the journeys between us. A coherent, useful cognitive theory will have to meet a large number of constraints. Rarely, however, do good solutions to problems which involve meeting multiple constraints emerge by decomposing the problem via the constraints themselves. Knowing the least expensive restaurant in town, the one with the best food, and the one with the healthiest menu is not particularly helpful in allowing one to find a good square meal at a fair price.

The purpose of this chapter is to identify a set of necessary conditions for psychological models capable of supporting the design of environments to promote skillful and effective human activity, that is, cognitive engineering (Fischoff, Slovic, & Lichtenstein, 1978; Norman, 1986; Rasmussen, 1986; Woods & Roth, 1988). This effort is motivated by my own limited success in attempting to apply the products of cognitive science to cognitive engineering. My experiences have led me to believe that the central problem that needs to be overcome to make the products of cognitive science more relevant to design is identifying a more productive set of dimensions along which modeling efforts can be decomposed. To support cognitive engineering, the decomposition must be derived from an overall framework capable of ensuring that the resulting research products can be reassembled into a coherent theory useful for design.

A description of a solution to any problem, even if expressed only as a set of necessary conditions, plays a crucial role in formulating a problem decomposition capable of ensuring that the subproblem solutions can be effectively integrated. We have to know where we are going if we want to get there. In terms of the previous analogy, we have to know that our goal is a good square meal at a fair price in order to determine how to decompose the problem of finding an appropriate restaurant. A necessary step toward a more applicable cognitive science, therefore, is a statement of the set of constraints that must be met if a psychological model is to support design.

There may be no good reason to expect that the set of constraints that must be satisfied to support design are identical to the set of constraints cognitive scientists normally use to guide their scientific explorations. In fact, I will suggest that the necessary conditions for an acceptable psychological model in cognitive science are quite different than the necessary conditions for a psychological model capable of guiding design. Many of the difficulties involved with trying to apply cognitive science modeling arise out of this mismatch. Cognitive science has simply decomposed its central problem in a manner that is very unfortunate for the cognitive engineer.

As Carroll (1991) has noted in regard to the failure of psychology to meaningfully contribute to understanding the problem of human-computer interaction (HCI), the realization that the products of a basic science do not provide effective resources for application can provide important lessons for the basic science itself (also see Flach, 1990b; Gibson, 1967/1982; Neisser, 1976). The solution to the problem of creating a scientific basis for cognitive engineering is not merely one of improving the designer's access to research findings (e.g., Meister, 1989), moving research into naturalistic or operational contexts (e.g., Klein, 1989), or improving generalizablity from experimental results (e.g., Hammond, Hamm, & Grassia, 1986), although each of these goals is surely important. Rather, I am convinced that the solution must lie in a reformulation of the questions posed by basic psychological research itself: a reformulation driven by an understanding of the psychological nature of the design product and the knowledge that is required to create it.

4.1   The Psychological Nature of the Design Product

A standard modeling approach in cognitive psychology is to hold a task environment relatively fixed and to create a description of the cognitive activities underlying a person's behavior in that environment. The designer, on the other hand, is faced with the reverse challenge of creating an environment to elicit a desired behavior, with the ultimate goal being the creation of a design that is maximally consistent with the principles underlying how people skillfully and effectively interact with the world. In problem solving terms, the solution space for the scientist is a set of plausible cognitive theories, whereas the solution space for the designer is a set of technologically feasible environments. We can thus characterize the scientist's problem as a search among possible cognitive "solutions" to a given task and the designer's problem as a search among possible environments to obtain a given cognitive solution. These are symmetrical psychological problems of comparable subtlety and difficulty, requiring equally sophisticated empirical and theoretical methods. In this sense, a theoretical/applied dichotomy does not appear to be a faithful way of portraying the difference between the practices of cognitive psychology and environmental design.

The primary reason for the perpetuation of the theoretical/applied distinction is the lack of appreciation for the psychological nature of the design product. Although the scientist creates theories of cognitive function, it is assumed that the designer creates not theories but merely environments: a mix of hardware and software that is best conceived in technological rather than in psychological terms. But this perspective is based on an overly restrictive view of what the environment is, from the standpoint of understanding human behavior. Although a design product may be implemented in hardware and software, this is the wrong level at which to view the relevant features of that product, just as it would be wrong to look for the relevant features of a psychological theory in the software or ink in which it is realized.

As Carroll and Campbell (1989) noted, each design product is actually an instantiation of the designer's theories of how the environment influences how people behave, think, and skillfully perform, however rudimentary and fragmentary these theories may be. Although the (good) psychological theory is only implicit in the design of a (good) VCR interface, for example, it is nonetheless real in exactly the same sense that the physical theories and electrical engineering principles implicit in the design of the inner workings of the VCR are real. If one wants to understand or predict the functionality of the VCR, you had better know the operative physical theories underlying its design. Similarly, if you want to understand or predict human interaction with the VCR, one had better know the psychological theory underlying its design as well. Although the VCR can of course be looked at as an assemblage of physical matter, this is the wrong level at which to look for the relevant structure of the machine, either for understanding electromechanical function or for understanding user interaction. Both the electrical engineer and cognitive engineer structure the physical matter using organizing principles derived from theories within their own disciplines. It is the adequacy of these theories, rather than any facts solely about the physical form of the machine, that determine whether the VCR will play and whether the user can play it.

Each instance of human interaction with any artifact is thus a psychological experiment testing the assumptions embodied in the environmental design (cf. Wise's 1985, construal of an architectural design in terms of scientific hypotheses). Although it may be fashionable within the cognitive engineering community to bemoan how little guidance modern psychology provides the designer, the psychological nature of the design product is inescapable. The correct response to the current and unfortunate lack of applicable psychological research is not to attempt to do psychology-free design (because this is impossible — the design will not be apsychological but instead reflect the designer's "folk" psychological theory), but rather to ask what kinds of psychological models are needed to support cognitive engineering, and to begin the long range empirical and theoretical work necessary to realize them.

This is not to imply, of course, that cognitive engineers can wait for a more applicable psychology to emerge before making design commitments, or that they should turn all attention to modeling and away from the design of prototypes and expanding technological opportunities. It may be, as Braitenberg (1984) has suggested, that human capacities for synthesis far exceed capacities for analysis, in the sense that our creative products reflect a degree of implicit or tacit knowledge that is far more elaborate and rich than the knowledge we can explicitly state and formalize. The direct manipulation interface was not deduced in any interesting sense from psychological theory — in fact, a case could be made that we still have no unified psychological theory that would predict the profound superiority of direct manipulation over command interfaces for various tasks. It just so happened that in this case the designer had an implicit understanding of how people naturally interact with the world which was closer to the truth than any explicit and formal psychological theory available at that time. Although cognitive psychologists may be able to identify why command-line interfaces are inefficient in various ways, it was the designers and not psychological research that pointed toward environments that support more efficient interaction, a finding that should probably inform psychological research itself.

There is a catch, however, to this design-as-research strategy. Assuming a particular prototype of a design concept is successful, any useful generalizations which emerge from creating the prototype will be at the level of the psychological assumptions underlying the design, rather than at the level of the particular technologies used to implement the design. To return to a point made earlier, the hardware and software implementation is the wrong level to look for the relevant features of the design product. Especially in HCI, a vast amount of research effort has been expended trying to answer questions comparing various interface technologies, for example, design options such as scrolling windows, hypermedia, and so on. This research is of dubious value (see also Vicente, this volume), because the "it depends" answers produced by such efforts will only lead to a never ending series of technology-specific design principles, rather than a stable and generative theoretical account of human-environment interaction that can guide design in novel situations.

To profit from the design-as-research strategy, then, it is incumbent upon the researcher to make explicit the psychological assumptions that contributed to the success of the prototype system. A successful system demonstrates nothing other than its own success, unless the possibly implicit psychological theory underlying the design is articulated. Although forcing the researcher to articulate theoretical assumptions prior to environmental design (as would be demanded by traditional experimental psychology methodology) may actually impede progress – synthesis may be more efficient than analysis; the hope for *generalizable* conclusions from such demonstrations surely rides on whether the researcher can subsequently identify the psychological hypotheses that were validated by the success of the prototype. There is probably no alternative to traditional experimental methodology for this purpose. A research program using the design-as-research strategy must include both an initial synthesis phase followed by an analysis phase in which the implicit psychological theory guiding synthesis is made explicit, tested, and communicated.

4.2   Modeling to Support Design

I argued that a good psychological theory is an essential aid to design by discussing the psychological status of the design product and also by showing that the problem faced by the designer is not one of mere application, but is instead itself a theoretical problem comparable to that faced by the scientist. A search for environments to promote a particular mode of cognitive activity and behavior (the designer's task) is no more an applied endeavor than is a search for accounts of cognitive activity and behavior that are promoted by particular environments (the scientist's task). There is, however, an important difference between these two problems. The designer and scientist search in opposite directions; one reasons over environmental models, whereas the other reasons over cognitive models. It should be expected, therefore, that different types of heuristic guidance will be necessary to direct search in the two cases. As a result, the theories that best provide heuristic guidance to the scientist will have different properties than the theories that best provide heuristic guidance to the cognitive engineer. I now examine how the theories that would best support reasoning over environments might differ from the theories that would best support reasoning across cognitive activities.

Much of the current understanding of cognitive-level, human-environment interaction consists of a set of somewhat independent environment-process-behavior triples, each of which provides a psychological model of how a person might achieve a particular behavior in a specified environment (cf. Hancock & Warm, 1989). When the difficult but important job of integrating this knowledge into coherent theory is attempted, efforts typically focus on integrating across the process and behavior dimensions rather than across the environment dimension. The result is that understanding the environmental contribution to behavior is a largely ignored component in the theoretical unification. One approach, for example, to achieving theoretical unification of this set of triples is to integrate across the behavior dimension. The results here are powerful, typically hybrid "cognitive architectures" (Card & Newell, 1989). These general purpose cognitive frameworks have the processing resources to produce a wide variety of behaviors, from simple motor responses to complex problem solving and planning. Yet, another approach is to integrate primarily across the process dimension in an attempt to show that the functionality of a wide variety of existing models can be subsumed under a single process modeling formalism. Cognitive models demonstrating how symbolic processing techniques can be implemented using neural network or connectionist formalisms are good examples of partial theoretical unification along the process dimension.

Theoretical integrations along the environment dimension, however, are hardly ever attempted but are critically needed to support the cross-environmental reasoning inherent in design. It should not come as a surprise that most cognitive psychologists are not overly concerned with this type of theoretical unification, because an acceptable scientific product is a model of behavior in a specified environment, and rarely is reasoning backward from cognitive theory to environment required. Except perhaps in experimental design itself, rarely is the cognitive psychologist forced to reason across environments in order to activate specified cognitive modes. Significant exceptions (i.e., attempts at theoretical integration across environmental influences on cognition and behavior) are Rasmussen's (1986) theory of multilevel environmental representation as reflected in the "abstraction hierarchy," and Hammond, Hamm, Grassia, and Pearson's (1987) efforts to obtain a rich set of environmental and task descriptors so that the cognitive mode underlying judgment behavior (e.g., analytical, intuitive) can be predicted and promoted through environmental manipulation. Only a unified theory of the environmental influences on cognition can guide the designer's search for environments to activate specified cognitive processes.

But what would a unified theory of the environment look like, and what types of guidance would it provide? Such a theory would integrate diverse knowledge of what the psychologically relevant aspects of the environment are for the purpose of trying to understand or predict human behavior and performance. One must, for example, determine when it will be appropriate to understand the environment in terms of stimuli and reinforcements as in behaviorism; cues, criterion, and feedback as in models of judgment; options, chance nodes, choice nodes, and probability distributions as in decision theory; initial states, goal states, and operators as in Newell and Simon's (1972) problem-solving theory; affordances or constraints on action as in Gibson's (1979) ecological theory; system state variables and differential equations as in manual control theory; and so on.

Each of the above forms of environmental description has its place. No single representation of environmental structure will do justice to understanding the many different forms of cognition and behavior observed in complex human-environment interaction (Rasmussen, 1986). The reflection of environmental structure in behavior is manifest in various ways, and each way is suggestive of a different model that best describes the structure of the environment to which productive behavior must be sensitive. A large part of design activity, in fact, can be viewed as the selection of appropriate environmental descriptions. In some cases, the cognitive engineer faces the problem of selecting an environmental description for an existing candidate design which will assist in predicting the cognitive activity and behavior the environmental design will promote. In other cases the cognitive engineer can operate earlier in the design cycle, and the central problem will be to create a design concept, expressed as an environmental model, that promotes a specified mode of cognitive activity maximally consistent with the demands of a task. Guidance for both of these cognitive engineering activities can only come from theoretical frameworks that support the designer's reasoning over alternative environmental models.

Psychology may already have the rudiments of such a theory, but perhaps oddly this knowledge is expressed not so much in existing models of cognitive activity, but rather in the process of designing experiments capable of successfully activating those cognitive activities for scientific study. That is, it is the often tacit and unformalized knowledge guiding experimental design in studies of cognition that approaches the type of understanding needed to reason effectively over environmental models. The knowledge underlying the experimenter's ability to promote a particular mode of cognitive activity is quite similar to the type of knowledge necessary to guide system design. Much has been made of the inability of basic experimental psychology research to guide design (e.g., Meister, 1984; Rouse, 1987), but perhaps the fundamental difficulty is that the knowledge the designer needs goes beyond the experimental findings, and it may approach the knowledge needed to have actually designed these experiments.

4.3   Issues in Environmental Modeling

Cognitive engineering thus demands techniques for environmental modeling with a strong theoretical basis, in which the resulting environmental models are as explicit, formal, and precise as the models used to describe internal cognitive activity. A cognitive psychology capable of predicting environmentally situated behavior and of supporting design will therefore have to be concerned as much with the environment as with internal cognitive activity (e.g., Anderson, 1990; Brunswik, 1952; Gibson, 1979). When one looks at the types of models produced by current cognitive psychology, however, rarely does the environmental model receive close to the amount of attention the internal cognitive model receives. There are at least three reasons for this state of affairs.

First, experimental psychologists often feel the need to simplify their environments for purposes of control and are thus able to get by with highly simplistic and impoverished environmental models (compare the length of the stimulus description — the environmental model — with the length of the description of the internal psychological model in most papers in the cognitive experimental literature). Second, and especially in research within the cognitive science orientation, often no distinction is even made between the description of the external environment and the subject's internal representation of the environment. Although it may indeed be the case that interesting issues can be addressed with this approach (e.g., differences in the types of internal representations used by expert and novice problem solvers), these accounts start so far downstream that they fail to capture any influences of the external problem representation on the efficiency of problem-solving activity (but see Larkin & Simon, 1987). Finally, rarely is it the case that researchers working within an established paradigm are forced to reason across widely varying environmental conditions, with the result that assumptions about environmental descriptions can remain implicit within a given research program. There may be no pressure to unconfound the environmental from the internal constraints on cognition and behavior when environmental manipulations are made over a very narrow range.

Thus, the open problem for cognitive engineering is to determine under what environmental conditions various cognitive activities will be activated and required for effective task performance. To evaluate a candidate design, the issue is not only to understand cognitive processes such as problem solving, decision making, and working memory, but also to determine *what* problems will have to be solved, *what* decisions will have to be made, and *what* working memory demands will be, given various design concepts for a particular task.

In the following section I discuss two types of constraints on acceptable psychological models that arise due to the need to represent both internal and external influences on cognition and behavior. The first set of constraints are structural. I argue that a model's structure must be capable of representing both cognitive and environmental organization in a single, unified format; (i.e., that the appropriate unit of analysis and modeling must be the human-environment system, rather than the human alone). The second set of constraints concern the content of acceptable models. Cognitive engineering is most in need of environmental models that assist in understanding fluent, skilled human interaction with the world, rather than environmental models that rationalize detached intellectual activity. In most cases the design goal is (or should be) to create a design that promotes fluent and skilled activity, rather than a design that promotes cognitively intensive control of behavior. We require environmental models that capture the features of environments that promote effective, skilled performance in order to define a design target and also to identify the causes of error-prone cognitive activity in current systems.

4.3.1   Modeling the Integrated Human-Environment System

One of the earliest attempts to model human-machine interaction concerned manual control behavior, such as steering a car or flying an aircraft. Engineers familiar with the design of electromechanical feedback control systems turned their attention to modeling the human as a feedback control system so that vehicles could be designed with control demands within the range of human capabilities. Control theory has a well specified language for environmental modeling. The controlled "plant" (airplane, automobile) can be described in terms of a transfer function that relates system inputs (steering adjustments) to system outputs (heading). The human as a feedback controller can be described in similar terms. In this case the input might be the heading of the automobile and the output would be a steering command. As Flach (1990a) noted, the goal in this endeavor was to discover the human transfer function; that is, a description of the function relating stimuli to response during manual control behavior. At this schematic level of description, much current psychological modeling shares this goal of finding invariance in the mapping from stimuli to response, rather than at the level of the human-environment system.

These engineers were in for a rude awakening, however, as empirical results indicated that there was no single human transfer function. Rather, the human transfer function appeared to adjust to changes in the dynamics of the controlled system. As Birmingham and Taylor (1954) noted, the ability of the human to adjust to the environmental transfer function was so great as to suggest "that 'the human transfer function' is a scientific will-o'-the-wisp which can lure the control system designer into a fruitless and interminable quest" (p. 1752). Subsequent modeling attempts (McRuer & Jex, 1967) were only successful once the search for invariance in behavior shifted to the level of the human-machine system, rather than in human behavior alone. The crossover model of human manual control behavior developed by McRuer and his colleagues is a statement of behavioral invariance at the level of the human-environment system.

Why should this finding concerning human perceptual-motor behavior inform our discussion of cognitive-level human-environment interaction? The answer is that there appears to be little reason to expect that cognitive-level behavior will be any less adaptive to environmental structure than is perceptual-motor activity. In fact, there are a variety of reasons to believe that just the opposite is the case, that human cognitive interaction with the world is even less constrained, and thus more flexible, than is perceptual-motor interaction. Note also that the correct response to this situation, and the one pursued by these manual control researchers, is to describe both human and environment as an integrated unit and to use this unified human-environment model as a tool in the search for behavioral invariance. Pursuing such a strategy requires formalisms capable of expressing both internal and external constraints on behavior in the same language, such as the transfer function representations used to model both the manual controller and the controlled system.

Although rarely used, this approach has been successfully applied to understanding cognitive-level behavior. The Lens Model framework for the description and analysis of human judgment (Brunswik, 1952, 1956; Hammond, 1955) is a unified description of both the human judge and the environment. As such, it has been a fruitful tool in understanding both environmental and cognitive constraints on judgment abilities, and has been enlightening as to a number of consistencies in judgment performance that would likely not have surfaced without some mechanism for partialing out the environmental contribution to behavior (e.g., see Brehmer & Joyce, 1988). The recent book by Anderson (1990) describing the Rational Analysis framework also represents a step in the direction toward integrated human-environment system modeling. This framework provides resources to address the question of how both internal "computational" constraints and external environmental structure combine to determine the processes that will be engaged to perform a particular task.

When we turn to the problem of understanding the kinds of environmentally "situated" (Suchman, 1987; Whiteside, Bennett, & Holtzblatt, 1988) activity typical of behavior in modern human-machine systems, it is clear that much work remains to be done before an integrated human-environment modeling approach will be possible. However, the kinds of psychological descriptions already being proposed for describing dynamic human interaction with technological systems indicate a clear shift toward understanding how both environmental and cognitive structure contribute mutually to the production of skilled behavior.

In a pair of penetrating analyses of the cognitive-level ecology of human-machine systems, Hutchins (1988, 1991) suggested that human cognition and behavior cannot be understood apart from the external devices in the environment that have been designed to perform cognitive functions. In ship navigation (Hutchins, 1988), for example, human interaction with notepads, checklists, and calculators can sometimes be used in lieu of memorial, procedural, and computational operations; and in modern aircraft (Hutchins, 1991), much of the cognitive burden for memory of intended and current speeds has been allocated to external memory structures within the cockpit. In such environments, the entire cognitive function is distributed across both person and environment. It is not surprising, then, that understanding these integrated systems requires describing both internal and external cognitive functions in mutually compatible terms. We have come full circle: The computer metaphor that gave rise to a description of human cognition in terms of information processing has been turned back upon the world, as seen in the description of the environment as external memories, external problem representations, and the like.

The importance of these environmental aids to thought and behavior cannot be overestimated. Much of modern psychological research paints a rather dismal picture of human cognitive abilities and limitations, leaving some of us in a state of wonder over how it can even be possible for human cognition to have resulted in its modern achievements. But rarely does even the scientist work in isolation from external cognitive tools, as Donald (1991) noted.

"For example, there is no internal wiring schema to support the kind of synthesis made possible by a scientific diagram; the synthesis is out there, in the diagram itself. The theoretician depends heavily upon a huge variety of external cognitive props — mathematical notations, curves, plots, histograms, analog measurements, and technical jargon — to arrive at a theory. Without these things, thoughts of this kind would simply not be possible, because the end-state or "conclusion" reached by the mind is driven directly by the external representation itself. The locus of a process like theoretical synthesis would thus be difficult to attribute to any single part of the internal-external network that makes up such a system." (pp. 378–379)

The same comments would also apply, and perhaps in even greater force, to understanding the mechanisms underlying skilled activity, such as flying an airplane, driving a car, or performing the many routine tasks we find in daily life. Skilled activity is often accompanied by a heightened level of intimacy with the world rather than by increased detachment, an observation that leads to the hypothesis that intensive exploitation of environmental structure plays a key role in productive behavior. As Norman (1993) has noted:

> *"With a disembodied intellect, isolated from the world, intelligent behavior requires a considerable amount of knowledge, lots of deep planning and decision making, and efficient memory storage and retrieval. When the intellect is tightly coupled to the world, decision making and action can take place within the context established by the physical environment, where the structures can often aid as a distributed intelligence, taking some of the memory and computational burden off the human."* (Chapter. 10, p. 6)

The human-environment system must serve as the unit of analysis and modeling to allow the internal cognitive activity necessary for productive behavior to be predicted as a function of environmental design and also to identify how necessary cognitive activity can be engineered through environmental manipulation.

4.3.2 The Need for Models of Fluent Interaction with the World

Because the goal of the cognitive engineer is often (but not always) to create an environmental design that promotes fluent and effective skilled behavior, the features of environments that support fluent as opposed to cognitively intensive behavior need to be identified and described. In many existing human-machine systems, the reason that complex cognitive processing is necessary for effective performance is that the environments in which the operators work are quite unlike those environments in which human psychological abilities evolved. As a result, the acquisition of fluent modes of behavior is impeded, and the end state is one of only partially effective adaptation. The problem of "situation awareness" (e.g., Sarter & Woods, 1991) in the modern commercial aircraft cockpit is a prime example. Edwards (1988) went so far as to describe the cockpit as an "opaque veil," (pp. 3-25) and Bohlman (1979, cited in Weiner and Nagel, 1988) suggested that the difficulty of maintaining an active understanding of the aircraft and airspace from cockpit displays is so great that it is appropriate to speak of crews as constructing "theories" of their situations.

As one who tries to make a living constructing theories, I find it most unsettling to think that theoretical abilities are sometimes necessary to ensure safe flight. What kind of psychological theory would provide the most leverage for remedying the situation awareness problem? Because cognitively intensive activities such as inductive inference, hypothesis generation, and mental modeling are observed in current systems, it seems only natural that better accounts of these activities are the key to enhancing interaction. Such accounts could presumably guide the design of aids to assist flight crews in their theoretical tasks, or the design of training methods to make crews better theoretical thinkers. It is natural to view such attempts with suspicion, however, because problem-solving aids have the potential to create their own set of human-machine interaction problems (Woods & Roth, 1988), and training to make people more "rational" problem solvers or decision makers has yet to be proven effective.

The alternative solution, of course, is to design environments more consistent with the principles underlying skilled, dynamic human interaction with the world. Pursuing this strategy, however, requires techniques for environmental modeling capable of representing the features of task environments that both promote and inhibit the acquisition of fluent modes of behavior. Only models of productive, skilled behavior can provide the resources for a task-analytic approach capable of identifying features of an environmental design that are both consistent and inconsistent with the principles underlying skilled activity. The problem of identifying demands for complex cognitive activities such as problem solving, planning, and decision making posed by a given environmental design chiefly requires models of skilled, fluent behavior, not models of problem solving, planning, or decision-making activities.

Why is this the case? Due to their roots in either economic theory or artificial intelligence, rational action models such as those mentioned earlier are more concerned with sufficiency considerations than they are with necessity considerations. The great appeal of such models is their ability to describe and often prescribe behavior in a huge variety of situations. Nearly any, and perhaps all, behavior can be rationalized as being the result of some cognitively intensive process such as search through a problem space, hypothetico-deductive inference, or the comparative evaluation of options with respect to a goal structure or utility function. No empirical evidence could ever be brought to bear on limiting the sufficiency of these rational methods for action selection. However, identifying when these sorts of complex cognitive activities will actually be necessary for successful performance requires models capable of indicating when such activities are not necessary.

My observations of skilled human behavior in complex systems have led me to the working hypothesis that cognitively intensive methods for action selection are used only as a last resort, that is, when effective perception-action solutions are not readily available. Predicting cognitive demands thus requires modeling approaches capable of defining when effective perception-action solutions will not be available, and this knowledge can best be provided by a theory of perception-action skill. I realize that the claim that skilled performers will typically opt for perception-action solutions to cognitive tasks may strike the cognitive psychologist as being counter intuitive. However, intuitions based mainly on laboratory findings may be distorted by the fact that experiments on cognition are typically carefully designed to preclude the availability of perception-action shortcuts for meeting task demands. Although my own intuitions are largely based on observations of behavior in operational settings, even in the laboratory I am continually amazed at the cleverness of subjects who are able to short-circuit demands for complex cognitive activity by cuing off the whir of a disk drive or an aberration in the graphics software. I have ceased to be surprised and frustrated by such cleverness and have begun to view the tendency toward the perceptual selection of action as a fundamental aspect of skilled behavior. There is no doubt, however, that more empirical research is needed to clarify this issue. However, the necessary experiments must provide rich enough environmental conditions and sufficient practice so that both cognitively intensive and perception-action task solutions are made available. Such laboratory experiments are rarely conducted.

These "experiments," however, are performed every day in both complex human-machine systems and in more everyday work settings. As an example, over the past 2 years I have made fairly extensive observations of the behavior of short-order cooks working busy rush periods at a local area grill. My interest in this behavioral situation arose because skilled performance in this setting appeared to possess many of the same properties I have observed in my more limited observational studies in complex operational settings, and also because 24 hour access to this environment can be readily secured for the price of a cup of coffee. And by making a well timed food order or by initiating conversation with the cook, one even has a (albeit limited) capacity for intervention and control over task demands.

In the environment I have studied, the cook uses an assortment of automated devices such as fryers and ovens, combined with substantial manual activity at the grill to coordinate the preparation of the many items within each order, while preparing multiple orders simultaneously. Describe this task in any formalism for rational action and the task demands appear overwhelming. Observe this type of skilled human-environment interaction, though, and I believe the following will be apparent. First, there is an intensive degree of intimacy in the cook's perception-action interaction with the environment. The cook maintains tight perceptual contact with the world and always seems to be taking some sort of action. Rarely if ever does the cook appear to engage in detached, contemplative cognitive activity. Task demands are uncertain and arrive dynamically, and ongoing behavior must be sensitive to a number of unpredictable events.

What allows this perceptually intensive mode of interaction to be productive? Note that the cook's environment is highly structured, but nearly all of this structure is visible. The most efficient "problem representation" for the cook to use is an external one: the grill area itself. Action selection based on the external environment has considerable economies as compared to action selection based on internal representations of the environment. The environment considered as a problem representation serves as an external memory capable of being perceptually accessed, updates itself automatically and in parallel, serves as an external memory store, is internally consistent, and is always veridical (also see Reitman, Nado, & Wilcox, 1978). The world takes care of its own "truth maintenance."

When uncertainties do occur using such an external representation (i.e., perceptually available information underspecifies constraints on activity), these uncertainties can often be resolved through perception-action rather than accessing stored knowledge. How well cooked is the underside of a steak? Flip it and see. The cook not only uses the structure already present in the environment, he or she can dynamically create structure in order to make perception-action solutions available and thereby reduce cognitive burdens. For example, the cook may organize the placement of meats in order of doneness, may lay out dishes or plates to serve as a temporary external memory of orders to be prepared (also see Beach, 1988, pp. 342-346), and may even generate new information "displays" by introducing constraint in the controlled environment, causing a hidden variable to covary with a visible one. For example, the cook may adopt the strategy of continually flipping meats so that the doneness of the top side can always be used as a reliable indicator of the doneness of the underside. In a very real sense, the cook is both performer and online interface designer.

Skilled human-environment interaction is thus a response to

environmental structure as well as a source of environmental structure to be subsequently exploited. The environmental structure created by the cook's own "tricks" and routinized strategies plays a role similar to the structure created through the environmental design process itself in promoting cognitive efficiencies. The former structuring merely happens "on-line" and is thus short lived, whereas the latter happened during the design of the grill area and is thus reflected in the static and permanent organization of the design. But both forms of structure, whether contributed at one point in time by the designer or continually by the cook, result in cognitive economies through the enablement of perception-action solutions to the task. For example, the external memories and displays dynamically created by the cook play a similar cognitive role to the timing mechanisms used in toasters and ovens to offload memory demands to the world. Because of the existence of self–produced environmental structure, the acquisition of situated skills will always resist faithful description solely in terms of increasingly efficient mechanisms sensitive to a fixed set of environmental information. A model of skill acquisition in dynamic human-environment interaction would also have to describe how the actor's external environment becomes increasingly structured by activity itself, and thus increasingly informative to the actor, over the course of skill development.

Much of the responsibility for dynamic human-environment interaction lies in the perception-action mechanisms at the interface between the performer and the world. The development of skilled, dynamic interaction relies on abilities to exploit environmental structure to obtain perception-action solutions to tasks and, where none naturally exist, to create additional environmental structure in such a way as to enable perception-action solutions. If such structure is not provided by the designer, the performer will seek to create it through activity that introduces new forms of structure. The productivity of this mode of behavior requires the availability of sources of information to specify the environmental constraints to which behavior must be sensitive in order to be effective and the availability of actions capable of both changing the environment and of creating additional sources of information to further enable the perceptual guidance of activity. These are, I believe, features common to nearly all environments in which the acquisition of fluent, dynamic interaction is observed. They are also features lacking in the many technological environments of interest to cognitive engineering, due largely to interfaces that highly restrict perception-action access to the controlled system. The absence of such features is one major cause of the difficulty of acquiring skills in such systems and the reason that the end state of learning is often one of only partially effective adaptation.

4.4   Toward an Ecological Perspective

The previous discussion has centered on identifying a number of necessary conditions for psychological models to support environmental design. It is time now to turn toward outlining a methodological strategy with the potential to address some of the gaps in our knowledge discussed earlier. Many of the necessary features for psychological models that have been identified are suggestive of the possibility that an ecological approach to human-environment interaction may yield fruitful tools for cognitive engineering. The ecological approach was pioneered by Brunswik's (1952) and Gibson's (1966, 1979) theories of how knowledge of environmental structure can provide important constraints on psychological explanations. In particular, Brunswik's emphasis on taking the human-environment system as the unit of analysis and modeling, perhaps best represented in the Lens model framework (Brunswik, 1952, 1956; Hammond, 1955), and Gibson's (1979) focus on how fluent interaction can be described as perceptual specification of environmental constraints on activity, blend nicely with the claims that cognitive engineering is most in need of models of skilled interaction with the world and models that take the human-environment system as the unit of analysis.

  In the following discussion I take some initial steps toward identifying opportunities the ecological approach might offer for cognitive modeling to support design. However, and for readers already familiar with Gibson's views especially, it is important to first discuss what an ecological approach to cognitive engineering does not require. First, it does not require that we conceive of all human–environment interaction as purely perceptually guided activity. Direct perceptual guidance of action, as discussed by Gibson, might surely be possible, although it is likely that it is specific to those information-rich environments in which perception evolved or to artifactual environments designed to mimic such environments. There is no reason to expect that evolution anticipated the modern aircraft cockpit or the word processor. In such environments the need for postperceptual processes such as problem solving and decision making is quite likely. The ecological and information processing approaches need not always be considered to be at odds, but may instead both contribute to a more complete understanding of human-environment interaction.

Second, the adoption of an ecological approach does not necessarily imply a commitment to studying fluent behavior in the natural environment. Gibson rallied against the use of abstract information displays for the study of visual perception, the types of displays often found in existing human-machine systems. But in a larger sense, Gibson, like Brunswik before him, was arguing for using environmental conditions as the basis of scientific study which are representative of the conditions in which a target behavior of interest occurs. And, for better or worse, a cockpit or control room looks much more like a laboratory than it does the natural terrestrial environment. These are the target environments of interest to the cognitive engineer. For this reason, these environments, or carefully made abstractions of them, are the places in which the ecological approach to cognitive engineering should be carried out.

4.4.1 Resources for Cognitive Modeling

Brunswik (1952, pp. 1-102) offered the Lens model as a description of how the human and environment could be described in an integrated fashion, using the principle of parallel concepts (e.g., see Hammond, Steward, Brehmer, & Steinmann, 1975). As shown in Figure 4.1, the Lens model is a symmetrical framework that represents how both environmental and cognitive structure mutually contribute to judgment performance. The organism has available a set of cues (x_i's) which bear specified relations ($r_{e,i}$'s) to an environmental criterion to be judged (e.g., a medical diagnosis). The relations between the cues and the criterion may take various forms and vary in ecological validity. Similarly, the ways in which the organism makes use of the cues ($r_{s,i}$'s) to arrive at a judgment may take various forms and vary in cue utilization. The framework is an expression of the principle of parallel concepts in that each concept on one side of the model has a counterpart on the other side. This framework has a number of attractive properties which result from representing the organism and environment in compatible terms.

Perhaps, most importantly, the Lens model framework allows the modeler to measure the degree to which the environmental structure which relates the cues to the criterion is reflected in the manner in which the cues are cognitively structured to produce a judgment. High levels of achievement are an indication of a highly *adaptive* cognitive

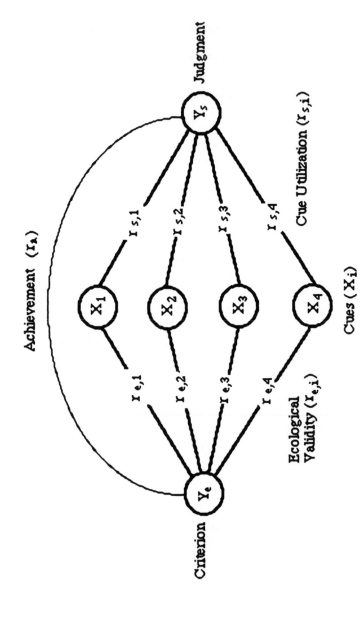

Figure 4.1. *The Lens model of Brunswik.*

organization; that is, a cognitive strategy that mirrors the environmental structure to which behavior must be sensitive. As many of the previous comments in this chapter suggest, some sort of adaptivity-oriented view of cognitive activity is likely to be required in order to understand skilled human-environment interaction as well. In addition, the Lens model allows one to localize the causes of less than fully productive behavior to either the environmental structure, the cognitive structure, or both. Weather forecasters, for example, frequently err in their predictions, but only an analysis of both the ecological validity of the cue structure and their policies for cue utilization can yield an understanding of the reasons for these errors. We often have similar interests in the design and analysis of human-machine systems. Did a particular error result from an operator making incorrect use of displayed information, or was the error the result of a well-adapted operator confronted with not fully diagnostic information? Quite different types of remedial action can (and should) result depending on the answers to questions such as these.

Thus, the Lens model framework offers a good starting point for developing an approach for representing skilled human-environment interaction. However, my previous comments suggest that we require models that not only take the human-environment system as the unit of analysis, but we also need models capable of representing fluent, skilled behavior in order to identify demands for more complex cognitive activity. From this perspective, the Lens model has two important deficiencies. First, action itself is not explicitly represented. The Lens model is an epistemological framework for modeling judgments about the state of the world, not for representing activity or the selection of action (see also Brehmer, 1986). It is interesting to note, however, that early formulations of the Lens model (Brunswik, 1952; Hammond, 1966), as well as its conceptual precursor presented by Tolman and Brunswik, (1935, especially Figure 1), were indeed concerned with both the input (stimulus) and output (response) sides of the organism. That the original model had this dual emphasis can be seen in Brunswik's (1952) statement:

> *"The probability character of intra-environmental relationships, their limited "ecological validity," becomes of concern in two regional contexts: on the reception or stimulus side as the equivocality of relationships between distal physical or social objects and proximal sensory stimuli or cues, and on the effection or reaction side as the equivocality of relationships between proximal outgoing behavioral*

responses, or means, and their remote distal results and effects."
(p. 22)

Although Brunswik did recognize this symmetry between the problems of perceiving or judging and acting, more modern Brunswikian research has been mainly limited to issues of perception, multiple-cue probability learning, and policy capturing (Brehmer, 1986). As noted by Brehmer, one possible reason that the Brunswikian framework has not been fruitful in explicating the psychology of action is that it conceptualizes action mainly as a process of choice.

Conceiving action as a product of judgment and choice may be fruitful for certain activities, but probably not for understanding fluent, skilled behavior. As suggested by Hammond et al. (1975), judgment is a "cognitive activity of last resort" (p. 272). Judgmental abilities will only be called on when the available information only probabilistically specifies a criterion, and actions capable of manipulating environmental variables to gain more diagnostic information are not available. Note that these environmental properties are exactly those features of task environments I have previously described as being the major impediments to the development of fluent, perceptually guided interaction. Like formalisms for rational action, then, it may be quite possible to interpret the skilled selection of action within the Lens model framework. However, such a model is not likely to capture those special features of task environments that allow for the acquisition of fluent human-environment interaction.

The need to explicitly represent fluent interaction, as well as those features of the world that promote it, suggest that we must consider the problem of how the environment of the skilled performer should be described. Gibson, with his ecological physics and theory of affordances (1979), proposed an action-oriented environmental description in order to understand how perception may orient behavior to environmental opportunities for action. Such an approach results in a description of the environment in terms of the opportunities for action it presents the performer. The resulting description can be called an *affordance space*, akin to the decision space descriptions resulting from decision theory, the problem space descriptions resulting from problem-solving theory, or the cue space descriptions resulting from theories of probabilistic judgment. In most cases, an affordance space will be a dynamic description of the environment, as both the environmental structure and the performer's resources for action will change over time, and a dynamic affordance structure will result.

Note that creating an affordance space environmental description does not commit one to any particular position concerning how affordances may be detected to guide activity. Gibson was most concerned with those situations in which perceptual information is available to specify affordances, and in such cases interaction can be described as the perceptual detection of information capable of orienting behavior to action opportunities. However, in other situations perception-action access to the environment may be restricted or impoverished, or information other than that specifying the immediately present affordance structure must be taken into account for behavior to be productive. In such cases, performers may have to engage in more elaborate cognitive activity in order to detect the affordance structure or to combine information specifying affordances with other information in order to select actions. Regardless of the type of either perceptual or postperceptual activity required to orient behavior to an affordance structure, a description of the world in terms of affordances is still a valuable tool in understanding how environmental structure is reflected in cognition and behavior.

For our purposes, the concept of an affordance space is especially important because it provides resources to compensate for the two deficiencies of the Lens model identified earlier. First, an affordance space can in some cases play the role of the criterion in the Lens model, thus shifting the emphasis from passive judgment to the identification and selection of opportunities for action. Second, the possibility that perceptual information is capable of fully specifying the affordance space (i.e., direct perception) suggests that we must relax the a priori assumption underlying the Lens model that the available information is only probabilistically related to the criterion. The question of determining the relationship between the available information and the environmental affordance structure is an empirical one. The ecological task analysis framework presented next is an attempt to integrate Brunswik's Lens model and Gibson's affordance theory into a unified framework for modeling skilled human-environment interaction.

4.4.2 The Framework for Ecological Task Analysis

Integrating concepts from Brunswik's Lens model and Gibson's affordance theory results in the *ecological task analysis* framework depicted in Figure 4.2. Like the Lens model, the proposed framework is a symmetrical arrangement that represents the integrated human-environment system. Not only does ecological task analysis exploit the

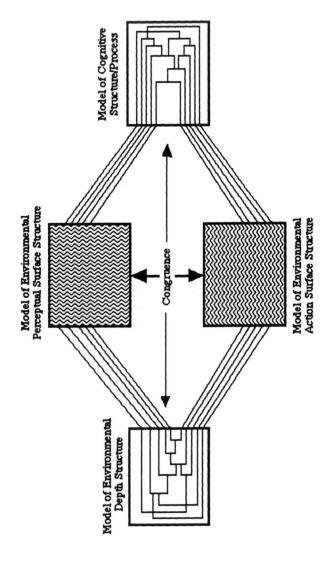

Figure 4.2. *The Ecological task analysis framework. Like the Lens model framework for human judgment, the Ecological task analysis framework represents the integrated human-environment system in a symmetrical arrangement. Matches between the perceptually oriented and action-oriented environmental differentiations indicate opportunities for the fluent perceptual guidance of activity. Structural mismatches between the models of surface perceptual structure and surface action structure are indicative of demands for cognitive activity and corresponding models of the environmental depth structure.*

principle of parallel concepts to capture certain symmetries between cognitive and environmental structure, the proposed framework also uses a principle of parallel concepts to suggest certain symmetries between perception and action. This latter symmetry is evident in the relationship between the upper model of environmental perceptual structure and the lower model of environmental action structure.

Ecological task analysis begins with the creation of two complementary descriptions of the surface structure of the environment. As in the Lens model framework, we rely on a distinction between environmental surface structure which exists at the interface between the performer and the world and the environmental depth structure which exists remotely, behind the surface structure so to speak. The two descriptions of surface structure, shown in the middle of Figure 4.2, are models of environmental perceptual structure and environmental action structure. The description of surface perceptual structure in ecological task analysis plays a similar role to the cue description in the Lens model. The model of surface perceptual structure describes the environmentally available information. In human-machine systems, one can think of surface perceptual structure as the information available from interface displays about the state of the controlled system. Unlike the Lens model framework which captures only the surface perceptual structure of the environment, however, ecological task analysis also requires a description of the surface *action structure* of the environment. In a human-machine system, one can think of surface action structure as the actions made available by interface controls.

In the Lens model framework, the relation between surface perceptual structure and depth structure is the relation between the readily available information and the environmental state a person is attempting to judge. Similarly, in the proposed framework the relation between surface action structure and depth structure is the relation between the readily available actions and the environmental change the person is attempting to effect. On the perceptual side, the surface/depth distinction reflects the difference between given and inferred. On the action side, the surface/depth distinction reflects the difference between readily available actions and required effects. The model of environmental depth structure on the left side of Figure 4.2 is a description of these potentially covert relationships, and the model of cognitive structure on the right side of the figure is a description of how environmental depth structure is reflected in cognition, and ultimately, behavior.

4.4.2.1 The Principle of Parallel Concepts: Perceptual and Action Structure

The description of environmental surface structure in terms of both perception and action is necessary to capture how the skilled perceptual guidance of activity is possible. The possibility for perceptual guidance of activity relies on the availability of perceptual information capable of fully specifying environmental affordances. In terms of the proposed framework, perceptual guidance of action is not to be described as the perceptual detection of surface information to infer some covert or remote depth property of the environment. Rather, the perceptual guidance of action is to be described as the use of one form of surface structure, namely perceptual, to specify another form, namely, the surface action structure. As described in detail later, the ecological task analysis process proceeds by examining the models of perceptual and action surface structure to identify the congruence between these two forms of environmental description. Matches between these two models indicate opportunities for fluent, perceptually guided activity. Mismatches between these two models are indicative of demands for cognitive activity to overcome the perceptual nonspecification of action. Various forms of mismatch are possible: Each form of mismatch is suggestive of a different type of necessary cognitive activity and a different type of remedial interface design solution.

Before describing the analysis process, the symmetrical nature of the models of perceptual and action structure must be discussed. The model of surface perceptual structure is an environmental description using a performer's perceptual capacities as a frame of reference in which the environmental structure is described. This description is relational in the sense that the resulting environmental model reflects both the perceptual capacities of a performer and the environmental structure. It is easy to overlook that such descriptions are actually relational in nature, as we often speak as if a perceptually generated differentiation of the environment is purely a function of the environmental structure and not a function of perceptual capacities. However, the fact that some forms of environmental structure are seen as objects whereas others are not, or the fact that some aspects of the environment are seen as being blue whereas others are red, are as much facts about the perceptual system as they are of the environment. A perceptually oriented environmental model is a relational construct created by using perceptual capacities as a frame of reference in which environmental structure is measured and described.

The second model of environmental surface structure required for ecological task analysis is an action-oriented description of the environment. The model of surface action structure is an environmental description using a performer's action capacities as a frame of reference in which the environmental structure is described. This description is relational in the sense that the resulting environmental model reflects both the action capacities of a performer and the environmental structure. This action-oriented environmental model represents the world in terms of its opportunities for action. This environmental model thus generates a differentiation of the world in terms of the degree to which various spatiotemporal environmental regions are consistent in various degrees with the taking of various actions. An action-oriented environmental model is a relational construct created by using action capacities as a frame of reference in which environmental structure is measured and described.

The symmetrical nature of the perceptually oriented and action-oriented environmental descriptions required for ecological task analysis should be apparent. These two environmental models differ only in that one uses perceptual functionality as the yardstick for environmental measurement, whereas the other uses action functionality as the yardstick. One can view Gibson's affordance theory as springing from the recognition of a certain symmetry between perception and action, with the resulting insight that either the perception or action interfaces between the organism and the environment can be used as a measurement reference frame. The two resulting models thus reflect two different primitive differentiations of the environment; one generated by using perceptual capacities to understand how the world is carved up with respect to perception, and the other generated by using action capacities to understand how the world is carved up with respect to action. As Barwise and Perry (1983, p. 11 emphasis added) have suggested:

> "The emphasis is on how the organism differentiates its environment, on the sorts of uniformities it recognizes across situations. Different organisms can rip the same reality apart in different ways, ways that are appropriate to their own needs, their own perceptual abilities and their own capacities for action. This interdependence between the structure the environment displays to the organism and the structure of the organism with respect to the environment is extremely important. For while reality is there, independent of the organism's

*individuative activity, the structure it displays to an organism reflects
properties of the organism itself. "*

Neither the perceptually oriented environmental description nor the
action-oriented environmental description results in a more primitive,
privileged, or objective ontological picture of the world. However, the
claim is sometimes made that an action-oriented description of the
world in terms of affordances is scientifically illegitimate, because it is
relativized to the actor and is thus in some sense subjective. Note that
the correct response to this possible criticism is not to adopt the heroic
position that affordances are in some sense independent of the
performer. They are not; they arise from using the performer's action
capacities as a frame of reference for environmental description. Rather,
to counter this argument one must merely emphasize that the
supposedly scientifically legitimate perceptually oriented environmental
descriptions which portray a world of objects and properties are just as
relativized to the capacities of the performer as are action-oriented
environmental descriptions. One merely takes perceptual functionality
as the frame of reference for environmental description, whereas the
other takes action functionality as the frame of reference.

4.4.2.2   The Process of Ecological Task Analysis

I call a process whereby environmental models of surface perceptual
and action structure are created, and mismatches between the
environmental differentiations represented in the two models are
identified and described, an *ecological task analysis* of a human-
environment system. What I believe to be the central contribution of an
ecological approach to cognitive modeling can now be stated quite
simply. A preliminary ecological task analysis of a human-environment
system is required to identify the degree to which an interface (natural
or artificial) between the human and environment is consistent with the
principles underlying fluent interaction, and by doing so such an
analysis helps specify what cognitive processes will be necessary for
effective behavior. An ecological task analysis is thus similar in spirit to
Marr's (1982) computational-level theory that attempts to define the
necessary functionality of vision models, and Anderson's (1990) rational
analysis that attempts to define the necessary functionality of a variety
of cognitive models. Before outlining the analysis process, a few
comments comparing the goals of ecological task analysis and
Anderson's approach in particular may be valuable.

The goal of ecological task analysis is to define the necessary functionality of any cognitive processes that may be required to support effective human-environment interaction. Like Anderson's rational analysis, then, the present approach defers the question of how any necessary cognitive activities might be carried out. In this way the two approaches share a common perspective. However, ecological task analysis takes human-environment interaction to be its primary concern, whereas the current formulation of rational analysis is concerned with cognitive activities such as memory, categorization, and problem solving. But as Anderson(1990) himself noted, all these cognitive abilities are useless if they do not in some way serve the goal of action selection, and for Anderson, action selection is always to be understood as the result of a problem-solving or decision making exercise (p. 192). Perhaps it doesn't much matter what words Anderson uses to describe the processes underlying action selection, but what does matter is the nature of the environmental models that result from such a choice (decision trees and problem state spaces).

As our previous comments suggest, all action selection, whether resulting from cognitively intensive processes or from more efficient perception and action, can be rationalized into one of these frameworks for explaining productive behavior. Modeling to support design, however, must focus not on the sufficiency of these frameworks but rather on limiting the conditions of their necessity. Ecological task analysis, therefore, starts not by assuming action selection is governed in any particular manner, but rather has as one if its goals to define what sorts of governing mechanisms will be necessary for modeling any instance of human-environment interaction. Ecological task analysis attempts to meet this goal through the use of environmental models that allow for a description of action selection in terms of the perceptual guidance of activity. Only when an examination of these environmental models indicates mismatches in perception-action environmental structure will ecological task analysis result in a construal of action selection in terms of decision making, problem solving, or any other cognitively intensive activity.

4.4.3   The Principle of Parallel Concepts: Environmental Depth and Cognitive Structure

As mentioned earlier, ecological task analysis begins with the creation of the models of perceptual surface structure and action surface structure which provide the bridge between the human and the environment. The analysis process proceeds by then examining the congruence between these two models in terms of the manner in which they differentiate the environment with respect to perception and action. The process of examining congruence will be discussed in detail later. The result of this exercise, however, is the specification of a model of the environmental depth structure and a complementary model of internal cognitive structure and process. It is very important to note that ecological task analysis requires that modeling the environmental depth structure comes after creating the models of perception and action structure. That is, we do not begin by assuming that the task environment possesses a certain intrinsic depth structure (e.g., problem state space, decision tree, linear cue-criterion function) and thereby a corresponding structure to cognitive activities (e.g., heuristic search, comparative evaluation of alternatives, linear cue combination rules). Rather, we let the examination of the surface perceptual and action structure of the environment guide the selection of the model of depth structure and thereby the corresponding structure for cognitive activities.

Herein lies the major difference between the proposed approach and many current modeling approaches in cognitive science: Considerations of perception-action functionality define the necessary functionality for cognitive processes, rather than defining perception-action functionality by an a priori cognitive model of action selection. The shift is away from viewing perception and action as servants of cognitive activities, providing input information for these activities and executing any action commands that are handed down. The shift is toward viewing cognition as a servant of perception and action. Cognition is thus viewed as a gun for hire, that is called on to meet only those task demands that cannot be met with perception and action mechanisms. The role cognitive activities will play in any given case can only be defined by first identifying mismatches in the perception and action structure of the environment.

I have tried to make the case for sequencing the analysis in this fashion at a variety of previous points in this chapter. We are obviously looking for a model of environmental depth structure that corresponds to, and helps make sense of, the cognitive activities that will actually be engaged to serve action selection. Which cognitive activities will be necessary for productive behavior, however, can only be determined by a detailed analysis of the degree to which perceptual information is available to specify productive action. Different interface (control display) designs for the same task environment can differ radically in terms of the cognitive demands they make on the performer. Even for a simple tracking task, a poor interface may require a performer to undergo considerable problem solving search, while a good interface may allow a performer to operate in a fluent perception-action mode, for meeting the same task demands. As a result, different environmental models (e.g., problem spaces vs. transfer functions) will be needed in the two cases to describe the different ways in which environmental structure is reflected in cognition and behavior. Note, however, once a model of environmental depth structure and a corresponding model of internal cognitive activities are selected, as in the Lens model framework the congruence between the environmental and cognitive models can be examined in order to identify any ways in which cognitive limitations or biases place constraints upon productive behavior.

This is obviously a highly schematic description of the models necessary for ecological task analysis. Yet, I wonder if it is possible to get more precise about the content of these descriptions without doing a potential injustice to the richness of the perceptual and action structure of any realistically complex behavioral situation. One great allure of environmental models which rationalize action selection as decision making or problem solving is that relatively low-dimensional environmental descriptions can be used. Such low-dimensional representations are advantageous in that they can be easily applied across a variety of contexts and are thus suggestive of how the psychological processes underlying action selection might be organized in a context-free, general purpose format. Describing the world in terms of the interaction of "raw" environmental structure with perceptual and action capacities, on the other hand, has the potential to create environmental models of almost unlimited dimension.

Nevertheless, my own view is that any method capable of identifying the mechanisms underlying skilled human interaction in a setting of any reasonable complexity requires models capable of preserving many of the fine details at the interface between performer and environment. The richness of the world's perceptual and action structure, seemingly necessary for the fluent operation of the perception-action system, severely overtaxes our highly limited cognitive-linguistic resources for environmental description. I doubt that at the current time a task analysis technique capable of guiding human-environment interaction modeling can possibly be much more than a charge to the modeler to undertake the arduous process of identifying and describing the overwhelmingly rich interface between the performer and the world.

For this reason, the following description of the analysis process will consist of an abstract discussion of the possible results of an ecological task analysis, along with a concrete example of how such an analysis can be performed for a particular behavioral situation. Figure 4.3 depicts four possible results of an ecological task analysis in terms of the congruence of the resulting perceptually oriented and action-oriented environmental models. The grid lines in each of the schematic environmental models indicate the manner in which the environment is spatiotemporally differentiated with respect to either perceptual capacities or action capacities. The four cases, each describing a different mapping between perceptual and action structure (many-to-one, one-to-many, one-to-one, and many-to-many) are described separately.

4.4.3.1 Case I: Perceptual overspecification of action

In the first case shown in Figure 4.3, the perceived environment is overdifferentiated with respect to the environmental differentiation in terms of opportunities for productive action. Many different perceptually distinct situations all point to a single opportunity for productive action. Object or configural displays are one type of design solution available for coping with perceptual overspecification of action. Object displays are an attempt to reduce the dimensionality of the perceptual space so it becomes aligned with the lower dimensional action space. These displays perform this function by organizing the originally overdifferentiated perceptual information in such a way that perceptually salient relational features emerge that are differentiated in a manner identical to the differentiation reflected in the action space. When a display-based solution is not used, however, the performer will have to develop some ability to overcome perceptual overspecification. Perceptual pattern recognition is one process that could potentially result in an alignment of the perception and action spaces, although some naturally occurring relational properties must be perceptually available to enable this solution. When the possibility for pattern recognition is neither naturally supported nor supported through

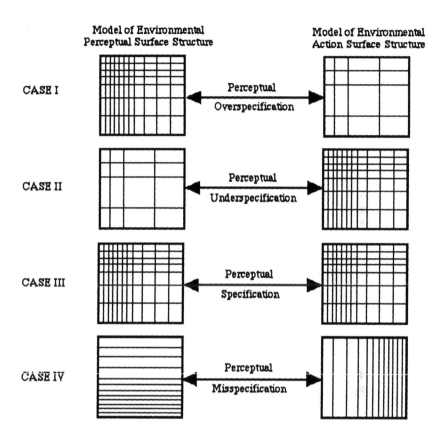

Figure 4.3. *Four possible results of an ecological task analysis. Each case reflects a different type of match or mismatch in environmental perception and action structure, and is thus indicative of a different form of cognitive demands, and a different type of remedial interface design solution.*

configural display design, it is likely that a significant amount of categorical or instance-based learning may be required in order to identify consistencies in the mapping from perception to action. Note, however, that there is still the possibility of fully productive performance in all these cases because perception merely overspecifies action, not misspecifies action, as in Case IV described later.

4.4.3.2 Case II: Perceptual underspecification of action

In this case, the perceived environment is underdifferentiated with respect to the environmental differentiation in terms of productive action. There is simply not enough perceptually available information in order to uniquely specify the appropriate action alternative. One likely cause of perceptual underspecification of action is that there is hidden-state information in the environment; i.e., the performer must know something about the previous history of the environment, keep this information in memory, and then integrate this memorial information with the perceptual information in order to uniquely specify the appropriate action alternative. Building memory into the displayed interface using trend or historical displays is one strategy for aiding performance in such cases. Another cause of perceptual underspecification is that information about future state, rather than past state, is not perceptually available. Perception identifies a number of action candidates, but the selection of the appropriate action requires knowledge of the downstream effects of an action, and these effects are not perceptually apparent. Predictor displays or fast-time simulations are two approaches for making this information available to the performer. Unaided performance, however, will require considerable learning before skill can be acquired. Internalization of environmental dynamics in the form of an internal model may be necessary in order to gain access to past or future state information. The problem solving models of Newell and Simon can be viewed as descriptions of the cognitive activities that may be necessary when the downstream effects of an action must be taken into consideration. Exploratory behavior, or physically "trying out" solutions, is a method available for taking into account information about the future state without the use of an internal model, although this form of activity may not always be possible. A form of perceptual learning which is available to overcome initial perceptual underspecification is perceptual differentiation (e.g., Gibson, 1969). Here, the perceptual capacities of the performer change in order to increase the dimensionality of the perceptual space to bring the perceptual differentiation of the environment into alignment with the action-oriented environmental differentiation. Perceptual learning of this type, however, requires (perhaps initially subtle) dimensions of stimulation to which perception can eventually become sensitive.

4.4.3.3 Case III: Perceptual specification of action

In this situation, fluent performance can be expected to develop without significant cognitive demands or conceptual learning. The performer's preestablished perceptual competencies provide the ability for unique specification of productive action. No rule-based information integration is required. Neither is information about the history nor future of the environment necessary. Therefore, no internalization of environmental dynamics is required to provide these forms of information. If it were the case that the designer could always be certain that the perceptually-oriented and action-oriented environmental models of the human-environment system were correct, interfaces that support the perceptual specification of action would be our undeniable design target. However, if unanticipated changes occur in either the diagnosticity of the displayed information, the functionality of the interface controls, or the environmental dynamics, a fluent, informationally encapsulated perception-action mode of control may carry on without the performer paying heed to these environmental disturbances.

4.4.3.4 Case IV: Perceptual misspecification of action

In the final case, the mapping between the perceptually available information and productive action is many to many. Behavior might well be productive in this situation, but not because the currently perceived situation is particularly informative. Models of behavior in such situations typically endow the performer with a considerable amount of knowledge to overcome perceptual misspecification, or else they give up hope for a deterministic account of action selection and instead opt for finding invariance in aggregate performance through the construction of probabilistic cognitive and environmental models. In fact, the Lens model of human judgement can be considered to be a special case of the ecological task analysis framework under the assumptions that perception misspecifies action (judgment), and that the environmental depth structure can be described with cue and cue criterion correlations indicating the covert relationships among these variables.

I have little to say about this case because, frankly, it is something of a catch-all. However, I do think it is important to emphasize that when the modeler finds what appears to be a case of perceptual misspecification, yet observes productive action selection with a heavy reliance on perceptual information, it is likely that the performer knows some things that the modeler does not. I suspect that such cases are more frequent than we may care to admit. The long history of findings on the context-sensitivity of reasoning, decision making (Kahneman & Tversky, 1979; Tversky & Kahneman, 1981), and problem solving (Kotovsky, Hayes, & Simon, 1985, pp. 248-294) demonstrate that people pay considerably more attention to the concrete presentation of a problem situation than do many abstract cognitive models.

There is, of course, really no environmental stuff that is "context" as opposed to a "relevant structure." Context is always defined with respect to a model: It is simply those aspects of the environment that a given model fails to represent, and as a result, those aspects that are rendered incapable of producing behavioral variance. Findings that demonstrate the intensive context sensitivity of cognition and behavior can be seen to be, in part, a reflection of the fact that many current environmental models are either overly abstract or perhaps even cut across the grain of the perceptually oriented and action-oriented environmental differentiations which are the basis of ecological task analysis. I suspect that in some cases the apparently unruly mappings which give rise to complex or probabilistic accounts of cognition can be straightened out as much by increased attention to environmental modeling as by increasingly elaborate cognitive modeling.

4.4.4   An Example of Ecological Task Analysis

Because the goal of ecological task analysis is to identify the cognitive demands imposed by a task as a function of the environmental surface structure as revealed through an interface, the results of the analysis are useful for motivating both interface design and process models of interaction. A study demonstrating the utility of ecological task analysis in motivating interface design to facilitate the acquisition of skilled interaction in a dynamic decision and control task is presented in Kossack (1992). In this section, we present an example illustrating the utility of the approach for functionally specifying cognitive demands as a precursor to psychological process modeling.

We have performed modeling of human-environment interaction in a dynamic microworld in order to advance approaches that could provide resources for interface design (Kirlik, Miller, & Jagacinski, 1993). At a concrete level, the experimental apparatus consisted of rich, graphically displayed sources of information and a mixture of both continuous and discrete controls, similar to the kind of interface technology typical of many modern human-machine systems. At an abstract level, the task required subjects to engage in both manual control and supervisory control (Sheridan, 1987) of a set of semiautonomous craft operating in a simulated world.

Subjects used a joystick to pilot a scout vehicle within a partially forested world shown on a dynamic, colored, graphical map or situation display showing the entire 100-square mile area to which activity was confined. The scout's major activity was to discover hidden objects (cargo and enemy craft) within the world. The scout was therefore equipped with a 1.5 mile radius radar for this purpose. Subjects used four additional craft primarily to act on the discovered objects, i.e., to engage both stationary and mobile enemy craft and to load cargo and unload it at a home base. Subjects also had to attend to a number of resource management constraints (e.g., fuel, missiles, cargo capacity) in order to successfully complete each 30 minute experimental session.

The task was quite complex, and many hours of practice were required to achieve mastery. However, at skilled levels of performance the selection of action was quite rapid, and a fluent and often seamless mode of dynamic interaction characteristic of much skilled behavior was observed. The *apparent* economy of behavior in this environment led to the hypothesis that subjects were relying heavily on rich graphical information as an external problem representation, with some of the attendant advantages of this processing mode as discussed earlier in relation to our example of the short-order cook. However, it seemed unlikely that a perception-action mode of control was possible in all cases, because some of the constraints on productive action were not easily identifiable from the displayed information, and actions were not always available to resolve uncertainties associated with nonspecific perceptual information. An ecological task analysis of this human-environment system was performed in order to identify situations in which the control-display interface supported a perception-action processing mode, as well as those situations in which the interface design may have required subjects to use a more cognitively intensive mode of action selection. The results of the task analysis were used to motivate the design of a process model capable of successfully mimicking subject behavior.

4.4.5   An Ecological Task Analysis of Search Behavior

The present example in this section concerns modeling the selection of continuous search paths for the scout through the simulated world. A more complete description of this model as well as a description of environmental and cognitive modeling for dynamic discrete action selection in the laboratory task can be found in Kirlik, et al. (1993). Figure 4.4. is a depiction of a world configuration as it was displayed to subjects. The open regions, indicated in white, were displayed in light brown. The lightly forested and heavily forested regions, shown as light gray and dark gray, were displayed as light green and dark green, respectively. Only home base (the unfilled circle) and the initial location of the scout are shown in the figure.

Searching the world for cargo and enemy craft requires consideration of two capacities for action: scout locomotion and

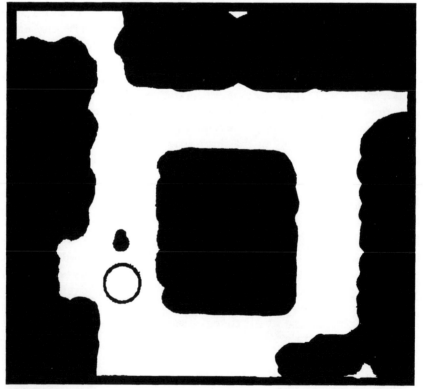

Figure 4.4. *Map display showing home base, the scout, and forested areas.*

sighting objects with scout radar. Locomotion was most efficiently performed in open rather than forested regions due to the need to navigate around trees. Sighting objects, on the other hand, was more efficiently performed in lightly forested areas because objects were considered to be more densely located in forests. In addition, the fuelrange constraints influenced search-path selection because fuel expenditure rates were designed so that scout had to refuel at home base at some point during the middle third of the experimental session.

Figure 4.5 shows search paths created by two different subjects. Both began by traveling north along the boundary of the inner forest, turning east to follow the top boundary of this forest, looping back west along the boundary of the upper forest, turning south along the border of the left forest, visiting home base for refueling, then departing to the east and then the north, at which point the session terminated. This boundary hugging behavior resulted from the interaction of the two major criteria for search-path selection: searching forests with the 1.5-mile radar to discover objects, and locomoting through open terrain to cover as much area as possible. What is the most faithful description of the cognitive activities underlying search behavior? What is an appropriate model for the depth structure of this environment? We of course would like to find a model of depth structure that captures those environmental features to which cognition and behavior was sensitive. One could of course formulate this process as a constrained optimization problem and use a generate-and-test procedure to create alternative paths and then evaluate them with respect to an objective function. However, there are an infinite number of possible paths, and the

 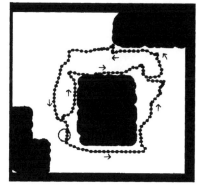

Figure 4.5. *Search paths produced by two different subjects in the same world configuration.*

computational demands appear overwhelming. One could also attempt to describe this process in rule-based terms, using an environmental description in terms of perceptually salient objects such as forests and their borders. With this model, the cognitive processes underlying search behavior would be described in terms of the manipulation of symbols standing for discrete aspects of environmental structure.

Ecological task analysis, on the other hand, suggests that we delay assumptions about modeling environmental depth structure and cognitive processes until after the models of environmental perceptual structure and action structure have been constructed. An initial cut at constructing the perceptually oriented environmental model required for ecological task analysis would be to describe the objects and properties perceptually apparent on this map display. However, as will be seen later, ecological task analysis suggests that we iterate and refine this initial perceptually oriented model after constructing the action-oriented model by using the latter to help identify any initially overlooked perceptual information capable of specifying the action-oriented structure of this world. The information used as the basis for action selection may be considerably more subtle and rich than the information preserved when using perception in a purely descriptive capacity (e.g., Bridgeman, 1991; Neisser, 1989; Shebilske, 1991).

Constructing the action-oriented environmental model requires the use of action capacities as a frame of reference for environmental description. Because two action capacities underlie search behavior (locomotion and sighting objects with radar), we must describe the action-oriented structure, or affordances, of this environment for both locomotion and sighting objects. We consider search affordances to be a simple combination of the locomotion and sighting affordances. Figure 4.6 shows the distribution of locomoting, sighting, and searching affordances as maps of the world in which the paths shown in Figure 4.5 were generated. Figure 4.6a shows the world as it appeared on the map display. Figure 4.6b shows the locomoting affordance, calculated by assigning a value of zero for open regions, a value of -1.5 to lightly forested regions, and a value of -2.0 to heavily forested regions. These values were assigned to reflect the difficulty of rapidly flying the scout through these regions differing in tree density. Darker regions on the maps indicate higher affordance values.

The affordance values were selected by attempting to construct an objective measure of the degree to which relevant actions could be performed as a function of environmental structure. For example, Fig4.6c shows the world sighting affordance structure. To construct this

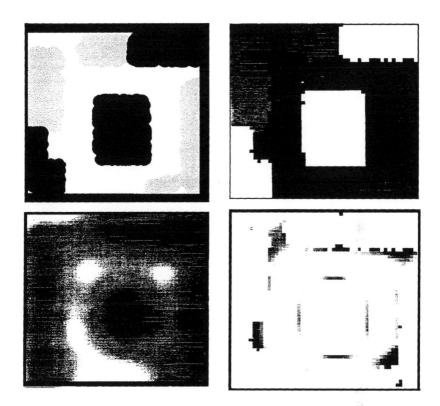

Figure 4.6. *Four representations of the same world. (a) Displayed representation. (b) Locomotion affordance representation. (c) Sighting affordance representation. (d) Search affordance representation scaled to emphasize local optima in the affordance structure.*

map, a four-dimensional vector was associated with each world location to indicate the percentage of area that would be covered by scout radar centered at the location that was open region, lightly forested region, heavily forested region, and area beyond the world boundaries. For each point, the inner product of this vector and a sighting affordance vector was taken to determine the sighting affordance of a particular world location. The sighting affordance vector was the same for each world location and indicated the density of cargo and enemy craft within each of the four types of regions. The sighting affordance vector had a value of zero for open regions and area beyond world boundaries, and a value of 1.0 for lightly and heavily forested regions. A maximal sighting affordance would exist, therefore, in cases in which the entire scout radar range covered a forested region, and a minimal sighting affordance would exist when the entire scout radar range covered either an open region or area beyond the world boundary. The graded structure of the sighting affordance distribution results from the complex interaction between the circular radar capabilities of the scout and the irregularly shaped open and forested regions that determined object density, or more generally, the interaction between the subjects' action capacities and the environmental structure.

Figure 4.6d shows the search affordance structure created by simply summing the values of the locomoting and sighting affordances at each world location (i.e., a location affords search if it affords both locomoting and sighting objects). This map has been rescaled to clearly indicate local optima in the search affordance structure. Considered three-dimensionally this map indicates peaks and ridges of high search affordance and valleys and holes of low search affordance. The peak areas indicate the best compromise between the conflicting demands for locomotion through open regions and sighting objects in forested regions.

In order to define opportunities for fluent perception-action solutions to this task, and also to define models of environmental depth structure and cognitive process, we now examine the congruence between the perceptually oriented and action-oriented environmental models. Note that our original perceptually oriented model differentiates the world differently than does the action-oriented model. The two models are apparently out of alignment. We now ask the question, what information contained within the perceptually-oriented model is available to specify the action-oriented structure, or affordances, in this the world? Using knowledge of human perceptual capacities together with knowledge of the displayed environmental structure, we attempt to find a way in which perception could possibly measure the displayed world in a manner that specifies the action structure to the most faithful extent possible. What perceptual information would be necessary to specify the search affordance structure?

First, note that the perceptually oriented model differentiates the world in an isomorphic manner to the differentiation provided by locomotion affordances. If perception can identify whether a given location is an open region, light forest, or heavy forest, then information is perceptually available to fully specify locomotion affordances. To fully specify search affordances, however, perception would also have to be able to measure the sighting affordance structure. Given the manner in which sighting affordances were constructed, we can define the nature of the necessary perceptual information in this case. Specifically, when foveating at a particular world location, perception would have to supply a measure of the amount of forested area within a circular area defined by the 1.5 mile radar radius of the scout. Although psychophysical experiments are surely needed to assess the degree to which this is possible (and could be straightforwardly conducted), in this case I simply assume that such perceptual judgments are possible, although we may expect certain forms of systematic measurement errors. The result of such experiments would be an empirically based, perceptually-oriented model of the world which indicates the information available to specify the search affordance structure. Constructing the action-oriented environmental model, however, was necessary to first identify what kinds of psychophysical experiments to conduct.

Assume, for the sake of this exercise, that the results of such experiments suggested that people did have the perceptual ability to specify the search affordance structure. (If the results indicated that the subjects could not reliably estimate the search affordance structure, we would then be able to define the kinds of cognitive activities necessary to do so for behavior to be productive, i.e., to be in alignment with search affordances.) What would search behavior look like if subjects were simply allowing search behavior to be governed by the perceptually detected search affordances? We would perhaps expect in this case that the scout would be flown up the steepest gradient in the search affordance structure from its present position. But although the local organization in search paths may be describable in this fashion, search paths also have a global organization that is not well captured by this simple search model.

This mismatch between observed behavior and the behavior that would result from a simple perception-action solution to this task is suggestive of what kinds of additional cognitive demands this task makes on the performer. We construe long-range or global path planning as the selection of a sequence of way points to be visited, in which each way point is a peak or ridge in the search affordance structure. We still however, can construe short-range or local navigation in terms of the simple perception-action model discussed above that results in search affordance gradient ascent. The selection of an appropriate sequence of waypoints is constrained by the need to avoid backtracking through previously searched regions, and also the need to return to home base at some point in the middle third of the mission. There is no readily available perceptual information capable of specifying these constraints on productive activity. Thus, we have a case of perceptual underspecification of action, in which the subject must apparently try out a number of alternative solutions to assess the downstream effects (i.e., consistency with backtracking and fuel constraints) of each, prior to selecting a global search path.

The resulting model is shown in Figure 4.7. At the start of a session, the model identified the peak areas in the search affordance map as candidate waypoints to visit during a mission. These peak areas were submitted to a generate-and-test mechanism which attempted to order the way points to acceptably meet backtracking and fuel constraints. Note that although a cognitively intensive process was needed for this purpose, a relatively small number of waypoints are considered, because the search affordance structure could be used to obtain a relatively low-dimensional representation of the world for long-range path planning (i.e., the set of local optima). Many of the fine details of local search affordance structure can be ignored during this process. The output of this process was an ordered sequence of waypoints. The first waypoint was then selected as a destination and was thus considered to possess an affordance for visiting. The scout did not fly in a linear path to the way point, however, because scout motion was determined not only by the visiting affordance but also by the local search affordance structure in the vicinity of the scout. Detailed motion commands for the scout were created by considering the perceptually measurable search affordance structure to operate on the scout as an attractive force field, which when combined with the force exerted by the visiting affordance of the current way point determined the direction of motion on a second-by-second basis. Large weights on the local search affordance values relative to the weighting on the visiting affordance provided by the current waypoint, resulted in meandering motion that was very sensitive to search affordance structure. In contrast, a large weight on the visiting affordance relative to the local search affordances resulted in a direct path to the current way point which largely disregarded the local search affordance structure. This search model was one component in an empirically validated process model of skilled human-environment interaction in the laboratory task. A discussion of the entire process model, including both continuous and discrete action components, can be found in Kirlik et al. (1993).

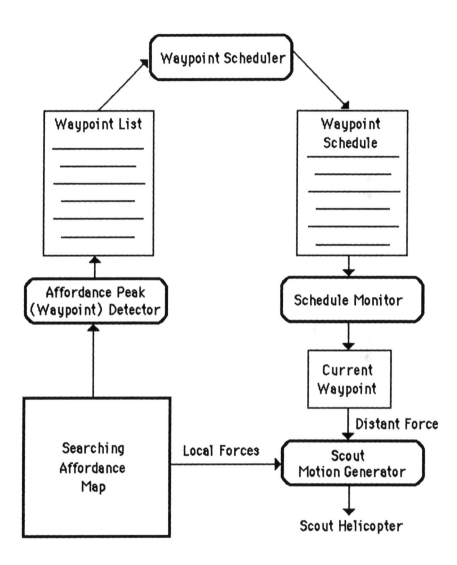

Figure 4.7. *Process model of search path generation. Local optima in the search affordance structure are organized into a series of waypoints via a generate-and-test scheduler. Detailed scout motion is produced by a combination of attractive forces from both the current waypoint and the local search affordance structure in the vicinity of the scout.*

4.5   Conclusion

Psychological models must meet special constraints in order to effectively support cognitive engineering. Current psychology methodology does not typically result in models that meet these constraints, largely because the environmental influences on cognition are investigated in piecemeal fashion. Although this manner of decomposing the cognitive modeling problem may result in research products that meet current scientific standards of merit, a progression toward the type of theoretical unification capable of guiding design is not evident in modern research. In addition, the current conception of the relation between cognitive modeling and design as one of theory versus practice does little to foster the development of either more design-relevant psychology or more psychologically-relevant design. Such a conception only serves to license the widening gap between psychological modeling, which abstracts away much of the complexity of the environment, and design practice, guided mainly by expert knowledge of particular environments but little psychological theory.

Surely, knowledge of both the internal and environmental constraints on cognition and behavior are necessary to understand any instance of human-environment interaction. However, modeling often proceeds as if the proper focus is on what happens internal to the performer, between eye and hand so to speak, and where environmental particulars can be tacked on as input parameters or variables. This approach does not do justice to the many ways in which environmental structure is reflected in cognition and behavior and does not result in models that can be used for identifying invariances in the behavior of the integrated human-environment system.

To overcome this problem, two requirements for psychological models to support design were identified. First, the human-environment system must serve as the unit of analysis in order for a model to serve as a tool for identifying invariances in human-environment interaction and for systematically describing how cognition and behavior are influenced by environmental design. Models without rich environmental components cannot guide design because they provide no descriptive resources for representing the many ways in which cognition and behavior are sensitive to environmental manipulation.

Second, models capable of guiding design must provide resources for describing the activities underlying fluent, skilled human-environment interaction, because this is often the design target. Models that conceptualize all activity as being guided by processes such as problem-space search or the comparative evaluation of options with respect to a utility function may be impressive from a sufficiency standpoint. However, these models are incapable of reflecting the distinction between the cognitively intensive activities that are well described in such terms, and more fluent activities that can surely be rationalized into these frameworks but are actually controlled in an economical perception-action mode. If models rationalize all activity as having a problem solving or decision-making basis, for example, how can such models prescribe design manipulations that *eliminate* demands for problem-solving or decision making? Only modeling approaches that admit the possibility that activity can be controlled in a fluent perception-action mode can motivate such design manipulations.

Guided by these requirements for psychological models to support design, ecological task analysis was proposed as a theoretical framework in which to identify the cognitive demands promoted by environmental designs and to suggest how interfaces can be manipulated to alleviate certain cognitive demands. The framework owes a heavy debt to Brunswik's approach for representing the human-environment system as an integrated unit and to Gibson's approach for understanding fluent interaction as guided by the perceptual specification of action opportunities. In ecological task analysis, two symmetrical models of the surface structure of the environment are constructed to reflect how the environment is differentiated with respect to both perception and action. These models are then used to identify matches between the perceptual and action structure of the environment, which reflect opportunities for the perceptual guidance of action and various forms of mismatch between perception and action structure. Each form of mismatch is reflective of a specified: (a) type of cognitive demand necessary to compensate for the mismatch (b) type of learning necessary to overcome the mismatch and (c) type of remedial design manipulation necessary to eliminate the mismatch and thus the associated cognitive demand. An example of ecological task analysis was described in which the framework was used to identify and model the perception-action and cognitive processes underlying route planning with a graphical map display.

Acknowledgments

I am grateful to Bill Gaver, Ken Hammond, Kevin Hodge, Art Ryan and the editors for valuable comments on a previous draft of this chapter, and to Peter Lenn and the kind folks at the Majestic Diner in Atlanta for assistance during the cook studies. This research was supported by NASA Ames Grant NAG2-656 to the Georgia Institute of Technology, Robert J. Shively, technical monitor.

4.6   References

Anderson, J. R. (1990). *The adaptive character of thought.* Hillsdale, NJ: Lawrence Erlbaum Associates.

Barwise, J., & Perry, J. (1983). *Situations and attitudes.* Cambridge, MA: The MIT Press.

Beach, K. D. (1988). The role of external mnemonic symbols in acquiring an occupation. In M. M. Gruneberg, P. E. Morris, and R. N. Sykes, (Eds.), *Practical aspects of memory: Current research and issues, (Vol. I.* pp. 342-346). New York: Wiley.

Birmingham, H. P., & Taylor, F. V. (1954). A design philosophy for man-machine control systems. *Proceedings of the I.R.E.,* XLII, 1748–1758.

Bohlman, L. (1979). Aircraft accidents and the theory of the situation. Resource management the flight deck. *Proceedings of a NASA/Industry Workshop. NASA Conference Proceedings 2120.*

Braitenberg, V. (1984). *Vehicles: Experiments in synthetic psychology.* Cambridge, MA: The MIT Press.

Brehmer, B. (1986). *New directions in Brunswikian research: From the study of knowing to the study of action.* Unpublished Manuscript, Uppsala University, Sweden.

Brehmer, B., & Joyce, C. R. B. (Eds.), (1988). *Human judgment: The SJT view.* New York: North Holland.

Bridgeman, B. (1991). Separate visual representations for perception and for visually guided behavior. (pp. 317–327). In S. R. Ellis, M. K. Kaiser, and A.J. Grunwald (Eds.), *Pictorial communication in virtual and reality environments.* New York: Taylor & Francis.

Brunswik, E. (1952). The conceptual framework of psychology. In, *International Encyclopedia of Unified Science .* (Vol. 1, No. 10). Chicago, IL: University of Chicago Press.

Brunswik, E. (1956). *Perception and the representative design of psychological experiments.* Berkeley: University of California Press.

Card, S. K., & Newell, A. (1989). Cognitive architectures. In J. I. Elkind, S. K. Card, J. Hochberg, and B. Messick Huey, (Eds.), *Human performance models for computer-aided engineering.* (pp. 173-179). Washington, D.C.: National Academy Press.

Carroll, J. M. (1991). Introduction: The Kittle House manifesto. In J. M. Carroll, (Ed.) *Designing interaction.* (pp. 1-16), New York: Cambridge University Press.

Carroll, J. M., & Campbell, R. L. (1989). Artifacts as psychological theories: The case of human-computer interaction. *Behaviour and Information Technology, 8,* 247–256.

Claxton, G. (1988). *Growth points in cognition.* New York: Routledge.

Donald, M. (1991). *Origins of the modern mind.* Cambridge, MA: Harvard University Press.

Edwards, E. (1988). Introductory overview. In E. L. Wiener and D. C. Nagel (Eds.), *Human Factors in aviation.* (pp. 3-25), New York: Academic Press.

Fischoff, B., Slovic, P., & Lichtenstein, S. (1978). Fault trees: Sensitivity of estimated failure probabilities to problem representation. *Journal of Experimental Psychology: Human Perception and Performance, 4,* 330–344.

Flach, J. M. (1990a). Control with an eye for perception: Precursors to an active psychophysics. *Ecological Psychology, 2,* 83–112.

Flach, J. M. (1990b). The ecology of human-machine systems I: Introduction. *Ecological Psychology, 2,* 191-205.

Gibson, E. J. (1969). *Principles of perceptual learning and development.* New York: Appleton-Century-Crofts.

Gibson, J. J. (1982). James J. Gibson autobiography. In E. Reed and R. Jones. (Eds), *Reasons for realism.* (pp. 7–22), Hillsdale, NJ: Lawrence Erlbaum Associates. (Original work published 1967).

Gibson, J. J. (1979). *The ecological approach to visual perception.* Boston: Houghton-Mifflin.

Gibson, J. J. (1966). *The senses considered as perceptual systems.* Boston: Houghton-Mifflin.

Hancock, P. A., & Warm, J. S. (1989). A dynamic model of stress and sustained attention. *Human Factors, 31,* 519–537.

Hammond, K. R. (1955). Probabilistic functioning and the clinical method. *Psychological Review, 62,* 255–262.

Hammond, K. R. (1966). *The psychology of Egon Brunswik.* New York: Holt, Rinehart, and Winston.

Hammond, K. R., Hamm, R. M., & Grassia, J. (1986). Generalizing over conditions by combining the multitrait-multimethod matrix and the representative design of experiments. *Psychological Bulletin, 100,* 257–269.

Hammond, K. R., Hamm, R.M., Grassia, J., & Pearson, T. (1987). Direct comparison of analytical and intuitive cognition in expert judgment. *IEEE Transactions on Systems, Man, and Cybernetics, SMC-17,* 753-770.

Hammond, K. R., Stewart, T. R., Brehmer, B., & Steinmann, D. O. (1975). *Social judgment theory.* (Center for Research on Judgment and Policy Report No. 176), Boulder: University of Colorado Institute for Behavioral Science.

Hutchins, E. (1991, August). *How a cockpit remembers its speeds.* Paper presented at the Second NASA Ames Aviation Safety / Automation Program Investigators Meeting.

Hutchins, E. (1988). *The technology of team navigation* (Institute for Cognitive Science Rep. 8804). La Jolla, CA: University of California, San Diego.

Kahneman, D., & Tversky, A. (1979). Prospect theory: An analysis of decisions under risk. *Econometrica, 47,* 263–291.

Kirlik, A., Miller, R. A., & Jagacinski, R. J. (1993). Supervisory control in a dynamic uncertain environment: A process model of skilled human-environment interaction. *IEEE Transactions on Systems, Man, and Cybernetics. SMC-23,* 929-952.

Klein, G. A. (1989). Recognition-primed decisions. In W. B. Rouse (Ed.), *Advances in man-machine systems research.* (pp. 47-92), Greenwich, CT: JAI Press.

Kossack, M. (1992). *Ecological task analysis: A method for display enhancement.* Unpublished Master's Thesis, School of Industrial & Systems Engineering, Georgia Institute of Technology, Atlanta, GA.

Kotovsky, K., Hayes, J. R., & Simon, H. A. (1985). Why are some problems hard? Evidence from the tower of Hanoi. *Cognitive Psychology, 17,* 248-294.

Larkin, J. H., & Simon, H. A. (1987). Why a diagram is (sometimes) worth a thousand words. *Cognitive Science, 11,* 65–99.

Marr, D. (1982). *Vision.* New York: Freeman.

McRuer, D. T., & Jex, H. R. (1967). A review of quasi-linear pilot models. *IEEE Transactions on Human Factors in Electronics, HFE-8,* 231–249.

Meister, D. (1989). *Conceptual aspects of human factors.* Baltimore, MD: The Johns Hopkins University Press.

Meister, D. (1984). Letter. *Human Factors Society Bulletin, 27*, 2.

Neisser, U. (1989, August). *Direct perception and recognition as distinct perceptual systems.* Address presented to the Cognitive Science Society.

Neisser, U. (1976). *Cognition and reality.* New York: W.H. Freeman.

Newell, A., & Simon, H. A. (1972). *Human problem solving.* Englewood Cliffs, NJ: Prentice-Hall.

Norman, D. A. (1986). Cognitive engineering. In D. A. Norman and S. W. Draper (Eds.), *User centered system design* (pp. 31-61). Hillsdale, NJ: Lawrence Erlbaum Associates.

Norman, D. A. (1993). *Things that make us smart.* Reading, MA: Addison-Wesley.

Rasmussen, J. (1986). *Information processing and human-machine interaction: An approach to cognitive engineering.* New York: North-Holland.

Reitman, W., Nado, R., & Wilcox, B. (1978). Machine perception: What makes it so hard for computers to see? In C. W. Savage (Ed.), *Perception and Cognition Issues in the Foundations of Psychology, Minnesota Studies in the Philosophy of Science, Volume IX.* (pp. 65–88), Minneapolis: University of Minnesota Press.

Rouse, W. B. (1987). Much ado about data. *Human Factors Society Bulletin, 30*, 1–3.

Sarter, N. B., & Woods, D. D. (1991). Situation awareness: A critical but ill-defined phenomenon. *International Journal of Aviation Psychology, 1*, 45–57.

Shebilske, W. L. (1991). Visuomotor modularity, ontogeny and training high-performance skills with spatial instruments. In S. R. Ellis, M. K. Kaiser, and A. J. Grunwald, (Eds.), *Pictorial communication in virtual and real environments.* (pp. 305–315), New York: Taylor & Francis.

Sheridan, T. B. (1987). Supervisory control. In G. Salvendy (Ed.), *Handbook of human factors* (pp. 1243–1268). New York: John Wiley and Sons .

Suchman, L. (1987). *Plans and situated actions: The problem of human-machine communication.* New York: Cambridge University Press.

Tolman, E. C., & Brunswik, E. (1935). The organism and the causal texture of the environment. *Psychological Review, 42*, 43–77.

Tversky, A., & Kahneman, D. (1981). The framing of decisions and the psychology of choice. *Science, 211*, 453–458.

Vicente, K. J. (1995). A few implications of an ecological approach to human factors. In J. M. Flach., P. A. Hancock., J. K. Caird., and K. J. Vicente, (Eds.), *Global perspectives on the ecology of human-machine systems*. Hillsdale, N.J.: Lawrence Erlbaum Associates.

Wason, P. C., & Johnson-Laird, P. N. (1972). *Psychology of reasoning: Structure and content*. Cambridge, MA: Harvard University Press.

Wiener, E. L., & Nagel, D. C. (Eds.) (1988). *Human factors and aviation*. New York: Academic Press.

Whiteside, J., Bennett, J., & Holtzblatt, K. (1988). Usability engineering: Our experience and evolution. In M. Helander (Ed.), *Handbook of human-computer interaction*. (pp. 791–817). New York: North-Holland.

Wise, J. A. (1985). Decisions in design: Analyzing and aiding the art of synthesis. In G. Wright (Ed.), *Behavioral decision making* (pp. 283-308). New York: Plenum.

Woods, D. D., & Roth, E. M. (1988). Cognitive systems engineering. In M. Helander (Ed.), *Handbook of human-computer interaction* (pp. 3–43). New York: North-Holland.

Chapter 5

Virtual Ecology of Work

Jens Rasmussen and Annelise Mark Pejtersen

Risø National Laboratory

5.0   Introduction

This chapter examines the design of information systems to support work in a period of rapid change of technology, market conditions, and company policies. Under such conditions, effective systems can no longer evolve by incremental empirical adjustment. Design must be based on a predictive conceptual framework.

For many modern work systems, stable work procedures are not the norm. Many tasks are discretionary. Explicit consideration of goals and constraints and an exploration of the boundaries of acceptable performance are required. For this reason, the object of modeling can no longer be the "task," but must include all the features of the work environment and the interpretation of these features by the actors. The interaction of task and operator constraints creates the task ad hoc.

Basically, human actors are goal directed, adaptive mechanisms. Great diversity in behavioral patterns is found among the members of an organization. No two individuals are occupied by the same activity; nor will a task be performed in exactly the same way twice. The variety of options with respect to "what to do when and how" in many work situations is immense. In order to predict why a particular piece of behavior is chosen instead of another possible pattern, we have to understand how the degrees of freedom (i.e., the action alternatives in a particular situation) are eliminated so that one unique sequence of behavior can manifest itself. As long as action alternatives remain, behavior is indeterminate until a choice is made. In other words, we have to identify the constraints which shape behavior by guiding the choices taken by the individual together with the subjective performance criteria that are applied by the individual actors to resolve the remaining degrees of freedom.

The framework must serve to represent the characteristics of both the physical work environment and the "situational" interpretation of this environment by the actors involved, depending on their skills and values. In order to bridge from a description of the behavior shaping constraints in work domain terms to a description of human resource profiles and subjective preferences, several different perspectives of analysis and languages of representation are necessary (see Figure 5.1).

It is necessary to adopt a strategy of analysis that converges rapidly by eliminating the degrees of freedom in the sets of behavior shaping constraints represented within the different dimensions:

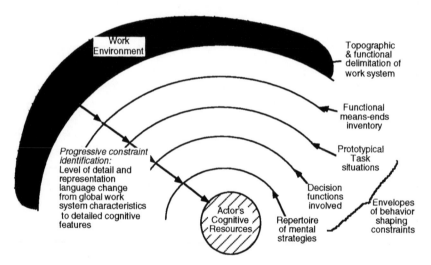

Figure 5.1. *Shift in language to relate descriptions of work and of cognition.*

1. A topographic delimitation of the work space should be found and an explicit identification of the goals, constraints, and means for action which are available to an actor.
2. A delimitation in time to determine the task situation will be made, followed by
3. A delimitation and shift in representation language to describe the decision task. The following step then involves a focus on
4. The mental activities and a related shift in language in order to have a description compatible with a representation of the actor's cognitive resource profile and performance criteria.

This framework supports a stepwise narrowing down of the degrees of freedom faced by an actor and, in addition, the necessary shifts in the language of description when going from the context of the work domain, the task situation, the decision and information processing task, onto human cognitive and emotional factors. In other words, the transformations serve to link descriptions of

1. The work or *problem space*
2. The situational *dynamics*
3. The *information* required for coping
4. The actor's state of *knowing*.

To this end, several different dimensions of analysis are necessary. The subsequent sections present a framework for cognitive work analysis that has emerged from our field studies (Rasmussen, Pejtersen, & Schmidt, 1990).

5.1   Identifying Behavior Shaping Constraints

5.1.1   Work Domain Constraints

A representation of the territory of work should identify the entire network of means-ends relations, that is, in Ashby's (1962) terms, the world of "possibilities" or "the requisite variety" that is necessary to cope with all the requirements and situations which may appear during work. For example, changes caused by the introduction of information systems are likely to propagate to all coupled activities. An overview of the means-ends network of an entire work organization can be useful to judge the likely propagation.

This domain representation defines the functional territory within which the actors will navigate or, in ecological terms, the *affordance space*. The means-ends representation is structured in several levels of abstraction as follows: The lowest level of abstraction represents the physical anatomy of the system and the appearance of its elements, that is, its material configuration. The next higher level describes the physical activities and processes of the various elements in a language related to their specific material properties (e.g., physical, mechanical, electrical, or chemical processes). At the level above this, work functions are represented by more general concepts without reference to the physical processes or parts by which the functions are implemented. At the level of abstract function the functional implications are found that are used to set priorities and coordinate resource allocation to the various general work functions and to compare their results with the goals and constraints formulated at the upper level. This level of abstract functions represents functionality/intentionality in terms of flow of values for which laws of conservation are valid, such as monetary values, energy, material, people, and so on. (For a detailed discussion of the means-ends hierarchy, see, e.g., Rasmussen; 1986, Rasmussen, Pejtersen, & Schmidt, 1990). An important feature of this complex means-ends network is the many-to-many mapping found among the levels. If this was not the case, there would be no room or need for human decision or choice.

At the lower levels, elements in the description represent the material properties of the system. When moving from one level of abstraction to the next higher level, the change in system properties represented is not merely a removal of detailed information about physical or material properties, but information added on higher level principles governing the co-functioning of the various elements at the lower level. In manmade systems, these higher level principles representing co-functions are derived from the purpose of the system (i.e., from the reasons and intentions behind the design, see Figure 5.2.

A change in the level of abstraction involves a shift both in the concepts and the structure of the representation. It is important to realize that the different levels represent information about the same physical world. The information used for representation at the various levels is chosen to serve as a set of links between the representation of the material work environment and its resources, on the one hand, and the representation of the ultimate human goals and objectives on the other. Thus, the means-ends hierarchy is formed by a progressive set of conceptual transformations.

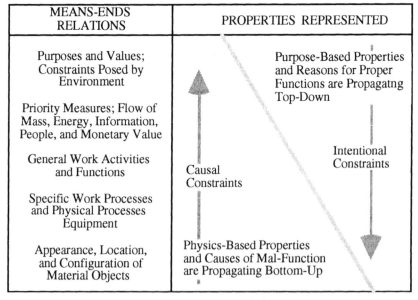

MEANS-ENDS RELATIONS	PROPERTIES REPRESENTED
Purposes and Values; Constraints Posed by Environment	Purpose-Based Properties and Reasons for Proper Functions are Propagatng Top-Down
Priority Measures; Flow of Mass, Energy, Information, People, and Monetary Value	
General Work Activities and Functions	Intentional Constraints
Specific Work Processes and Physical Processes Equipment	Causal Constraints
Appearance, Location, and Configuration of Material Objects	Physics-Based Properties and Causes of Mal-Function are Propagating Bottom-Up

Figure 5.2. *Any system can be described at several levels of abstraction.*

The levels of abstraction are illustrated in Figure 5.2 which also shows the different sources of regularity of system behavior underlying the representations at the various levels together with an indication of the directions of the propagation of changes. The higher levels of abstraction represent properties connected to the purposes and intentions governing the work system, whereas the lower levels represent the causal basis of its physical constituents. Consequently, perturbations of the system in terms of changes in company goals, management policy, or general market conditions will propagate downward through the levels, giving "rules of rightness" (Polanyi, 1958). In contrast, the effect of changes of the material resources, such as introduction of new tools and equipment or breakdown of major machinery will propagate upward, being causes of change.

Another way to look at the sources of regularity of the behavior of a goal oriented system is to consider them as behavior shaping constraints to be respected by actors working in the system. Any actor striving to control the state of affairs within a work system will have to operate on and through the internal constraints of the system. Control involves a change of the parameters of relational constraints in order to introduce a propagation of effects ultimately bringing the system into the intended goal or target state. This control involves operation on the causal constraints of the physical part of a system, or on the intentional constraints originating in the other actors of the system. Whether one or the other mode of control is appropriate depends on the task situation and the structure of the system.

In a tightly coupled technical system, such as a power plant, the intentional structure will be hardwired into a complex, automatic control and safety system, serving to maintain the functions of the entire system in accordance with the high-level, stable design goals (i.e., to produce power as requested by customers and as economically and safely as possible through the lifetime of the system). The physical processes of an automatic control system are only a medium for processing this intentional information and consequently will be of no significance except to the repair staff. The task of the operating staff is basically to make sure that the function of the automatic control actually represents the intentionality of the original design.

In contrast, in a loosely coupled workplace such as an office, a traditional manufacturing workshop, or a hospital, the functional coordination depends on the intentionality brought to work by the actors of the system. A city can be taken as an example of a loosely coupled work system and can be used to illustrate the role of the functional and intentional constraints represented by the means ends hierarchy of Figure 5.2. At the bottom level of the means-ends hierarchy, the representation of the material objects present in a city is found. The intentional element at this level is only represented in the selection and configuration of objects necessary to supply the requisite variety for the higher level functions. In a city, people and houses, furniture and tools, cars and street lamps are plentiful; whereas large boulders, wild animals, and village ponds are avoided or, at least, kept at controlled locations. The next level represents the processes of a city, such as moving goods and people, sleeping, feeding, shaping and assembling products, and chemical and physical production processes. The intentional aspect is the control of the configuration and boundary conditions of the material world so as to support the required processes. One level up, the functions of a city are found such as transport, trade, health care, administration, and public education. The intentional elements serve to coordinate the lower level processes so as to serve the various functions. At the level of abstract functions, the implications of the functions are represented in terms of values and resources absorbed. The intentional element represents the efforts to set priorities and to direct the flow of money, people, and goods so as to serve the higher level goals, such as the prosperity of the city, the well-being of the people, and the pride of the mayor.

The example illustrates how the functional aspects of the material world are represented up through all the levels, but in increasingly abstract and global terms. The intentional component is becoming increasingly influential at the higher levels and increasingly complex as more global functions are considered. It is also seen that the intentional part of the domain representation constitutes a hierarchical control function, serving to structure the functional implications of the material world at all levels. These functional implications propagate upward through the levels, whereas an explication of the intentional control criteria will propagate downward. Decision making by the actors can be required at any of the means-ends levels. Where decisions are made will depend on the level at which the actor perceives the discrepancy between the actual state of affairs and the intended target state. To visualize the ecology of work, therefore, it is necessary that the abstract representation of the functional implications matches the presentation of the intentional explications at each level.

For the individual decision maker, the intentional element is a very real part of the domain, being embedded in the form of behavior-shaping constraints present in the institutions and the accepted rules of conduct of the work system. In this way, the intentional system of a city is well established. It is, however, still flexible, and public debates, political parties, and efforts of the local government serve to modify the structure in response to the results of experiments made by political activists, social scientists, people violating rules of conduct (e.g., by stressing tax laws), and so on.

The properties of different work domains represent a continuum from tightly coupled, technical systems in which the intentional properties of the work domain are embedded in the functions of a control system, through systems in which intentionality is represented by social rules and values, to systems in which the entire intentional structure depends on the actual users' subjective preferences. The relationship between the causal and intentional structuring of a work system and the degree to which the intentionality of the system is embedded in the system or brought to work by the individual actor is an important characteristic of a work domain to be considered in the design of ecological interfaces.

5.1.2   Work Situation Aspects

Activities in Domain Terms. The work domain description discussed in the previous section is a stationary, situation-independent representation of the options for choice among alternative means-ends relations. The representation is an inventory map of the options of the actors in all relevant work situations. It will, in general, be very complex, simply because all "possibilities" should be included, even if they are only implicitly present in the established work practice at the time of analysis.

Traditionally, the identification of work requirements in a particular situation is done in terms of a task analysis to produce a description in terms of sequences of actions. This is no longer an adequate approach when tasks are discretionary. Instead, activities must be decomposed and analyzed in terms of a set of problems to be solved or a set of prototypical work situations that represent recurrent, natural islands of activity which have reasonably well-defined boundaries. That is, the activities are labeled in work domain terms, (e.g., manufacturing, libraries, or hospitals). For each of these work situations, the relevant subset of the means-ends relations (i.e., the "actualities") will be instantiated and will control the behavior of the involved staff.

The identification of the subset of behavior-shaping functional constraints which are relevant in a particular situation is rather straightforward. This is not the case for the intentional constraints. In most loosely coupled work systems, many degrees of freedom are found at the intermediate levels. Institutional intentional constraints, propagating top-down, will interact with local, situation-, and person–dependent criteria. Therefore, identification by an actor of the local, intentional behavior shaping constraints depends on the ability to identify the state of cooperating agents' work intentions. In many cases, actors judge the intentions of cooperators through informal communication channels (e.g., body language and other nonverbal cue see Rasmussen, Pejtersen, & Schmidt, 1990).

Activities in Decision-Making Terms.   In the previous section, activities were analyzed in work-domain terms. At a certain level of analysis, however, it is necessary to switch to a description in decision–making terms so that the information requirements of a task can be determined. This representation can be used to analyze possible mental strategies and their match with the competence and cognitive resources of individual actors.

Decision making is more manageable when decomposed into subroutines connected by more or less standardized key nodes. Such prototypical nodes represent states of knowledge about different features of a task and are very useful for linking different processes, for bringing results from previously successful subroutines into use in new situations, and for communicating with cooperators in decision making. This is important because a complex decision task may be shared, not only by a number of cooperating team members, but also across time with procedure designers, system programmers, and computers.

This level of analysis as well as the following consideration of mental strategies are focused on information aspects. The basic categories of decision functions to be considered for a more focused identification of behavior shaping constraints include information retrieval, situation analysis and diagnosis, priority judgment and goal evaluation, planning, execution, and monitoring. These decision functions involve different information processes. The identification of these processes is necessary to map onto a description of the mental strategies that are likely to be chosen by an actor in a particular work situation and that will influence an actor's situational interpretation of the constraints found in the work domain.

The "objective" behavior shaping constraints as identified by analysis of the work domain and task situations are transformed by the interpretation of the individual actor through the choice of mental strategy. This transformation not only depends on the level of representation of the work system, the actor considers, but also on the implicit representation of the constraints in the actor's repertoire of heuristics derived from previous encounters. The perception of the work environment changes as constraints become implicit in work strategies. Thus, perception is embedded in a tacit and subjective representation of the work context. Consequently, it is necessary to study the different analytical and empirical strategies brought to work by actors at various levels of expertise.

5.1.3   Aspects of Mental Strategies

This dimension of analysis serves to identify the mental strategies that can be used and are likely to be chosen by an actor. The context of reasoning, that is, the causal and/or intentional field within which analytic reasoning and perceptual judgment will take place, depends on the mental strategies applied. The choice of strategy depends on the competence and performance criteria of the individual actor and, therefore, determines the behavior-shaping constraints at the detailed level.

It is well known that several, basically different mental strategies can be used for a decision task. This has been studied for diagnosis (Rasmussen, 1993) and libraries (Pejtersen, 1979). Analysis of diagnostic strategies (Rasmussen, 1991) demonstrates that the context in which diagnostic reasoning takes place in the various strategies can be of a very different nature. In search by recognition, the context is only implicitly defined in terms of the pool of episodic experience possessed by the diagnostician. In decision table search, the diagnostic context is a hierarchically ordered set of categories found by induction (typical for medicine) or deduced from a model of the functional structure of the system (as typical for operating instructions for technical systems). For diagnosis by hypothesis and test, the diagnostic context is a representation of the functional structure of the system used for deduction of symptoms from a postulated cause to be matched with the observed symptoms. In topographic search for the location of a fault, the diagnostic context is a representation of the physical or functional anatomy of the system.

Consequently, the manifestation of the properties of a particular system in terms of behavior shaping constraints to be considered for modeling behavior and for system design will vary significantly as a function of the mental strategy taken by the individual actor. This problem is even more complicated, because the strategies, their context, and the perspective taken by the actor will change dynamically during a decision task.

A *mental strategy* is an abstraction describing one consistent reasoning approach characterized by a certain kind of mental representation and a particular interpretation of observations. The concept of a *formal* strategy serves to formalize the information processes, their information requirements, and the cognitive load involved. The *natural* strategies applied in actual work situations, for several different reasons, involve very frequent shifts among such formal strategies caused by very subjective and situation-dependent factors. One reason to shift between strategies is their very different resource requirements with respect to time taken, type of mental model available, number of observations necessary, and so on. Shift in strategy is a very effective way to circumvent local difficulties along the path of work.

Another reason for a shift of strategy in a particular work scenario is that the diagnostic objective will change during a session. Initially a medical doctor or a process operator will be concerned with the question of whether he or she is confronted with a need for rapid compensatory action, that is, the user is concerned with the potential consequence of the present state. Next, the operator will be concerned with the choice of a function to control for compensation. Then he or she will be concerned with the correction of the present abnormality and, finally, may be concerned with prevention of a repetition. This means that a diagnostician may need to shift between strategies and thus change perspective in the interpretation of constraints several times during a task. An important implication of this discussion is that an information system should support all the strategies relevant for a task to allow users to shed mental work load by shifts among strategies. Forcing users to work through problems using only one strategy, preferred by the designer, rather than to pass around them by another strategy will strongly influence acceptance of a system. It is, therefore, important to study how the choices of and shifts among mental strategies influence the interpretation of behavior-shaping constraints. Understanding this interpretation is essential for the design of proper representations for interfaces to the work domains.

To sum up, the means-ends map represents the problem space, that is, the structure of the work system and the conceptual levels at which the functional and intentional constraints can be formulated (more or less) objectively, as seen from the system's point of view. Analysis of task situation characteristics of the work system serves to select those constraints that are active at certain times and, in a way, focuses attention on the dynamics and relevant paths open in the space for certain work functions. An identification of decision functions serves to transfer the information processing domain. The analysis of mental strategies that can be used serves to identify the subjective interpretation of the behavior shaping constraints related to different subjective performance criteria and cognitive resource profiles, that is, the actor's knowing discussed in the next section.

5.1.4   Aspects of Cognitive Resources

This dimension of analysis serves to uncover the competence and cognitive resources of the actors and their subjective performance criteria. Humans have different modes of control of their interaction with the environment. During familiar circumstances, interaction is based on a real-time, multivariable, and synchronous coordination of physical movements with a dynamic environment. Quantitative, time–space signals are continuously controlling movements. The automated patterns are activated and chained by cues perceived as signs; no choice among alternatives is required. During skill-based performance, higher level conscious control may be concerned with rehearsal of previous scenarios to prepare the world model for future demands by updating the state of the dynamic world model and preparing it for the proper response when time comes. This anticipatory updating of the internal model is equivalent to Gibson's (1966) attunement of the neural system underlying direct perception of invariants of the environment in terms of "affordances."

When this intuitive reaction to the context is no longer effective, a mismatch will be experienced between the state of affairs in the environment and the intuitive expectations of an actor. In this case, a skilled professional will normally perceive a small number of alternatives for action, and efforts will be focused on search of information that can resolve the ambiguity, that is, performance depends on active perception often calling for exploratory actions. The important point to consider here is that an expert will need no more information than is necessary to resolve the choice between the perceived action alternatives. If only two alternatives are perceived to be present, only one bit of information is needed as long as the actor is embedded in the context, even when it is a complex work environment (Rasmussen, 1991).

An important implication for interface design is that decision makers are not subject to "information input" from an environment that has to be analyzed; they are asking very specific questions. They are, in a Gibsonian sense, able to consult "invariants" in the environment by direct perception if the environment is well structured and transparent. In this way, performance in a familiar environment can be represented by a cue-response hierarchy, based on the human ability to read off cues at several levels of abstraction in the means-ends representation.

5.2   Design of Ecological Interfaces

Support of work in a dynamic environment cannot be aimed only at tools for preplanned tasks. Exploration of ends and means, opportunities and constraints present in a work situation, is necessary for adaptation to new requirements. An information system that presents for decision makers a complex, rich information context as a scenario for direct perception as well as for "thought experiments" can be an effective support for natural decision making. Complexity in itself is not a problem, given meaningful information is presented in a coherent, structured context. However, decision makers can be overloaded by presentation of many separate data when interface designers fail to provide a coherent, structured context. Professionals in work are not subject to information input; they are actively asking questions to the environment based on their perception of the context. Therefore, a rich support of perception in context is a key design issue. It is our experience that outside designers and consultants who are not thoroughly familiar with a work domain often overestimate the amount of information required for action by a specialist who is effectively adapted to the work and engaged in its context. At the same time, these designers underestimate the complexity of a meaningful display acceptable to such experts who are able to selectively tune to the cues necessary for action.

Design of ecological information systems attempts to exploit the large capacity of the human perceptual and sensor--motor system as it is found in natural environments by supplying a complex, but transparent information environment. Direct perception becomes possible if the representation is appropriate both for the level in the means-ends structure most relevant for the task and for the level of cognitive control at which the user chooses to perform. In Gibson's terms, the designer must create a virtual ecology which maps the relational invariants of the work system onto the interface in such a way that the user can read the relevant affordances to actions. In other words, an effective way to support actors in a dynamic work environment in coping with new requirements is to present a map of the territory of work in which actors have to navigate. The ecological system approach can be summarized in three principles (Rasmussen & Vicente, 1989):

1. *Skilled routines*: To support interaction via time-space signals, the user should be able to act directly from and on the display, and the structure of the displayed information should be isomorphic to the part-whole structure of the repertoire of automated sensori-motor patterns.

2. *Rule-based know-how*: Provide a consistent one-to-one mapping between the work-domain constraints and the cues or signs provided by the interface to release familiar actions.

3. *Knowledge-based problem solving*: Represent the work domain in the form of an abstraction hierarchy to serve as an externalized mental model that will support knowledge-based action planning.

5.3 Interface Design Guide: Maps of Design Territory

A complete cognitive work analysis following the framework reviewed in the previous sections is a time- and resource-consuming undertaking which cannot be spent on every system design. It is, therefore, important to be able to generalize from the analyses already made and to derive some general guidelines for design of ecological information systems. In the previous section it was argued that displaying the invariant structure and the affordances of the work domain is a more effective way to support actors during discretionary tasks in a dynamic work environment than procedural guides. In a world of dynamic requirements, a map supports navigation more effectively than route instructions. This argument can, by recursion, be applied to development of design guides, and it suggests generalization in terms of a description of the design basis of prototypical examples at different locations of a map of the general territory of system design.

It is the hypothesis underlying this approach that, by presenting conceptual maps of the territories relevant for information system design, it is possible to give a designer an improved intuition concerning similarities and differences between system designs relevant for different domain-task-user combinations. The following maps of design territories are considered for this purpose:

1. A map of prototypical work situations
2. A map of knowledge representations in design
3. A map for guiding display composition.

5.3.1 A Map of Prototypical Work Situations

The map in Figure 5.3 displays horizontally different work domains in a continuum. To the left are found loosely coupled systems constituted by a set of separate tools used by an autonomous user who defines the problem and composes the particular work domain on each occasion. To the right are found the highly structured and tightly coupled technical systems, such as process plants, and so on, typically served by dedicated operators who are paced by the system. Between these two extremes are found the autonomous actors in work environments constrained by policies, legislation, and other forms of regulations. The system design considerations between the two extremes are very different. The general human computer interaction (HCI) design guideline development traditionally is focused on the left part of the map, including such activities as Norman–Schneiderman approaches to "direct perception-direct manipulation" interfaces. Much computer supported Cooperative Work (CSCW) effort is focused at the center of the map, whereas the ecological interface approach typically is focused on the right-hand side of the map.

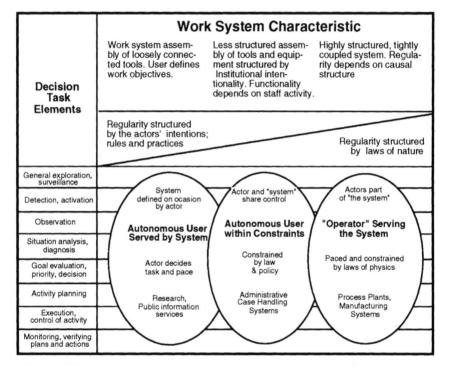

Figure. 5.3. *The map of prototypical work situations illustrates a continuum stretching from domains constituted by "natural environments" with loosely coupled assemblies of objects on the one hand to highly structured and tightly coupled technical systems at the other extreme.*

This map has three major dimensions; the typical work domain features, the related task situations, and the user profiles. Position on the map represents different prototypical WTU examples. Each prototype is intended to be an example of a class defined by a triplet WTU relation. Identifying the general class of an instance may allow generalization of design results across work situations which, on the surface at least, may seem to be very different from each other.

5.3.2   A Map of Knowledge Representations in Design

Figure 5.4 illustrates another conceptual map useful for design. This map gives an overview of the many different types of knowledge representations to be considered. The mental models of the ultimate user are different in content and form from the models developed by work analysis which, in turn, are different from the models used for interface design and from the formal models necessary for software development. The paths of transformation connecting these different models depend on the characteristics of the work domain. A map representing their relationship can be useful for generalization. This map is used in the examples discussed in the subsequent sections to illustrate the path of design in different work domains. The figure is organized into three vertical columns representing four categories of information.

Problem space: The center column shows the representations of the properties of the work system as described along the different dimensions of the framework for work analysis, such as

1. The means-ends structure of the work domain
2. The task situation in work terms
3. The decision task
4. The cognitive strategies that can be used.

The work analysis includes several levels of abstraction, representing the physical anatomy and topography, the processes in particular activities and tools, general functions, and the goals and value structures. This analysis defines the content of the knowledge base of the information system that must be available through the interface.

Knowing. The right-hand column of the Figure 5.4 shows the representations of the work domain which are found in textbooks, manuals, training material, and in the professional literature. Such material will influence the mental representations of system users. The figure also includes the influence from general information sources such as the media, television, news papers, and so on. The influence of these general information sources are particularly important for casual users. Such representations often have established general population stereotypes which can be exploited as metaphorical representations; for example, as in the multiple metaphor–multiple agent concept used for the Japanese "21st century personalized information environment" (Friend'21, 1991). These general sources can cause interference if not considered. Figure 5.4 illustrates that the choice of the form of display representation to some degree is independent of its content, and different forms can be chosen for representation to match the interpretation involved in skill-, rule-, and knowledge-based operations.

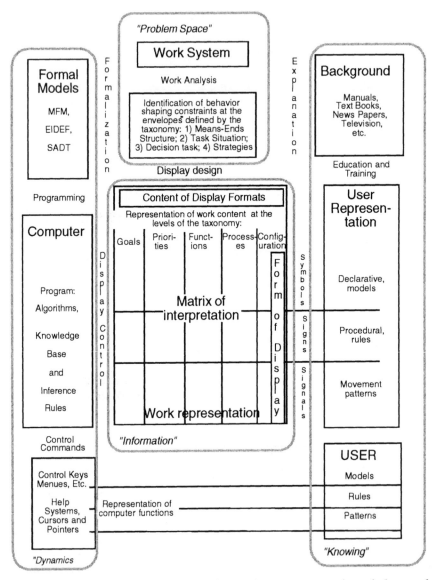

Figure 5.4. *A map showing how the work system, user knowledge, and information interact to determine knowledge representation.*

Information. The matrix in the center of the figure illustrates the complex matching problem to solve in interface design when the content of the interface should match a user's discretionary choice of means-ends level to work at. At the same time, the form of the display should match the need of the user's varying level of expertise.

Dynamics. The left-hand column shows that formalized, computational representations are necessary for driving the information systems. Normally, a design will aim at a direct communication between user and work through the center column. In some cases, however, control of parameters or functions of the program system is required, and the relevant program information must be available at the interface (see the bottom row of Figure 5.4). This communication path is useful in systems serving professional users who need to adapt the system to their particular needs.

5.3.3 Map of display composition and interpretation

The various modes of interpretation of information applied by users depends on their level of expertise and cognitive style and on the particular task situation. This interpretation will change with work domains, professions, cultures, and nations, and a conceptual map is important to guide design of prototypical graphic interfaces, icons, and metaphors.

The aim of ecological display design is to present a rich information environment that leaves the choice of level of interpretation and of cognitive control to the discretion of a user. For generalization it will be advantageous to identify some regularity in the relationship between the composition of complex displays and the structure of work requirements. In the present section some preliminary observations are discussed.

In most activities, a work sequence will involve shifts in the level of means-ends relations considered and in the span of attention of an actor. The examples given by Gibson (1966, 1979) of affordances and invariants which can be directly perceived in a natural environment show that the level of abstraction at which the environment will be perceived can be varied at will by the actor (see Rasmussen et al., 1990; Vicente & Rasmussen, 1990). The span of attention and the degree of perceptual chunking are discretionary and will be related closely to the level at which the actor expresses an intention to act. This, in turn, is determined by the complexity of the automated patterns of movements available to the actor for the particular activity.

This suggests that in order to facilitate skilled sensorimotor performance in complex work domains, a similar mapping between perceptual chunking and sensorimotor integration should be built into the interface. In other words, the interface composition should be isomorphic to the part-whole structure of an actor's hierarchical perception of task elements. This can be accomplished by revealing higher level information as aggregations of lower level information (e.g., through appropriate perceptual organization principles). In this way, multiple levels are visible at the same time in the interface, and the user is free to guide his attention to the level of interest, depending on his or her level of expertise and the current domain demands.

Low level display elements, therefore, should map closely to actions on physical items or components. Because such items will no doubt be found in a wide context, the form should match population or professional stereotypes. At a somewhat higher configural level, activity is related to functional relations for which stereotype representations (in some cases standards) for drawings and manuals are found within the involved profession. At the global level, activity is related to the constraint structure of the global task situation, and the representation for reasoning is specific for the kind of system, and the figural representation is based on the established conventions. This is, in particular, the case for the tightly coupled, technical systems illustrated in Figures 5.8 to 5.10 (see later in chapter). Also, Flach and Vicente (1989) advocated the use of the abstraction hierarchy as a framework for parsing the functional structure of work domains together with Gibson's concept of nested invariants as a framework for parsing the display composition in a way that will map naturally onto the abstraction hierarchy.

These considerations are focused on the structuring of the content of a composite display. The visual form to choose for display coding depends on the level of cognitive control for which it is intended. Figure 5.5 indicates the relevant sources and transformation paths.

For knowledge based reasoning, the visual form serves as a symbolic representation, that is, an externalized mental model of the relational network representing the work content at the proper level of the means-ends hierarchy. The form of displays should not be chosen to match the mental model of actors as found by analysis of an existing work setting, but so as to induce effective mental models for the tasks at hand in the new work ecology. For such representations, the established graphical designs used for teaching and found in textbooks are probably the best sources of design ideas, often having been used by system designers for decades.

In addition to the graphical representations which relate symbolically to the contents of the work domain, special consideration should be given to those elements of the display formats that basically serve as signs (i.e., cues for action at the rule-based level of performance). Symbolic, graphical representations used for knowledge–based reasoning will often also offer convenient cues for action. Being symbolically representative of system states, they will offer action cues based on defining attributes and, thus, serve to prevent errors due to underspecified actions.

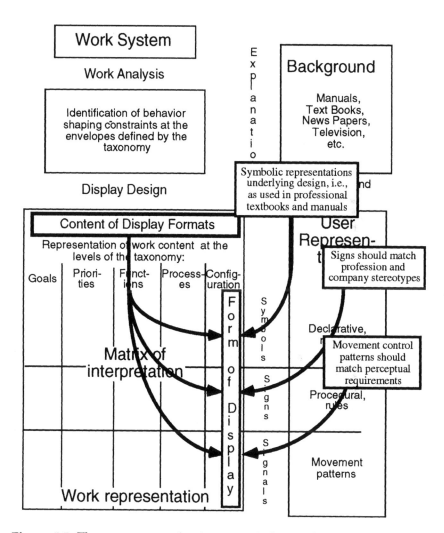

Figure 5.5. *The many sources that interact to influence the appropriate content and form for display formats.*

However, often special command icons are necessary to allow the user to manipulate the system in some way. Such icons are smaller pictorial entities on the display screen serving as signs that are related to actions by convention. They should be chosen so as to be easily understood and remembered (e.g., by their resemblance and/or suggestiveness regarding the related action).

For support of skill-based operation, the visual composition should be based on visual patterns that can be easily separated from the background and perceptively aggregated/decomposed in a way that matches the formation of patterns of movements at different levels of manual skills.

Navigation in the Information Systems. In general, the decomposition of the work domain in a particular situation is represented in a set, probably a hierarchy, of display formats. Therefore, special means are necessary to support users' navigation throughout the representation system in order to maintain their "cognitive momentum" throughout the accession process without "getting lost" (Woods, 1984). The problem is to create in the users a cognitive map of the display system including their present location, without constraining any possibility for them to develop and choose their own preferred work trajectories. Different approaches to this guidance are discussed for process plant and library systems in the subsequent sections.

5.4   WTU Examples

The aim of ecological information systems is to present to the user the relevant information about the state of affairs in the work environment structured according to its invariant properties. The regularity of behavior of a work domain comes from a combination of functional and intentional structures, and the invariants to be used in an interface, consequently, depend on the type of system considered. In the following examples, different sources of regularity to be considered for interface design in particular task situations are discussed for some characteristic WTU sets.

5.4.1   Operation of Tightly Coupled Technical Systems

Domain characteristics. The first example represents the tightly coupled technical systems located at the extreme right side of the spectrum in Figure 5. 3, such as power plants and chemical process plants. The basic source of regularity of the behavior of this category of work domains is the physical functionality governed by laws of nature. Functional invariants derived from laws of nature propagate upward through the means-ends hierarchy, because the entire system is *functionally integrated and tightly coupled* at all levels. At the physical level, functionality depends on the boundary conditions given to the physical processes by the material topography and anatomy; at the process level, functionality is governed by laws of nature constraining the particular electrical, mechanical, or chemical processes. At the level of general functions, invariant functional relations are governed by general laws of control theory, thermodynamics, information theory, and so on. The abstract functions relating to priority measures are governed by functional invariants defined by laws of conservation of values, matter, and energy.

When the functional degrees of freedom in a tightly coupled system are constrained by the physical anatomy and the resulting well defined functional structure, functional invariants can be defined by quantitative relationships at all the means-ends levels. These quantitative relationships can then be used to structure form invariants within ecological displays. The reasoning of the operating staff to explain the behavior of the process system will normally be based on the functional, relational structure of the process system.

The intentional structure defining the normal operational states of technical systems and the necessary protection against disturbances has normally been planned by the designer and embedded in an automatic control and protection system. The higher level goals with respect to production and safety are decomposed and implemented in terms of target states and boundaries of acceptable operation downward through the levels of a hierarchically organized control system that imposes target states and constraints on the quantitative relationships among functional variables. Because reasoning about the functioning of the control system, as mentioned, is based on the intentional design basis, it is important to make the design intentions accessible to the operating staff. In general, great effort is spent to supply the operating staff with factual data, whereas little is done to include design basis information in the information displays. This is a topic well suited for expert system technology. The subjective intentionality of operators is only relevant in boundary cases, for example, between whether to trust instructions or whether to rely on one's own judgment (see Figure 5.6).

Task Characteristics. For high hazard systems, such as industrial process plants, decision support systems are designed to support the task situations related to monitoring and disturbance control. Design priorities are given to supporting performance in the rare situations of disturbances that punctuate long periods of stable, automatic operation. In that case, the ecological approach to interface design aims at information systems and displays that can support skill- and rule–based performance during normal operation and, at the same time, can serve knowledge-based analysis during disturbances. Disturbance control involves diagnosis which, in a well structured system, can be effectively done by strategies comparing the actual state of the system with the normal, intended state in order to identify the location of the disturbance in the relational network. One important characteristic of process systems is, that the level of the means-ends hierarchy at which the diagnostic problem should be considered, varies with the phase of a disturbance control task and the related diagnostic perspective (Rasmussen, 1993).

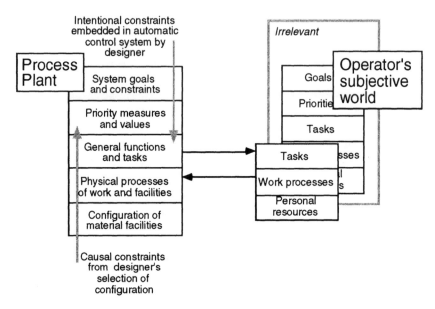

Figure 5.6. *In a tightly coupled, technical system, both the functional and the intentional structures are embedded in the physical configuration of the process and control system respectively.*

User Characteristics. In industrial process systems, the users are dedicated system operators, thoroughly familiar with the system. One important, operational role which is considered here is to monitor the function of the automatic control and safety systems and to control the operation during disturbances. The intentionality of actors is tied up by the system. The actors are dedicated system operators, and only during situations when an operator has to take on the role as a system designer (i.e., in situations requiring ad-hoc, goal-directed reconfiguration of the system) is some of the intentional structure dependent on the operators.

Content of Ecological Displays. For highly structured technical systems, the source of behavior shaping constraints to be represented in the display formats is exclusively defined by the work system because dedicated operators are serving the needs of the system according to the design intentions. In Figure 5.7, the map of knowledge representations is used to illustrate the path of analysis for identification of the content of information in displays.

Visual support of diagnosis in a system which has a stable and well–defined functional structure can be given by a display overlaying actual and intended states at different levels of representation. For direct perception diagnosis, that is, identification of the presence of a discrepancy between the actual and the intended states of operation, it is

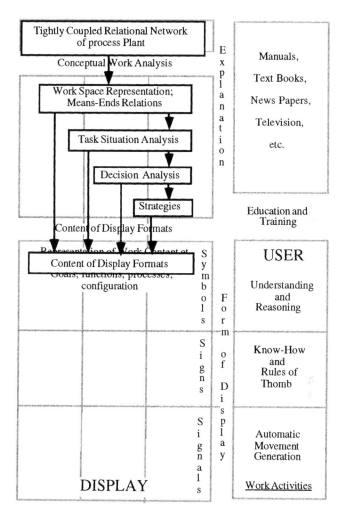

Figure 5.7. *In highly structured and tightly coupled technical systems, the elements as well as the structure of the domain representation are defined by the work system, and the source of relational structures of the display formats is entirely defined by the work system. The users are operators serving the needs of the system according to the design intentions.*

advantageous to represent the actual, functional state (the relationship among measured variables) and the intentional target and limit states by similar configural means. Consequently, the intentional structure (the intended operating states and the boundaries to unacceptable operation) should be represented by an overlay on the representation of the functional structure and states. Therefore, for this category of systems, the representation of invariants is normally in terms of diagrammatic representations of relationships among quantitative variables.

Form of Ecological Displays. The design of composite displays can draw advantage from the fact that the span of attention and shifts in abstraction are correlated in most task situations, and the trajectory in the abstraction-decomposition map is primarily along the diagonal as shown in Figure 5.8.

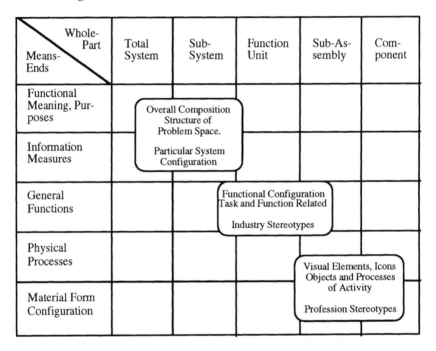

Means- Ends \ Whole- Part	Total System	Sub- System	Function Unit	Sub-As- sembly	Com- ponent
Functional Meaning, Pur- poses					
Information Measures					
General Functions					
Physical Processes					
Material Form Configuration					

Overall Composition Structure of Problem Space.

Particular System Configuration

Functional Configuration Task and Function Related

Industry Stereotypes

Visual Elements, Icons Objects and Processes of Activity

Profession Stereotypes

Figure 5.8. *In the design of composite displays, the source of regularity or the inspiration for design will have to be related to different sources of visual codes, depending on the level of aggregation.*

For an ecological display intended to give an overview of a technical process system to support monitoring and initial diagnosis of disturbances, therefore, the shift in level of abstraction of the representation should be coupled to the level of decomposition of the display elements. At the global level, display composition should reflect the plant structure and will be specific for the kind of system. Low level display elements, on the other hand, should map closely to actions on physical items or components. Because such items will no doubt be found in a wide context, the form should match population or professional stereotypes. At the intermediary, configural level, activity is related to functional relations for which stereotype representations (in some cases standards) are generally found within the involved profession. These representations are often found in textbooks in which they are used to illustrate the functional relations for students new to a profession.

An example of an ecological display for monitoring and diagnosing the operational state of the thermodynamic process of a power plant is shown in Figure 5.9 with an indication of the source of visual configurations as indicated by Figure 5.5 (for a detailed discussion, see Rasmussen & Vicente, 1992).

Heads-up Displays for Aircraft Piloting. Piloting of aircraft is another area for which the development of ecological displays has been explicitly considered.

Domain Characteristics. In this case, the work environment is characterized by the interaction of a vehicle with well-defined dynamic properties with a continuously changing topography, an interaction that can be characterized by a stable set of anatomical and configural elements which can be quantitatively represented.

Task Characteristics. Aircraft piloting depends heavily on skill and rule-based competence, and performance during normal as well as infrequent situations is important. Knowledge-based performance is typically found during rehearsal of critical scenarios ahead of missions for preparation of proper skill or rule-based responses (Amalberti & Deblon, 1992).

Ecological Display. Research on display design has focused on representation of those features of Gibson's "optical array" in a normal flight environment that supports perception of speed, height, and flight direction. Most "heads-up" displays integrate measured flight variables into a dynamic analog representation of the behavior-shaping features of the topography of the environment modulated by the movement of the craft. This representation can be generated by a representation of the quantitative relationships of the environment and the flight dynamics. Recently, designs have been introduced that present the pilots an envelope around the flight path recommended by the computer based fight management system. These recommendations are based on measured data and a model of the craft's flight dynamics and limiting properties. An example from Stokes and Wickens (1988) is shown in Figure 5.10. This kind of display is an implementation of Gibson's concept of "the field of safe travel" (Gibson & Crooks, 1938).

Navigation in Information Systems. In the previous sections, the discussion has been focused on a single, composite display for support of operations monitoring initial diagnosis of disturbed operations (i.e., faults). For support of the detailed diagnosis and resource management, more detailed displays are necessary at lower functional levels, and an ecological information system in an actual control room will include a

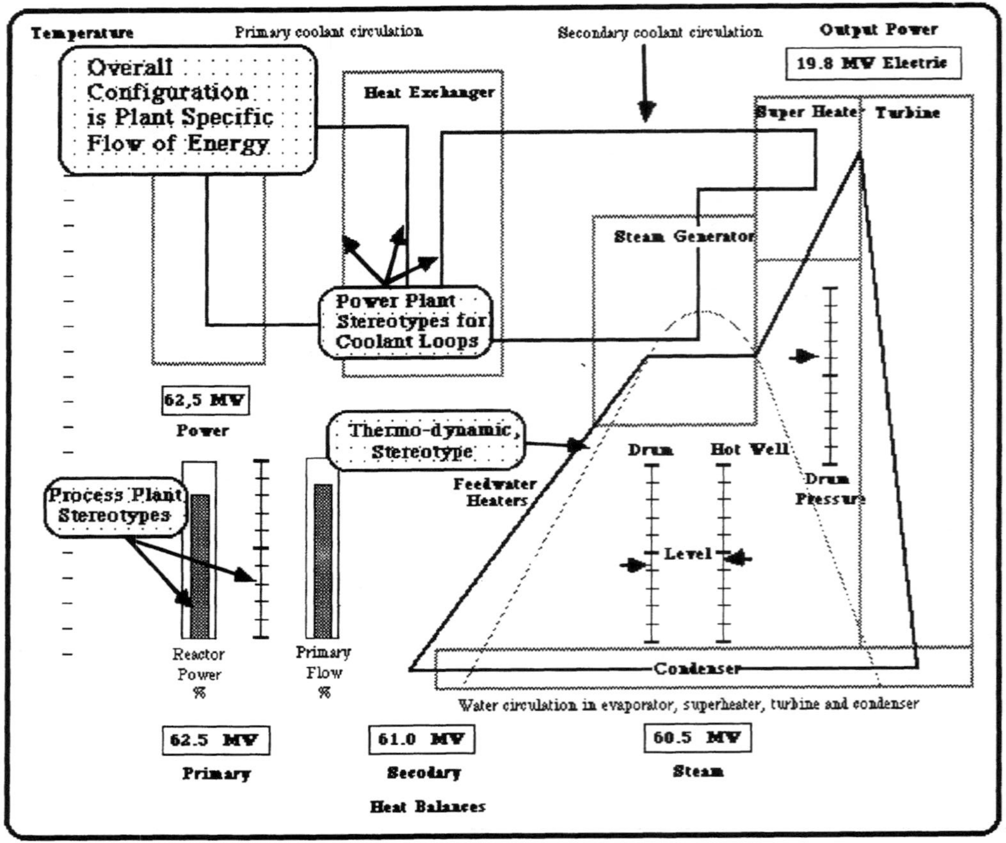

Figure 5.9. *An experimental overall display for a nuclear power reactor. The display is based on primary measuring data used to modulate the shape of graphical patterns that support perception of higher level functional features. Graphical patterns show the circulation paths of coolant and water. The temperature in different points of the paths can be read on the common scale. The patterns, therefore, immediately show the temperature rise and fall in core, heat exchanger, and so on, at the same time as temperature differences across heat exchangers can be read. Experiments with the dynamic display show that the propagation of a disturbance in the energy flow through the system can be directly perceived from the overall pattern. At the same time, primary data can be easily compared and the total thermodynamic loop can be visually decomposed in correspondence to the physical component that can be chosen for control. (Reproduced from Lindsay & Staffon, 1988).*

large set of displays. For a tightly coupled process system, the structure of the control tasks and, therefore, of the repertoire of the necessary display formats will match closely the structure of the means-ends hierarchy. Support of navigation in the display system, consequently, can be based on the means-end hierarchy (see Rasmussen and Goodstein, 1989).

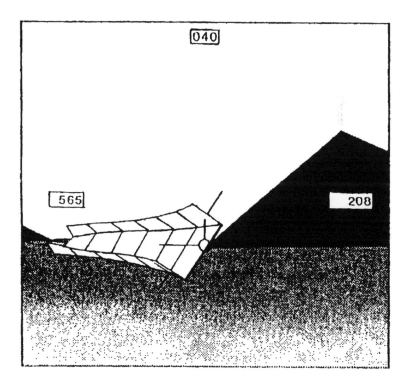

Figure 5.10. *A 'high-way-in-the-sky display showing the constraints around the flight path as defined by the flight-management system. It shows the recommended path that can be overruled by the pilot. (From "Aviation Displays," by A. E. Stokes & C. D. Wickens, 1988, in Human Factors in Aviation, Wiener, E., and Nagel, D (Eds.), (p. 392), New York: Academic Press. Copyright © 1988 by Academic Press. Reprinted by permission).*

5.4.2 Work in Loosely Coupled Systems

Work situations found within many other work domains are less constrained by the system configuration, because separate tools and equipment are used together in loosely coupled functional assemblies. Examples include manual assembling tasks in manufacturing workshops, maintenance tasks in process plants, work on construction sites and in offices, and so on. In the present discussion, a work situation in a public library will be considered as an example.

Domain Characteristics. In this category of work domains, invariants originating in causal laws will only be found at the lower means-ends levels, in the anatomy and processes of separate equipment and tools and the related work processes. Functionality at higher levels depends on human activities and the predominant source of regularity; that is, the invariants, related to the coordination of the work processes are only found in an intentional structure derived from institutional policies, legislation, socially established rules of conduct, and so on. In the example from a public library, the institutional intentionality is derived from financial constraints and legislation on the role of libraries in public education and their mediation of cultural values, etc. In the work situation considered, this intentionality is only reflected in the selection of books for the book stock and the instruction of librarians.

In this category of loosely coupled systems, functionality and intentionality is best represented in terms of the normal or intended connections of discrete, qualitative states, events, and acts, rather than in terms of quantitative, relational representation which is characteristic in cases of tightly coupled technical systems.

In work systems designed for support of the personal (private) needs of a user, the intentionality governing the use of the system within the envelope defined by the institutional policies and constraints are brought to work entirely by the user (see Figure 5.11). In consequence, the functionality of the work system must be designed so as to match the work processes of the users and the semantics of the functionality must match the user values and needs. How this can be realized depends on the degree to which the categories of users and their intentional structure can be analyzed and formulated. In the library example, this is possible to the extent that user needs in recreational reading can be identified, and their preferred search strategies in a book stock can be formulated and used for system design.

Task Situation Characteristics. Information systems for work in such domains will have to match the intentional structure brought to work by the user in a particular work situation, and an ecological interface representing also intentional invariants governing normal work patterns can only be identified when the task situation is defined and the user's intentional structure identified.

In the present case, the task situation considered for design of a support system is a user in a public library looking for fiction to read. In this task situation, the work domain is basically a collection of books and has no natural, functional structure. The intentional structure of a user

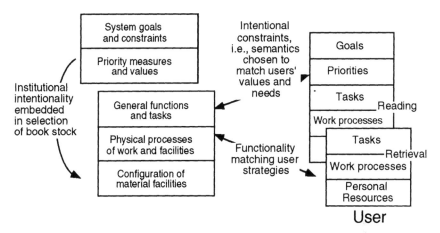

Figure 5.11. *In loosely coupled work systems, functionality and its propagation up through the levels depend on human activities. Intentional structure is defined dynamically by legislation, regulation, and company policies. The remaining degrees of freedom at the mid levels are closed by local and subjective intentional criteria. An ecological information system is organized so as to match the semantic dimensions of the information sources to the intentions of the users and the functionality of the system to the search strategies which are natural to the user for a given problem situation.*

looking for books must be identified by an analysis of user queries in the actual situation (Pejtersen, 1979, 1980, 1992a,b, 1993).

Design of Support System. The aim of the design of an ecological support system for this work situation, therefore, is to impose a functional structure on the retrieval system that matches the preferred search processes of the users. The semantics of the search to be represented should be the invariants and affordances which map onto the users' intentional structure, that is, their explicitly or implicitly formulated reading needs. For the library system discussed here, the functionality to build into the retrieval system has been identified by extensive analyses of the retrieval strategies that are preferred by users (Pejtersen, 1979) and an explicit formulation of the dimensions of the expressed reading needs. The dimensions of the user needs define the structure of the database, whereas the search strategies define the retrieval program. To sum up, the local functionality of the work system is derived from analysis of users' preferred search processes, whereas the indexing of the items of the data base match the users' query language. The transformation of representations involved is illustrated in Figure 5.12.

An overall structure of the information system which can be used to guide the user's navigation in the system and the form of the different displays cannot be derived from the local functionality corresponding to

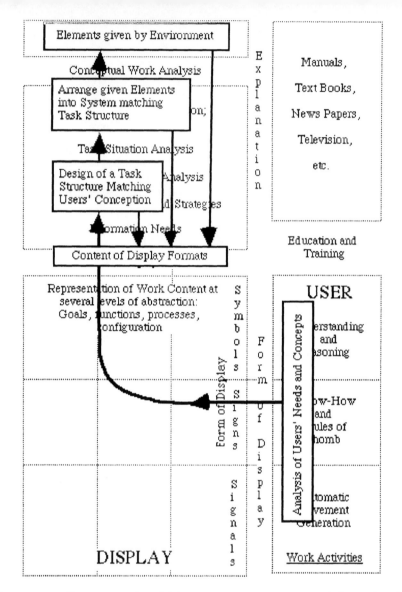

Figure 5.12. *For systems in which users are autonomous and determine the goals, task, and approach to work, the source of regularity to embed in the information system design and to represent in the content of interface formats is the problem space as defined by user needs. This does not imply that an information system cannot be highly structured by a formal concept, only that it should be based on an analysis of the relational structure of the users' needs folded onto the means for support given. This figure illustrates the use of the map of representations to define the source of the contents of the information displays and the necessary transformations during design.*

the individual search strategies. The users in a library represent a wide selection of the general public, and an overall organizing principle, therefore, cannot be related to a particular problem space and the related knowledge structure, as is the case for more structured systems such as process plants and manufacturing systems. Therefore, a high-level, organizing principle must be found that is familiar to the users from another context, that is, a metaphor should be used.

Because the functionality necessary for support of search strategies does not in principle constrain the context in which search is performed, a store house metaphor has been chosen for the overall functionality to exploit the general observation that "information is a question of where" (Miller, 1968; Semonides, 500 B.C.). The library system is described in detail elsewhere; in this section only a couple of display formats are shown to illustrate the approach.

Within the book house metaphor, separate "rooms" are available in support of different decision functions involved in retrieval. System functions supporting a particular search strategy are activated by choosing a workplace in a room of the book house (see Figure 5.13). Support is given to browsing (intuitive matching of semantic interpretation of icons to reading need), search by analogy (asking for a book similar to a known title), and analytic search (explicit analysis of needs).

Support of the user's explicit analysis of reading needs is given by a work desk presenting icons identifying the various dimensions of user needs as found from field studies (see Figure 5.14). By making the dimensions of the users' reading needs explicit, this display assists the users memory of aspects to consider in the analysis. At the same time, the icons function as command icons which allow specification of needs to be used for search by direct manipulation.

Support of the overall navigation in the information system is given by indication of the rooms that have been previously visited and that can be revisited by use of the command icons.

The library example illustrates the combined use of a representation of the invariant dimensions of the intentional structure governing users of an information system for a known task situation and a metaphorical principle for the overall organization of an information system. This approach is relevant for work support in task situations in the middle-to-left range of the domain map of Figure 5.3.

5.4.3.   Use of General Information Systems

For some information systems, such as lexicographic databases, information services for the general public, and so on. the task situation

Start Again   Select Other Database   Help

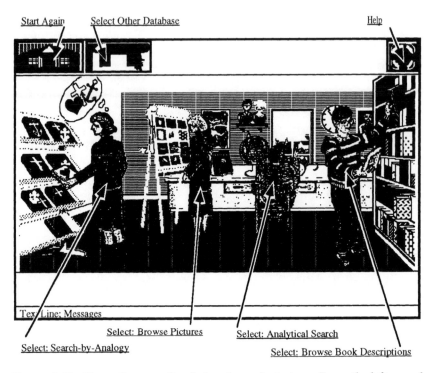

Text Line: Messages

Select: Browse Pictures   Select: Analytical Search

Select: Search-by-Analogy

Select: Browse Book Descriptions

Figure 5.13. *Shows the room for choice of search strategy. From the left: search by analogy, browsing pictures, analytical searches, browsing book descriptions. When the user selects a strategy, the system supports the search accordingly by a number of heuristics.*

and intentions of the casual users in general are considered to be unknown. The systems normally include a set of generally useful information sources and databases coming from many different public services, institutions, and commercial companies. No generally relevant intentional structure or task situation is normally formulated as a design basis to structure system function or interface design. The intentionality of the task situation is entirely defined on occasion by the actual user. Not even the higher level means-ends structure of a system is formulated except in very general terms of making information available to the public.

However, a structure must be imposed on the interface to assist the user to navigate and to remember the location of previously consulted items. In general, a cover story can be found that makes it possible for a user to transfer intuition and skill from a familiar situation in a previous context. Thus, a metaphor is used, such as the desktop metaphor for office systems, or the TV news, newspaper, and album metaphors of the Friend'21 personalized information environment (Friend'21, 1991).

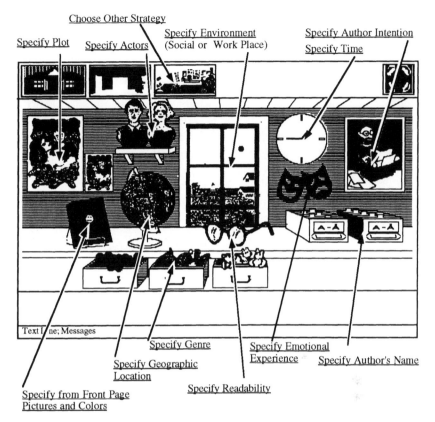

Choose Other Strategy

Specify Plot   Specify Actors   Specify Environment   Specify Author Intention
                                (Social or Work Place)   Specify Time

Text Line; Messages

Specify Genre   Specify Emotional
                       Experience   Specify Author's Name
Specify Geographic
Location
Specify from Front Page   Specify Readability
Pictures and Colors

Figure 5.14. *Shows the iconic display of the means-ends hierarchy of the work domain in terms of a multifaceted classification scheme based on an empirical analysis of users' search questions in libraries. When the user selects an icon, keywords are displayed, and the user can specify need by clicking on one or more keywords, and then a subsequent search is performed.*

Metaphors basically were used as a 'cover story' to couple the functionality built into a system to an intentional structure relevant to another domain. Metaphors are discussed by Clanton (1983). He classified them as functional (to reveal the relations between the system's functionality and the user's task perception—e.g., a spreadsheet), operational (these refer to actions in such a way that the user's natural expectations of the meaning and the results are confirmed—e.g., object manipulation), organizational (they show how notions on location and structure of information can be used for planning and application), and integrative (as described, these represent the integration of system components and subsystems in relation to the accomplishment of complex tasks).

5.5   Conclusion

It is, at present, an open question which calls for further research; of whether or not an ecological approach can be used also for general information services for the public. This approach requires an identification of the intentional structure of a set of prototypical user situations, of the structure of the involved task demands, the related reasoning strategies, and query dimensions. Considering the past problems with incompatible databases and retrieval strategies in general information services, an evolutionary approach to such services based on a growing set of carefully selected user situations will probably be fruitful. For further development of the ideas presented in this chapter, readers are referred to Rasmussen, Pejtersen and Goodstein (1994).

5.6   References

Amalberti, R., & Deblon, F. (1992). Cognitive modeling of fighter aircraft's process control: A step towards an intelligent on-board assistance system. *International Journal of Man-Machine Studies, 36*, 639-671.

Ashby, W. R. (1962). Principles of the self-organizing system. In H. Von Foerster, H. and G. W. Zopf (Eds.), *Principles of self-organization*. New York: Pergamon Press.

Clanton, C. (1983). The future of metaphor in man-computer systems. *Byte, 12*, 263-270.

Flach, J. M., & Vicente, K. J. (1989). *Complexity, difficulty, direct manipulation, and direct perception*. Technical Report EPRL-89-03. Urbana-Champaign: Engineering Psychology Research Laboratory, University of Illinois.

Friend'21. (1991). Proceedings of the International Symposium on Next Generation Human Interface Technologies, September November 1991, Tokyo. Institute for Personalized Information Environment.

Gibson, J. J. (1966). *The senses considered as perceptual systems*. Boston: Houghton-Mifflin.

Gibson, J. J. (1979). *The ecological approach to visual perception*. Boston: Houghton-Mifflin.

Gibson, J. J., & Crooks, L. E. (1938). A theoretical field-analysis of automobile driving. *American Journal of Psychology, 51*, 453-471.

Lindsey, R. W., & Staffon, J. D. (1988). A model based display system for the experimental breeder reactor II. Paper presented at the joint meeting of the American Nuclear Society and the European Nuclear Society. Washington, D.C.

Miller, G. A. (1968). Psychology and information. *American Documentation, 19,* 286-289.

Pejtersen, A. M. (1979). Design of a classification system for fiction based on an analysis of actual user-librarian communications; and use of the scheme for control of librarians' search strategies. In O. Harboe and L. Kajberg (Eds.), *Theory and application of information research.* (pp. 146–159). London: Mansell.

Pejtersen, A. M. (1992). The book house. An icon based database system for fiction retrieval in public libraries. In B. Cronin (Ed.), *The marketing of library and information services.* (pp. 572–591).

Pejtersen, A. M. (1979). Investigation of search strategies in fiction based on an analysis of 134 user-librarian conversations. In Henriksen, T. (Ed.), *IRFIS 3. Conference proceedings.* Statens Biblioteks-och Informations Hoegskole. (pp. 107-132). Olso, Norway.

Pejtersen, A. M. (1980). Design of a classification scheme for fiction based on an analysis of actual user-librarian communication, and use of the scheme for control of librarian's search strategies. In Harbo, O. and Kajberg, L. (Eds.), *Theory and Application of Information Research.* (pp. 167-183). London: Mansell.

Pejtersen, A. M. (1992a). *New model for multimedia interfaces to online public access catalogues.* The Electronic Library, 10, 359-365.

Pejtersen, A. M. (1992b). The book house. An icon based database system for fiction retrieval in public libraries. In Cronin, B., (Ed.), *The marketing of library and information services 2* (pp. 572-591). London: ASLIB.

Pejtersen, A. M. (1993). Designing hypermedia representations from work domain properties. In Fre, H. P., and Schauble, P. (Eds.), *Hypermedia '93.* Conference proceedings, 2273-289. Zurich: Springer-Verlag.

Polanyi, M. (1958). *Personal knowledge.* London: Routledge & Kegan Paul.

Rasmussen, J. (1986). *Information processing and human-machine interaction: An approach to cognitive engineering.* Amsterdam: North Holland.

Rasmussen, J. (1993). *Diagnostic reasoning in action.* IEEE Trans. SMC, 23, 981-993.

Rasmussen, J., & Goodstein, L. P. (1989). Information technology and work. In: M. Helander (Ed.), *Handbook of human-computer interaction.* (pp. 175-201) New York: North-Holland.

Rasmussen, J., Pejtersen, A. M., & Goodstein, L. P. (1994). *Cognitive systems engineering*. New York: Wiley.

Rasmussen, J., Pejtersen, A. M., & Schmidt, K. (1990, September). *A taxonomy for cognitive work analysis*. Technical Report Risø-M-2871.

Rasmussen, J., & Vicente, K. J. (1989). Coping with human errors through system design: Implications for ecological interface design. *International Journal of Man-Machine Studies, 31*, 517–534.

Rasmussen, J., & Vicente, K. J. (1990). Ecological interfaces: A technological imperative in high tech systems. *International Journal of Human Computer Interaction, 2*, 93–111.

Rasmussen, J., & Vicente, K. J. (1992). Ecological interface design: Theoretical foundations. *IEEE Transactions on Systems, Man, and Cybernetics. SMC 22*, 589-607.

Stokes, A. E., & Wickens C. (1988). Aviation displays. In E. L. Wiener and D. C. Nagel (Eds.), *Human Factors in Aviation*. (pp. 387-431). San Diego: Academic Press.

Vicente, K. J., & Rasmussen, J. (1990). The ecology of human-machine systems II: Mediating 'direct perception' in complex work domains. *Ecological Psychology, 2*, 207-249.

Woods, D. D. (1984). Visual momentum: A concept to improve the cognitive coupling of person and computer. *International Journal of Man-Machine Studies, 21*, 229-244.

Chapter 6

Toward a Theoretical Base for Representation Design in the Computer Medium: Ecological Perception and Aiding Human Cognition

David D. Woods
Cognitive Systems Engineering Laboratory, Ohio State University

6.0   Introduction

The technological potential for data gathering and data manipulation has expanded rapidly, but our ability to interpret this avalanche of data, that is, to extract meaning from this artificial data field, has expanded much more slowly, if at all. Significant advances in machine information processing seem to offer hope for advancing our interpretative capabilities, but in practice such systems become yet another voice in the data cacophony around us. The computer as a medium for supporting cognitive work is omnipresent. Yet, where is the theoretical base for understanding the impact of computerized information processing tools on cognitive work? Research in cognitive science has focused on grand questions of what is mind; artificial intelligence research continues to focus on building autonomous machine problem solvers; and the mainstream in both has assumed that cognition can be studied and simulated without detailed consideration of perception and action.

Ironically, the current state of research on computer support for human cognition has many interesting parallels to the state of perception research during the time when James J. Gibson developed the basis for ecological perception. The concepts and research program that Gibson introduced for perception also provide inspiration for needed concepts and research directions in aiding human cognition.

On the one hand, this chapter lays out the fundamental issues on aiding human cognition via the computer medium, drawing on and pointing to parallels to Gibsonian concepts. Examples include agent–environment mutuality which leads to the need for a new way to characterize problem solving habitats, the question of what is informative, the multiplicity of cues available in natural problem solving habitats, and the need to take into account the dynamism of real problems. If one takes the ecology of aided cognition seriously, it has fundamental implications for research on human-computer interaction, models of human performance, and the development of intelligent support systems.

The themes in this chapter, and, in fact, the title of this book, juxtapose concepts that are seen in the conventional wisdom of cognitive psychology and human factors as contradictory. Terms such as *representation, cognition, problem solving,* and *support for cognitive work* will be seen over and over again in the company of ideas that relate to perception, especially for this forum — ecological perception. This juxtaposition is not the result of momentary fashion; rather, it is indicative of a paradigm shift that has been underway for some time in human factors and human-machine systems (this shift can be seen in a variety of published works that began to appear in widely accessible forms starting in 1981, e.g.; Norman, 1981; Rasmussen & Rouse, 1981; Reason & Mycielska, 1982). The appeal of the analogy to ecological perception is, in part, based on the common ground of researchers experiencing the joys and pains of questioning the assumptions behind the current conventional wisdom.

6.1   Representation Aiding: The Role of Perception in Cognition

Representation aiding is one strategy for aiding human cognition and performance in complex tasks. The basic rationale is straightforward. Figure 6.1 depicts the pieces of the puzzle that combine to create this strategy. A fundamental finding which has emerged from cognitive science and related research is that the representation of the problem provided to a problem solver can affect his, her, or its task performance. There is a long tradition in problem solving research in that "solving a problem simply means representing it so as to make the solution transparent" (Simon, 1969, p. 77). I refer to this as the *problem representation principle*, and it is a way to summarize a widespread psychological result that the content and context of a problem can radically alter subjects' responses (examples can be seen in everything from text comprehension to deductive reasoning to organizational effects in visual search; cf., Norman, 1993 for a recent summary).

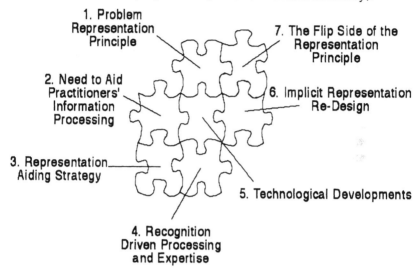

1. Problem
Representation
Principle

7. The Flip Side of the
Representation
Principle

2. Need to Aid
Practitioners'
Information
Processing

6. Implicit Representation
Re-Design

3. Representation
Aiding Strategy

5. Technological Developments

4. Recognition
Driven Processing
and Expertise

Figure 6.1. *The factors that contribute to the Representation Aiding approach for assisting cognitive work.*

A variety of researchers concerned with aiding human performance in complex, high consequence task domains (energized in part by responses to the Three Mile Island accident) seized this fundamental finding and turned it around (Goodstein, 1981; Rasmussen & Lind, 1981; Woods, Wise, & Hanes, 1981) — if the problem representation affects performance, then, when the goal is to improve human performance, one can develop new types of representations of the task domain to effectively support human problem solving — *representation aiding*.

This approach dovetails very neatly with another fundamental result that was gaining wider appreciation at about the same time. Studies of human performance at complex high consequence tasks (again, motivated in part by the need to improve safety and human performance in the aftermath of accidents) kept pointing to the critical role of situation assessment in expert performance (cf. e.g.,Woods, O'Brien, & Hanes, 1987, for a synthesis of results across several studies of operator decision making in both actual and simulated nuclear power plant emergencies). The results in a variety of studies examining practitioners in a variety of domains seemed to indicate that practitioners' behavior was based primarily on the ability to recognize the kind of situation evolving in front of them, selecting actions appropriate to those circumstances from a repertoire of doctrine about how to handle various situations and contingencies — recognition–driven processing (see Klein, Orasanu, Calderwood, & Zsambok, 1993 for an extensive treatment of this result and its implications). The critical role of *recognition-driven processing* in effective task performance points to the need for aiding practitioner situation assessment through improved representations of the task world.

In human performance this result has been called recognition-driven processing, but the underlying concept really has a longer history in cognitive psychology. The data that support the critical role of recognition-driven processing reemphasize the view that perception is a part of cognition rather than an independent front-end module that merely supplies input for cognition. Remember that Neisser (1976) called his fundamental principle of cognition the *perceptual cycle*, emphasizing the dynamical interplay and mutual dependence of perception, cognition, and action in contrast to linear information processing models in which independent modules interact via parameter passing. Perception and cognition guide action; acting (and the possibilities for action) and cognition direct perception; perceiving informs cognition (cf. Tenney, Adams, Pew, Huggins, & Rogers, 1992, for one explicit application of the perceptual cycle to understand the dynamics of human performance in one complex domain—pilot performance in commercial aviation).

Given the omnipresence of the representation principle in research on cognition, one may then ask how do representations affect human cognition; in other words, why do computer-based displays (representations) make a difference in human performance? There are several (perhaps overlapping) mechanisms that have been suggested to mediate the effect of external representation of a process or device or problem on the cognitive activities of the human problem solver:

(a) Problem structuring — the representation changes the nature of the problem and therefore the kinds of strategies that can be used to solve the problem;
(b) Overload/workload — a good representation shifts task-related cognitive activities to more mentally economical forms of cognitive processing, such as more parallel, more 'perceptual,' more automatic, drawing on different kinds of mental resources,

and so on. (note the inverse: a poor representation forces reliance on more deliberative, serial, resource-consuming cognitive processing);

(c) Control of attention — good representation supports attention related cognitive processes including switching attention, pre-attentive reference or peripheral access (Hutchins, 1991; Woods, 1992), divided attention, knowing where to look next;

(d) New secondary tasks — poor representation creates new secondary tasks that increase workload especially in high-criticality, high-tempo periods, that interrupt primary tasks, and that shift the focus of attention away from the actual task and to secondary interface and data management tasks (Woods et al., 1994; Woods, 1993b);

(e) Effort — the representation affects effort and therefore the effort-performance relationship for that individual and that task context (cf. Johnson & Payne, 1985; Payne et al., 1988).

The fourth piece of the puzzle is the technological developments that have been underway, providing new powers to develop representations via advances in the computer medium (computer-based graphics and intelligent data processing) and via increased penetration of the computer medium into places in which substantive cognitive work is performed. However, it is critical to note that representation aiding forces a reinterpretation of what it means to design human-computer interaction — the display of data through the medium of the computer should be considered in terms of how different types of representations vary in their effect on the problem solver's information processing activities and performance — *representation design in the computer medium* (Woods, 1991).

Although the technological advances increase the potential for attempts at representation aiding, their primary effect on research and development has been a kind of negative motivation. The technological changes that are now in motion with regard to information technology in complex, high-consequence applications are radically changing the kinds of problem representations available to practitioners (Cook, Woods, & Howie, 1990; Woods, Potter, Johannesen, & Holloway, 1991). Implicit representation re-design is widespread, accompanying the advances and increased penetration of the computer medium into domains of cognitive work. The implicit representation re-design going on should remind us that there is a flip side to the problem representation principle: *Poor representations will degrade task performance.*

The problem representation principle allows for no a priori neutral representations. The representations of the problem domain available to the practitioner can degrade or support information processing tasks and strategies related to task performance. Thus, there is an increasingly pressing need "with developing a theoretical base for creating meaningful artifacts and for understanding their use and effects" (Winograd, 1987, p. 10), in other words, to develop a theoretical base for representation design in the computer medium.

The concept for representation aiding as a means for computer-based decision support and some initial attempts to construct such systems predate 1979 (and the history of science, mathematics, and technology is replete with examples of non-computer-based representation aiding). However, the conjunction of several factors, beginning about 1979–1980, both pushed and pulled representation aiding as an approach to support cognitive work. The conjunction of need (failures in complex human-technological systems as energizer of work on aiding dynamic cognitive work), opportunity (advances and increasing penetration of computer-based graphic systems), and motive (the problem representation principle and recognition-based models of dynamic cognitive work) together have energized the development of representation aiding as a means for computer-based decision support.

6.2   The Computer as a Medium for Representation

The development of information displays within the computer medium is constrained and shaped by the properties of computer and display technologies as a medium for representation.

6.2.1   The Symbol Mapping Principle

The computer can be seen as a referential medium in which visual and other elements are signs or tokens that function as meaning carriers within a symbol system (Woods, 1991, in preparation). Peirce's (1903/1955) treatment of symbols points to representation as a three part relationship (Fig. 6.2) between (a) signs or tokens in the medium for representation, (b) the referent objects and environment, and (c) the interpretants or observers who extract meaning from the token-referent relationship given a larger context of goals and activities. What is unique is that Peirce includes the observer as part of the symbolic link itself.

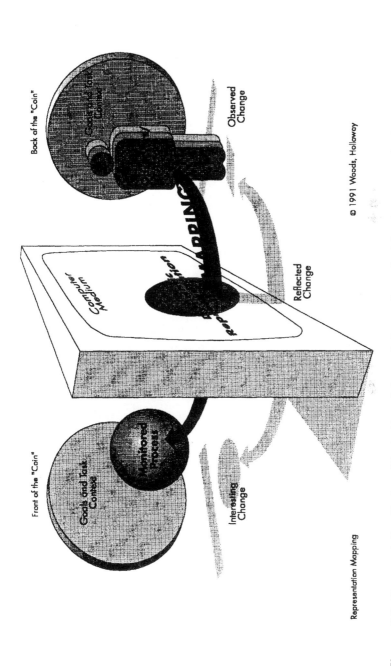

Figure 6.2. *Representation Design: Illustration of the symbol mapping principle. A computer display system functions as a representation of something for someone in some task/goal context.*

This is essentially what Gibson does in the mutuality assumption in which perceptual properties of the external world are specified with the observer/actor as a fundamental part of the variable. This idea changes the concept of information for a representation aiding approach, as it did for Gibson and ecological perception. Given the mutuality assumption, information is seen as a relation between the data, the world the data refer to, and the observer's expectations, intentions, and interests and not as a thing in itself (Woods, 1986). Results in cognitive psychology are replete with examples that reinforce this three-part relationship (e.g., Gonzalez & Kolers, 1982). For example, change the perceptual characteristics of a scene in ways that change properties of perceptual organization and visual search performance will be radically affected (e.g., the pop-out effect).

If the computer is a medium for representation, then we can think of representational forms as distinct from visual forms. The physical form of sign tokens within a medium for representation does not by itself indicate a mode of symbolizing. Pictures and words can both refer either propositionally or analogically (Gonzalez & Kolers, 1982; Woods, in preparation). How sign tokens represent depends on a three part relationship — how tokens map onto the structure and behavior of relevant objects in the referent domain for some agent or set of agents in some context (Figure 6.2). See Hutchins, 1991, for an analysis of this three part relationship for one particular referent (airspeed), set of representations (e.g., speed bugs), set of agents (flight crews), and goal/task context (descent phase of commercial air flights).

Thus, the fundamental principle for any discussion of types of displays from a representation point of view is the *symbol mapping principle:* computer-based displays *of* data function as a representation *of* something *for* someone *in* some task context. Or, more elaborately, representational form is defined in terms of how data on the state and behavior of the domain is mapped into the syntax and dynamics of visual forms in order to produce information transfer to the agent using the representation, given some task and goal context (Figure 6.2). The symbol mapping principle means that one cannot understand computer–based information displays in terms of purely visual characteristics; the critical properties relate to how data are mapped into the structure and behavior of the visual elements. The dynamic aspect of the mapping is critical. The symbol mapping principle means that to characterize or design a graphic form one must consider how the form behaves or changes as the state of the referent changes.

Given the symbol mapping principle, representational form in the computer medium is defined in terms of how data on the state and semantics of the domain is *mapped* into the syntax and dynamics of perceivable tokens/forms in order to produce information transfer to the agent using the representation (Woods, 1991, in preparation). Representational form cannot be assessed from visual appearance per se so that a single visual format such as a bar chart can be used to represent in different ways. Conversely, bar charts, trends, and other visual forms can be used to create the same representational form (or variants on a representational theme).

There are design challenges raised by the symbol mapping principle. One part of the design challenge is to set up the mapping between domain referents and visual tokens in the computer medium so that the representation captures task-meaningful semantics. What are interesting properties and changes in the monitored process or device? How are these properties and changes captured or reflected in the structure and behavior of tokens in the computer medium. Note that meeting this part of the mapping principle requires uncovering what are the task meaningful domain semantics to be mapped into the structure and behavior of the representation. This is the problem of cognitive task analysis (Woods, 1988): identifying what is informative and what are the *interesting* properties and changes of the domain referents given the goals and task context of the team of practitioners.

The second part of the design challenge for creating effective representations is to set up this mapping so that the observer can extract information about task-meaningful semantics. Becker and Cleveland (1987) call this the decoding problem, that is, domain data may be cleverly encoded into the attributes of a visual form, but, unless observers can effectively decode the representation to extract relevant information under the conditions of actual task performance (attention switching, time pressure, risk, uncertainty), the representation will fail to support the practitioner.

6.2.2   Virtual Perceptual Field

A fundamental property of the computer as a medium for representation is freedom from the physical constraints acting on the referent real-world objects/systems (see Hochberg, 1986, p. 22-2 – 22-3 for an elegant treatment of this property of the computer medium). In many media (e.g., most notably, cinema), the structure and constraints operating in the physical world will ensure that much of the appropriate information about relationships in the referent domain is preserved in the representation. On the other hand, in the computer medium, the designer of computer displays of data must do *all of the work to constrain or link attributes and behaviors of the representation to the attributes and behaviors of the referent domain.*

This property means that data displays in the computer medium can be thought of as a virtual perceptual field[1]. It is a perceivable set of stimuli, but it differs from a natural perceptual field and other media for representation, because there is nothing inherent in the computer medium that constrains the relationship between things represented and their representation. This freedom from the physical constraints acting on the referent real world objects is a double-edged sword in human-computer interaction (HCI), providing at the same time the potential for very poor representations (e.g., see Cook, Potter, Woods, & McDonald, 1991; Woods et al., 1991) and the potential for radically new and more effective representations.

Note the parallels to the Gibsonian revolt against tachistoscopic perception. For Gibson, the problem was to discover the higher order properties of the perceptual field (the optic array). Discovering potential sources of information in the stimulus array is a critical part of understanding how the perceptual systems functions. It is difficult to debate specific cognitive processing mechanisms without having some understanding of what it is the perceptual system needs to extract about its environment given its behavioral competencies.

The virtuality property of the computer medium creates a new twist on Gibsonian agenda. The designer creates the perceptual field and the mapping to the semantics of the domain. The designer's choices manipulate the properties of the virtual perceptual field at several levels to achieve an overall desired phenomenological effect/result (Woods, in preparation). One level I call workspace coordination — the set of viewports and classes of views that can be seen together in parallel or in series as a function of context (Woods, 1984). Another is the level of coherent process views where a kind of view is a coherent unit of representation of a portion of the underlying process or systems that the

---

[1]More properly, I should say that the computer medium can be thought of as an artificial perceptual field. I have chosen to use the word virtual to play off the current fashion in software. The use of virtual, as in virtual reality, is creating a new connotation for the word: virtual — giving the appearance or suggestion of a naturally occurring phenomenon whereby only approximating or even missing its essence; refers especially to computerized devices and systems.

observer could select for display in a viewport. A third level of analysis is the the level of graphics forms -- the forms, objects, or groups and their perceptual context. The designer's choices affect the kind of cognitive processing that the practitioner must bring to bear to do cognitive work through computer-based representations of the underlying system or process (e.g., mentally effortful, serially bottlenecked, deliberative processes versus mentally economical, effectively parallel, perceptual processes).

Another important property of the virtual perceptual field of computer-based display systems is that the viewport size (the windows/VDUs available) is very small relative to the large size of the artificial data space or number of data displays that potentially could be examined (Figure 6.3). In other words, the proportion of the virtual perceptual field that can be seen at the same time (physically in parallel) is very, very small. This property is often referred to as the *keyhole effect* (e.g., Woods, 1984). Given this property, shifting one's gaze within the virtual perceptual field is carried out by selecting another part of the

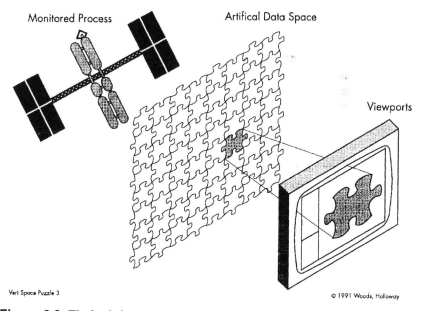

Monitored Process

Artifical Data Space

Viewports

Verl Space Puzzle 3

© 1991 Woods, Holloway

Figure 6.3. *The keyhole property of the computer medium as a perceptual field: The viewport size (the windows/VDUs available) is very small relative to the large size of the artificial data space or number of data displays that potentially could be examined.*

artificial data space and moving it into the limited viewport. But in the default design, the observer can see only one small portion of the total field at a time or a very small number of the potentially available displays (cf. Cook et al., 1990; and Woods et al., 1991, for examples). The consequences for the observer are exacerbated by the default tendency in interface design that places each piece of data in only one location within the virtual perceptual field (one "home").

However, how do we know where to look next in a virtual perceptual field like this (cf. Woods, 1984; Woods, Watts, & Potter 1993)? Meaningful tasks involve knowing where to look next in the data space available behind the limited viewports and extracting information across multiple views. Yet, the default tendency in interface design is to leave out any orienting cues that indicate in mentally economical ways whether something interesting may be going on in another part of the virtual perceptual field. Instead, the processes involved in directing where to look next are forced into a mentally effortful, high memory load, deliberative mode (in addition, the interface structure may create other cognitive problems in translating intentions into specific actions — Norman's (1988) Gulf of Execution). Observers must remember where the desired data is located, and they must remember and execute the actions necessary to bring that portion of the field into the viewport, given he knows what data are potentially interesting to examine next (cf. Woods et al., 1991, for a detailed case). One can see the potential problems that derive from this keyhole property by imagining what it would be like to function with no peripheral vision or without other orienting perceptual systems to help determine where to look next, that is, where to direct focal attention next.

Given this property of the computer medium and its implications for cognitive processing of the observer, it is ironic that discussions, guidelines, and empirical studies in HCI are stuck in the design of single isolated graphic forms/views. There is still virtually nothing available on the coordination of multiple views as a virtual workspace, virtually nothing on the information extraction problems that occur at the workspace level — keyhole effects, getting lost effects, and virtually nothing on the tradeoff between searching one display and searching within the larger network of displays (for one exception, see Henderson & Card, 1987).

But technological advances (both in power and in penetration of applications) proceed regardless of our understanding of representation design. Windowing capabilities have resulted in an undisciplined proliferation of displays that create navigation problems where users have difficulty finding related data, especially when complicated by hidden windows and complex menu structures. The proliferation of windows tends to fragment data across an increasingly complex structure of the virtual perceptual field. This forces serial access to highly interrelated data and increases the cognitive load in deciding where to look next (see Cook et al., 1990; Woods et al., 1991 for data and cases on poor window coordination at the workspace level of a display system). Users may possess great flexibility to tailor their workspace by manipulating the number, size, location, and other window parameters, but this flexibility creates extra data management burdens in event-driven task domains that increase practitioner workload often during high-tempo operations. Practitioner attention shifts to the interface (where is the desired data located in the display space?) and to interface control (how do I navigate to that location in the display space?) at the very times when his or her attention needs to be devoted most to assessing/managing the monitored process. This factor appears to be one source of the phenomenon that Wiener (1989) has termed *clumsy automation.* Clumsy automation or the clumsy use of technology is a form of poor coordination between the human and machine in which the benefits of the automation accrue during workload troughs, and the costs of automation occur during high-criticality or high-tempo operations.

Directing perceptual exploration to potentially interesting data in dynamic and event-driven domains is a fundamental competency of perceptual systems in natural perceptual fields. "The ability to look, listen, smell, taste, or feel requires an animal capable of orienting its body so that its eyes, ears, nose, mouth, or hands can be directed toward objects and relevant stimulation from objects. Lack of orientation to the ground or to the medium surrounding one, or to the earth below and the sky above, means inability to direct perceptual exploration in an adequate way (cf. Reed, 1988, p. 227, on Gibson and perceptual exploration in Gibson, 1966)." The problem of building representations in the computer medium that support rather than undermine this fundamental competency has barely been acknowledged, much less addressed (Woods, 1984).

6.2.3   Dynamic Reference

Another fundamental property of the computer as a medium for representation is that computer-based displays can behave and change. This is one property that distinguishes the computer from traditional media for visual expression. This property provides new potential and new challenges for developing representations. Previously, discussions of symbol-referent relationships were based at least implicitly on static symbols. But the computer medium creates the potential/challenge of dynamic reference — the behavior of the token is part of the process linking symbol to referent. Thus, to characterize or design a graphic form, one must consider how the form behaves or changes as the state of the referent changes.

In the current default in computer-based representations, the basic unit of display remains an individual datum usually represented as a digital value, for example, oxygen tank pressure is 297 psi (cf. Woods, 1991, or Woods et al., 1991, which contains examples of typical displays). No attempt is made in the design of the representation of the monitored process to capture or highlight operationally interesting events — behaviors of the monitored process over time, for example, the remaining cryogenics are deteriorating faster (for one exception, see Woods & Elias, 1988). Furthermore, this failure to develop representations that reveal change and highlight events in the monitored process has contributed to incidents in which practitioners using such opaque representations miss operationally significant events (e.g., Freund & Sharar, 1990; Moll van Charante et al, 1993).

In the most well known accident in which this representational deficiency contributed to the incident evolution (cf. Murray & Cox, 1989), the Apollo 13 mission, an explosion occurred in the oxygen portion of the cryogenics system (oxygen tank 2). The mission controller (electrical, environmental, and communication controller) monitoring this system was examining a screen filled with digital values (display CSM ECS CRYO TAB; Figure 6.4).

After other indications of trouble in the spacecraft, he noticed that oxygen tank 2 was depressurized (about 19 psi) as well as a host of other problems in the systems he monitored. It took a precious 54 minutes as a variety of hypotheses were pursued before the team realized that the command module was dying and that that an explosion in the oxygen portion of the cryogenics system was responsible. The digital display had hidden the critical event (2 digital values out of 54 changing digital numbers had changed anomalously; compare Figures 6.4, 6.5, and 6.6). So none of the three noticed the numbers for oxygen tank 2 during 4 particularly crucial seconds. At 55 hours, 54 minutes, and 44 seconds into the mission, the pressure stood at 996 psi–high but still within accepted limits. One second later, it peaked at 1,008 psi. By 55:54:48, it had fallen to 19 psi. If one of them had seen the pressure continue on

```
LM12839                    CSM ECS-CRYO TAB                           0613

CTE 055:46:51    (         )    GET 055:53:47    (         )    SITE
------LIFE SUPPORT--------                -----PRIMARY COOLANT-----
GF3571  LM CABIN P   PSIA|                CF0019  ACCUM QTY PCT      34.4
CF0001  CABIN  P     PSIA    5.1          CF0016  PUMP P    PSID     45.0
CF0012  SUIT   P     PSIA    4.3          SF0260  RAD IN T      F    73.8
CF0003  SUIT  Δ P   IN H20  -1.68
CF0015  COMP  Δ P   P PSID   0.30
CF0006  SURGE  P    P PSIA   891          CF0020  RAD OUT T     F    35
        SURGE QTY    LB      3.67         CF0181  EVAP IN T     F    45.7
   02  TK 1 CAP  Δ P PSID    21           CF0017  STEAM T       F    64.9
   02  TK 1 CAP  Δ P PSID    17           CF0034  STEAM P   PSIA |   .161
                                          CF0018  EVAP OUT T    F    44.2
CF0036  02 MAN P    PSIA    105
CF0035  02 FLOW     LB/HR   0.181
                                          SF0266  RAD VLV 1/2        ONE
CF0008  SUIT T       F      50.5          CF0157  GLY FLO LB/HR      215
CF0002  CABIN        F      65            ----SECONDARY COOLANT----
CF0005  CO2 PP      MMHG    1.5           CF0072  ACCUM QTY PCT      36.8
------------H20------------               CF0070  PUMP P    PSID |   9.3
CF0009  WASTE        PCT    24.4          SF0262  RAD IN T      F    76.5
        WASTE        LB     13.7          SF0263  RAD OUT T     F    44.6
CF0010  POTABLE      PCT   104.5          CF0073  STEAM P   PSIA     .2460
        POTABLE      LB     37.6          CF0071  EVAP OUT T  F      66.1
CF0460  URINE NOZ T  F      70            CF0120  H20-RES  PSIA      25.8
CF0461  H20 NOZ T    F      72            TOTAL PC CUR    AMPS
--------CRYO SUPPLY----------------02-1-----02-2-------H2-1----------H2-2---
SC0037-38-39-40 P       PSIA   876.5    906       225.7(03-1)    235.1
SC0032-33-30-31 QTY     PCT    77.63    O/S        73.24         74.03
SC0041-42-43-44-T       F     -189     -192       -417          -416
                QTY     LBS    251.1    260.0      20.61         20.83
```

Apollo 13, 906
1/22/92   © 1992 Woods and Holloway

Figure 6.4. *Partial reconstruction of the computer display (display CSM ECS CRYO TAB) monitored by the electrical, environmental, and communication controller (EECOM) at 55:54:44 mission time during the Apollo 13 mission.*

```
LM12839                    CSM ECS-CRYO TAB                           0613

CTE 055:54:45    (         )    GET 055:54:47    (         )    SITE
-------LIFE SUPPORT--------               -----PRIMARY COOLANT-----
GF3571  LM CABIN P   PSIA|                CF0019  ACCUM QTY PCT      34.4
CF0001  CABIN  P     PSIA    5.1          CF0016  PUMP P    PSID     45.0
CF0012  SUIT   P     PSIA    4.1          SF0260  RAD IN T      F    73.8
CF0003  SUIT  Δ P   IM H20  -1.68
CF0015  COMP  Δ P   P PSID   0.32
CF0006  SURGE  P    P PSIA   892          CF0020  RAD OUT T     F    35
        SURGE QTY    LB      3.68         CF0181  EVAP IN T     F    45.7
   02  TK 1 CAP  Δ P PSID    20           CF0017  STEAM T       F    64.9
   02  TK 1 CAP  Δ P PSID    15           CF0034  STEAM P   PSIA |   .161
                                          CF0018  EVAP OUT T    F    44.2
CF0036  02 MAN P    PSIA    105
CF0035  02 FLOW     LB/HR   0.163
                                          SF0266  RAD VLV 1/2        ONE
CF0008  SUIT T       F      50.2          CF0157  GLY FLO LB/HR      215
CF0002  CABIN        F      66            ----SECONDARY COOLANT----
CF0005  CO2 PP      MMHG    1.5           CF0072  ACCUM QTY PCT      36.8
------------H20------------               CF0070  PUMP P    PSID |   9.3
CF0009  WASTE        PCT    24.2          SF0262  RAD IN T      F    76.5
        WASTE        LB     14.2          SF0263  RAD OUT T     F    44.6
CF0010  POTABLE      PCT   104.5          CF0073  STEAM P   PSIA     .2460
        POTABLE      LB     37.6          CF0071  EVAP OUT T  F      66.1
CF0460  URINE NOZ T  F      71            CF0120  H20-RES  PSIA      25.8
CF0461  H20 NOZ T    F      72.1          TOTAL PC CUR    AMPS
--------CRYO SUPPLY----------------02-1-----02-2-------H2-1----------H2-2---
SC0037-38-39-40 P       PSIA   874.9   1008.3     225.7(03-1)    235.1
SC0032-33-30-31 QTY     PCT    75.45    60         73.24         74.03
SC0041-42-43-44-T       F     -190     -160       -417          -416
                QTY     LBS    251.1    O/S        20.61         20.83
```

Apollo 13, 1008.3
1/22/92   © 1992 Woods and Holloway

Figure 6.5. *Partial reconstruction of the computer display (display CSM ECS CRYO TAB) monitored by the electrical, environmental, and communication controller (EECOM) at 55:54:45 mission time during the Apollo 13 mission. Note oxygen tank 2 pressure showed a peak at this point of 1,008 psi.*

```
LM12839                 CSM ECS-CRYO TAB                        0613

CTE 055:57:02    (        )    GET 055:54:53    (        )       SITE
-------LIFE SUPPORT-----                -----PRIMARY COOLANT-----
GF3571  LM CABIN P   PSIA               CF0019  ACCUM QTY PCT    34.4
CF0001  CABIN    P   PSIA     5.6       CF0016  PUMP P    PSID   45.3
CF0012  SUIT     P   PSIA     4.13      SF0260  RAD IN T    F    73.8
CF0003  SUIT     P   IN H2O  -1.8    Δ
CF0015  COMP     P   P PSID   0.27   ↑
CF0006  SURGE    P   P PSIA  889        CF0020  RAD OUT T   F    35.2
        SURGE QTY    LB       3.9       CF0181  EVAP IN T   F    45.7
    O2  TK  1  CAP   P PSID  19         CF0017  STEAM T     F    64.7
    O2  TK  1  CAP   P PSID  17      ↑  CF0034  STEAM P   PSIA   .161
                                    ↑  CF0018  EVAP OUT T  F     44.6
CF0036  O2 MAN P     PSIA   108
CF0035  O2 FLOW      LB/HR    0.178
                                       SF0266  RAD VLV 1/2       ONE
CF0008  SUIT T       F       50.5      CF0157  GLY FLO LB/HR    215
CF0002  CABIN        F       65        ----SECONDARY COOLANT----
CF0005  CO2 PP       MMHG     1.6      CF0072  ACCUM QTY PCT    36.8
------------H2O------------            CF0070  PUMP P    PSID    9.3
CF0009  WASTE        PCT     28.9      SF0262  RAD IN T    F    76.5
        WASTE        LB      14.9      SF0263  RAD OUT T   F    45.1
CF0010  POTABLE      PCT    109.9      CF0073  STEAM P   PSIA  .2460
        POTABLE      LB      39.1      CF0071  EVAP OUT T  F    65.9
CF0460  URINE NOZ T  F      105        CF0120  H2O-RES  PSIA    26.2
CF0461  H2O NOZ T    F       78        TOTAL FC CUR    AMPS
--------CRYO SUPPLY-- ---------02-1-----02-2-------H2-1-----------H2-2---
SC0037-38-39-40 P         PSIA    872      19     225.7(03-1)    235.1
SC0032-33-30-31 QTY       PCT     72.3     01.17    73.24        74.03
SC0041-42-43-44-T         F      -189      O/S     -417          -416
                QTY       LBS    251.1     O/S      20.61        20.83
```

Apollo 13, 19
1/22/92   © 1992 Woods and Holloway

Figure 6. 6. *Partial reconstruction of the computer display (display CSM ECS CRYO TAB) monitored by the electrical, environmental, and communication controller (EECOM) at 55:54:48 mission time during the Apollo 13 mission. Note oxygen tank 2 is depressurized.*

through the outer limits, then plunge, he would have been able to deduce that oxygen tank 2 had exploded (Figure 6.7). It would have been a comparatively small leap to have put the whole puzzle of multiple disturbances across normally unconnected systems together (Murray & Cox, 1989, p. 406).

It is reported that the relevant controller experienced a continuing nightmare for 2 weeks following the incident, in which, when the astronauts reported a problem, "he looked at the screen only to see a mass of meaningless numbers." Finally, a new version of the dream came — he looked at the critical digitals "before the bang and saw the pressure rising. ... Then the tank blew, and he saw the pressure drop and told Flight exactly what had happened (Murray & Cox, 1989, p. 407)."

The poor representation could be compensated for through human adaptability and knowledge; in other words, as Norman (1988) likes to put it, knowledge-in-the-head can compensate for the absence of knowledge-in-the-world. However, what is the point of the computer as

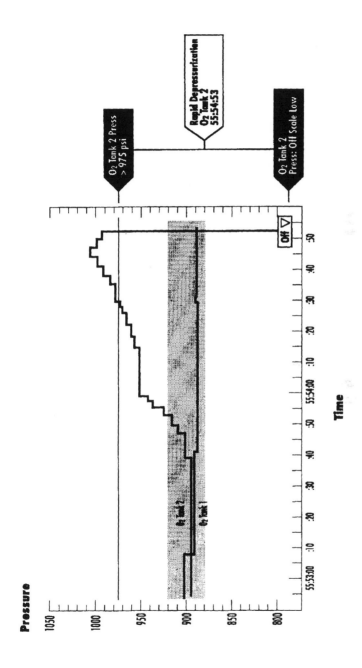

Figure 6.7. *Analogical representation based on the Significance Message graphic form (Woods & Elias, 1988) of the oxygen tank events during the critical seconds of the Apollo 13 mission.*

a medium for the display of data if it does not reduce practitioner memory loads. In fact, in computer system after computer system (e.g., Woods et al., 1991), we find that despite the availability of new computational and graphic power, the end result is an increase in demands on practitioner memory. The contrast cannot be greater with studies of successful, but often technologically simple, cognitive artifacts such as Hutchins (1991) which reveal how effective cognitive tools off load memory demands, support attentional control and support the coordination of cognitive work across multiple agents.

To begin to move toward better representations that do not obscure the perception of events in the underlying system, there are three inter-related critical criteria in representation design (Woods, in preparation):

1. **Put data into context:** (a) put a given datum into the context of related values; (b) collect and integrate data about important domain issues. Data are informative based on *relationships* to other data, relationships to larger frames of reference, and relationships to the interests and expectations of the observer. The challenge is the context-sensitivity problem — what is interesting depends on the context in which it occurs.

2. **Highlight changes and events.** Representations should highlight change/events and help reveal the dynamics of the monitored process. Events are temporally extended behaviors of the device or process involving some type of change in an object or set of objects. One key question is to determine what are 'operationally interesting' changes or sequences of behavior Examples include -- highlighting approach to a limit, highlighting movement and rate of change, emphasizing what event will happen next, and highlighting significant domain events (e.g., Woods & Elias, 1988). Representing change and events is critical because the computer medium affords the possibility of dynamic reference -- the behavior of the representation can refer to the structure and behavior of the referent objects and processes.

3. **Highlight contrasts.** Representations should highlight and support observer recognition of contrasts. Meaning lies in contrasts -- *some departure from a reference or expected course;* Representing contrast means that one indicates the relationship between the contrasting objects, states or behaviors. One shows how the actual course of behavior follows or departs from reference or expected sequence of behavior given the relevant context. Representing contrast signals both the the contrasting

states of behavior and their relationship (how behavior departs or conforms to the contrasting case). In this way, one is in effect, highlighting anomalies. By this, I mean one indicates what is anomalous (the contrast) as opposed to simply indicating that some unspecified thing is general is anomalous. For example, coding a number or icon red shows that some anomaly is present, but it does not show the contrast of what is anomolous relative to what (cf., Woods, 1992).

The relation to ecological perception is obvious: Event perception is at the heart of situation assessment and recognition-driven processing. As researchers in ecological perception have pointed out, events and sequence are more than a mere succession of static views. Ongoing events are dynamic patterns that directly specify the type of event. When the pattern is available (represented in the medium) there is no need to abstract or derive it; and the event structure is available even with discontinuous eye fixations (there is no minimal cognitive load in re-orienting after a glance away).

But given that the computer representation is free from the physical constraints acting on the referent objects, support for event perception in the computer medium requires the designer to actively identify operationally interesting changes or sequences of behavior and to actively develop representations that highlight these events to the observer given the actual task context. The default representations typically available do not make interesting events directly available for the practitioner to observe. Instead, the typical computer displays of data force the practitioner into a serial deliberative mode of cognition to abstract change and events from the displayed data (typically digital representations of sensed data).

6.3 Towards a Context-Bound Science of Human-Machine Systems

6.3.1 What is Informative?

The ubiquitous computerization of the modern world has tremendously advanced our ability to collect, transmit, and transform data. In all areas of human endeavor, we are bombarded with computer processed data, especially when anomalies occur. The problem of our day is data overload — computerization of process control centers generates huge networks of computer displays (a European computerized nuclear power control room containing over 16,000 displays is scheduled to come online in a few years; Easter, 1991), huge databases are generated about the performance of telecommunication networks, and so on. But our ability to digest and interpret these data has failed to keep pace with our capacity to generate and manipulate it. Practitioners get lost in large networks of computer-based display frames; they experience keyhole effects in trying to monitor dynamic systems through the narrow viewports of windowed computer screens; they are overwhelmed by the massive field of available data and fail to focus on the data critically important to a specific context.

Fundamentally, ecological perception sees the problem of perception in terms of understanding how meaning is specified and how we find what is meaningful or significant in the perceptual field (von Uexkull, 1934), for example, the perception of affordances. In the ecological view, perceptual systems are concerned with extracting meaningful information rather than a mere encoding and transmission front end for cognition. Similarly, the problem for human-machine systems is that the technological view emphasizes data encoding, manipulation, and transmission, not the processes involved in meaning extraction. Given the properties of the computer as a medium — narrow keyhole, the accumulation of ever larger amounts of data—the critical research issue is understanding how we focus in on what is interesting or significant (Doyle, Sellers, & Atkinson, 1989; Woods, 1984, 1986).

This is all based on how one answers the question of what is informative. Information is not a thing-in-itself, but is rather a relation between the data, the world the data refer to, and the observer's expectations, intentions, and interests. As a result, informativeness is not a property of the data field alone, but is a relation between the observer and the data field. The important point is that there is significant difference between the available data and the meaning or information that a person extracts from that data (e.g., Woods, 1986).

The available data are raw materials or evidence that the observer uses and evaluates to answer questions (questions that can be vague or well formed, general, or specific). The degree to which the data help answer those questions determines the informativeness or inferential value of the data. Thus, the meaning associated with a given datum depends on its relationship to the context or field of surrounding data, including its relationship to the objects or units of description of the domain (what object and state of the object is referred to), to the set of possible actions, and to perceived task goals (after Gibson, what that object/state *affords* the observer).

Thus, all issues in representation design (Woods, in preparation) revolve around putting data into context. The significance or meaning extracted from an observed datum depends on the relationship of that datum to a larger set of data about the state of the domain, goals, past experience, possible actions, and so on — the context in which the datum occurs. This means that processing of an observed value or change on one data channel (extracting meaning from such observations) depends on contact with a variety of other information derived from checking or being aware of or remembering or assuming the state of other data.

Not only does a datum gain significance in relation to a set of other data, but also what data belong in this relevance set will change as a function of the state of the domain and the state of the problem solving process. The above points — (a) a set of contextual data is needed to extract meaning from a datum, and (b) the data in this relevance set change with both system state and the state of the problem solving process — define the context-sensitivity problem (Woods, 1986, 1991). The amount of context-sensitivity present in a particular domain of application is a major cognitive demand factor that has profound implications for representation design and human performance. If the relevance sets are limited and do not change much with context, then many of the challenges in representation design are relaxed, that is, practitioners will be able to extract needed information, even from otherwise problematical representations. However, if the relevance sets are larger and are sensitive to changes in context (as in time-pressured, high-consequence applications), then the problem representation principle and its inverse become a particularly important contributor to human interaction with the computer display system and human performance at domain tasks. The quality of a representation depends on how it affects the cognitive processes involved in extracting meaning given the context-sensitivity problem. Poor representations present available data as individual signals without relating a signal to its context. The observer must acquire, remember, or build the context of interpretation datum by datum with the associated possibilities for incompleteness or error and with the associated mental workload (cf. Doyle et al., 1989, Woods & Elias, 1988, for examples of context–sensitive representations).

6.3.2   Agent-Environment Mutuality

Ecological perception also has pointed out that the problem in perception is not that the scene is underspecified requiring cognition to make up for an impoverished retinal image. Instead, if one considers actual visual scenes, the issue is that there are a multiplicity of cues available for perceptual systems, that is, the critical properties of the scene are overspecified (Cutting, 1986). Similarly, the problem in human-machine system performance is not a lack of data, but rather the difficulties associated with finding the right data at the right time out of a huge field of potentially relevant data (Woods, 1986).

The multiplicity of cues available in natural problem-solving domains creates the problem of deciding what counts as the effective stimulus. Psychologists cannot simply assert the grain of analysis or the specific driving stimuli in a particular behavioral situation without risking falling into the psychologist's fallacy of William James in which the psychologist's reality is confused with the psychological reality of the human in his or her problem-solving world. As Gibson saw, meeting this challenge required new ways and efforts to characterize the stimulus world based on an organism-environment mutuality. Meeting this challenge in person-machine systems also requires new ways and efforts to characterize the demands of problem solving domains based on an agent-environment mutuality (Rasmussen, 1986; Woods, 1988). Whereas Gibson called this ecological physics, in human-machine systems it has generally gone under the name *cognitive task analysis*. This is a set of empirically based investigations to model the mutual shaping between (a) the cognitive demands/constraints of the domain and (b) the strategies evolved by groups of practitioners given (c) the cognitive tools and resources available in that field of practice (for an outstanding example, see Hutchins, 1991).

Remember the design challenge for creating effective representations is to set up the mapping between domain referents and visual tokens in the computer medium so that the observer can extract information about task-meaningful semantics. One kind of error in representation design is failure to map task meaningful domain semantics into the structure and behavior of the representation. To avoid this, however, requires that designers directly confront the problem of cognitive task analysis (Rasmussen, 1986): identifying what is informative, what are the *interesting* properties and changes of the domain referents given the goals and task context of the team of practitioners.

Note the seeming paradox at the heart of the agent-environment mutuality assumption. In order to understand the psychology of human information processing, one must understand the nature of actual problems (i.e., information processing tasks) as it relates to the possibilities for information processing. For ecological perception, agent-environment mutuality meant the abandonment of the minimalist research strategy in perception based on excessive simplification of perceptual stimuli (e.g., tachistoscopic research) and a commitment to studying the properties of complex perceptual situations as they relate to the potential for perception by an organism. As ecological perception led to the need for a better understanding of the stimulus world-- ecological physics, progress in understanding aided cognition requires progress on understanding of the properties of its stimulus world -- an ecology of problems. To do this will require development of new methods and approaches to understand the dynamic interplay of people and technology (e.g., Hutchins, 1990).

Adopting the agent-environment mutuality assumption for human-machine systems forces a revolutionary shift with regard to normal disciplinary boundaries and research agendas. Technologists work on problems related to expanding the boundaries of what machines can do autonomously. However, data overload is a problem concerned with the interaction of people and technology in cognitive work. Experimental psychologists study human information processing, but almost always divorced from any technological context. They do not study how people create, use, and adapt cognitive tools to assist them in solving problems, or, what could be called, aided information processing. But problems like data overload or mode errors only exist at the intersection of people and technology.

6.3.3   The Context-Bound Approach to Human-Machine Systems

At the risk of oversimplifying, one can think of human factors being divided into two types. One is a context-free approach to study human-machine systems. It is characterized by studies of generic tasks; user modeling is a critical focus; results are organized by domain of application (aviation, consumer products, medical, etc.) or technological developments (e.g., hypertext). The research methods are laboratory–based hypothesis testing. The subjects tend to be naive and passive relative to the test tasks. The test tasks bear no or only a superficial relationship to the tasks and task contexts in the actual target domain of reference. The relationship between basic and applied research is seen as a pipeline in which basic work eventually flows toward applications.

The context-bound approach, on the other hand, is based on a commitment to study complex human-system interactions directly. Its methods, data, concepts, and theories are all bound to the situations under study, that is, to the context (Hutchins, in press; Woods, 1993a). In this approach it is axiomatic that one cannot separate the study of problem solving from analysis of the situations in which it occurs (Lave, 1988, p. 42). Hence, it has become popular to refer to the context-bound approach as situated cognition (cf. Suchman, 1987). This approach is characterized by studies of specific meaningful tasks as carried out by actual practitioners; models of errors and expertise in context are a critical focus; results are organized by cognitive characteristics of the interaction between people and technology (e.g., distributed or cooperative cognition, or how practitioners shape the information processing tools that they use). The research methods are based on field study techniques and detailed protocol analysis of the process of solving a problem. The study participants are active skilled practitioners. Understanding the cognitive task—practitioner strategy relationship and how practitioners adapt cognitive tools to aid them in their work— is fundamental. The relationship between basic and applied research is seen as complementary in which growing the research base and developing effective applications are mutually interdependent (Woods, 1993a). Recent examples of the context-bound approach are Mitchell's studies of satellite control centers (e.g., Mitchell & Saisi, 1987), my studies of nuclear power emergency operations and new support systems (e.g., Woods, O'Brien, & Hanes, 1987); Klein's studies of recognition-driven decision making in several domains (e.g., Klein et al., 1993); Cook's studies of computerized surgical operating room devices (e.g., Cook et al., 1990, 1991; Moll van Charante et al., 1993), Hutchins's studies of navigation (e.g., Hutchins, 1990; in press); P. Smith and J. Smith's studies revolving around blood matching decisions in immunohematology (e.g., Smith et al., 1991); and several studies of cognitive activities in commercial airline cockpits (Hutchins, 1991; Sarter & Woods, 1994).

To say that the study of human-machine systems can and should be context bound is not simply to call for more applied studies in particular domains. *"It is . . . the fundamental principle of cognition that the universal can be perceived only in the particular, while the particular can be thought of only in reference to the universal"* (Cassirer, 1923/1953, p. 86). As Hutchins puts it:

> *"There are powerful regularities to be described at a level of analysis that transcends the details of the specific domain. It is not possible to discover these regularities without understanding the details of the domain, but the regularities are not about the domain specific details, they are about the nature of human cognition in human activity"* (Hutchins, 1992, personal communication).

This reveals the proper complementarity between so called basic and applied work in which the experimenter functions as designer and the designer as experimenter (Woods, 1993a). "New technology is a kind of experimental investigation into fields of ongoing activity. If we truly understand cognitive systems, then we must be able to develop designs that enhance the performance of operational systems; if we are to enhance the performance of operational systems, we need conceptual looking glasses that enable us to see past the unending variety of technology and particular domains" (Woods & Sarter, 1993).

Each approach is subject to risks and dangers. The challenges for a context-bound science of human-machine systems are (a) methods and results for building cognitive task models (the ecology of problems), (b) methods and theory building to generate generalizable results from context bound studies (producing distilled results transportable across scenarios, participants, and domains rather than just diluted motherhood generalizations), and (c) methods and theories to stimulate critical growth of knowledge across context bound studies.

It is the context-bound approach that can draw on analogies to ecological perception. For example, what follows is a listing of the research program of a context-bound human factors expressed in terms parallel to the ecological perception and action research program (based on Pittenger, 1991):

1. Discovery of the events/information that serves to guide the observer's action in the domain/environment.
2. Specification of the information in the domain which supports perception/identification of events.
3. Design of a representation of the domain that supports perception and extraction of the information/events important to action.
4. Testing that the information/events are perceivable/extractable and that they are useful in guiding action.

6.3.4   The Adaptive Practitioner

In developing new computer-based information technology and automation, the conventional view seems to be that new technology makes for better ways of doing the same task activities. We often act as if the domain practitioner were a passive recipient of the resulting operator aids, the user of what the technologist provides for them.

However, this view overlooks the fact that the introduction of new technology represents a change from one way of doing things to another. The design of new technology is always an intervention into an ongoing world of activity. It alters what is already going on — the everyday practices and concerns of a community of people — and leads to a resettling into new practices (Flores, Graves, Hartfield, & Winograd, 1988, p. 154). Practitioners are not passive in this process of accommodation to change. Rather, they are an active adaptive element in the person-machine ensemble, usually the critical adaptive portion. Studies show that practitioners adapt information technology provided for them to the immediate tasks at hand in a locally pragmatic way, usually in ways not anticipated by the designers of the information technology (Cook et al., 1990; Roth, Bennett, & Woods, 1987; Flores et al., 1988; Hutchins, 1990). Tools are shaped by their users.

One of the forces that drive user adaptations is *clumsy automation* (Wiener, 1989). One of several forms of the clumsy use of technology occurs when the benefits of the automation accrue during workload troughs, and the costs of automation occur during high-criticality or high-tempo operations (Woods et al., 1994; Woods, 1993b).

Practitioners (commercial pilots, anesthesiologists, nuclear power operators, operators in space control centers) are responsible not just for device operation, but also for the larger system and performance goals of the overall system. Practitioners tailor their activities to insulate the larger system from device deficiencies and peculiarities of the technology. This occurs, in part, because practitioners inevitably are held accountable for failure to correctly operate equipment, diagnose faults, or respond to anomalies, even if the device setup, operation, and performance are ill suited to the demands of the environment. This creates the paradoxical situation in which practitioners' adaptive, coping responses often help to hide the corrosive effects of clumsy technology from designers.

Again there is a parallel to research in perception and especially, ecological perception. The human is an adaptive, active perceiver, not a passive element. We will begin to understand human-machine systems only when we begin to understand the adaptive interplay of practitioner and tools in the course of meeting task demands. Unfortunately, this demands, like Gibson demanded of the minimalist tachistoscopic school of work in perception, a paradigm shift for work on human-machine systems. It demands that researchers examine problem solving in situ — in complex settings, in which significant information processing tools are available to support the practitioner and in which domain–knowledgeable people are the appropriate study participants (Woods, 1993a).

6.4   Summary

The parallel to Gibson and ecological perception can be overdrawn with respect to the study of human-machine systems. The agent-environment mutuality assumption is (or should be) common to both endeavors. It is fairly easy to draw analogies between concepts in ecological perception and the ideas of some researchers in human-machine systems: the user as an active adaptive practitioner, the shift in the sense of what is informative, the search for meaning as a fundamental parameter in human-computer interaction, the need for a new way to characterize problem-solving habitats, the equivalent of an ecological physics, the multiplicity of cues available in natural problem-solving habitats, and the need to take into account the dynamism of real problems. But at another level the appeal of the term *ecology of human-machine systems* is based on the perceived need for a paradigm shift in human-machine systems — a parallel to the Gibsonian paradigm shift in research on perception and action. The paradigm shift is an abandonment of the context-free approach and methods in the study of human-machine systems and a commitment to the methods and agenda of a context-bound approach.

Acknowledgments

This work was supported by the Aerospace Human Factors Research Division of the NASA Ames Research Center under Grant NCC2-592 (Dr. Everett Palmer, technical monitor) and by the of the NASA Johnson Space Center (Dr. Jane Malin, technical monitor).

6.5   References

Becker, R. A., & Cleveland W. S. (1987). Brushing scatterplots. *Technometrics, 29,* 127–142.

Cassirer, E. (1953). *The philosophy of symbolic forms, Vol. 1: Language* (R. Manheim, Trans.). Yale University Press. New Haven, CT. (Original work published 1923).

Cook, R. I., Potter, S. D., Woods, D., & McDonald, J.S. (1991). Evaluating the human engineering of microprocessor controlled operating room devices. *Journal of Clinical Monitoring, 7,* 217–226.

Cook, R. I., Woods, D. D., & Howie, M. B. (1990). The natural history of introducing new information technology into a dynamic high-risk environment. *Proceedings of the Human Factors Society, 34th Annual Meeting.* Santa Monica, CA.

Cutting, J. E. (1986). *Perception with an eye for motion.* Cambridge MA: MIT Press.

Doyle, R., Sellers, S., & Atkinson, D. (1989). A focused, context sensitive approach to monitoring. *Proceedings of the Eleventh International Joint Conference on Artificial Intelligence.* IJCAI.

Easter, J. R. (1991). The role of the operator and control room design. In J. White and D. Lanning (Eds.), *European nuclear instrumentation and controls,* (Rep. PB92–100197). World Technology Evaluation Center, Loyola College, National Technical Information Service.

Flores, F., Graves, M., Hartfield, B., & Winograd, T. (1988). Computer systems and the design of organizational interaction. *ACM Transactions on Office Information Systems, 6,* 153–172 .

Freund, P. R., & Sharar, S. R. (1990). Hyperthermia alert caused by unrecognized temperature monitor malfunction. *Journal of Clinical Monitoring, 6,* 257-257.

Gibson, J. J. (1979). *The ecological approach to visual perception.* Boston: Houghton-Mifflin.

Gonzalez, E. G., & Kolers, P. A. (1982). Mental manipulation of arithmetic symbols. *Journal of Experimental Psychology: Learning, Memory, and Cognition, 8,* 308–319.

Goodstein, L. (1981). Discriminative display support for process operators. In J. Rasmussen and W. Rouse (Eds.), *Human detection and diagnosis of system failures*. New York: Plenum Press.

Henderson, A., & Card, S. (1987). Rooms: The use of multiple virtual workspaces to reduce space contention in a window-based graphical interface. *ACM Transactions on Graphics, 5*, 211–243.

Hochberg, J. (1986). Representation of motion and space in video and cinematic displays. In K. R. Boff, L. Kaufman, & J. P. Thomas, (Eds.), *Handbook of human perception and performance, I*. New York: Wiley.

Hutchins, E. (1980). *Culture and inference*. Cambridge, MA: Harvard University Press.

Hutchins, E. (1990). The technology of team navigation. In J. Galegher, R. Kraut, and C. Egid (Eds.), *Intellectual teamwork: Social and technological foundations of cooperative work*. Hillsdale, NJ: Lawrence Erlbaum Associates.

Hutchins, E. (1991). *How a cockpit remembers its speed* Technical Report. La Jolla: Distributed Cognition Laboratory, University of California, San Diego.

Hutchins, E. (in press). *Cognition in the wild*. Cambridge, MA: MIT Press.

Johnson, E., & Payne, J. W. (1985). Effort and accuracy in choice. *Management Science, 31*, 395–414.

Klein, G., Orasanu, J., Calderwood, R., and Zsambok, C. E. (1993). (Eds.), *Decision making in action: Models and methods*. Norwood, NJ: Ablex.

Lave, J. (1988). *Cognition in practice*. New York: Cambridge University Press.

Mitchell, C., & Saisi, D. (1987). Use of model-based qualitative icons and adaptive windows in workstations for supervisory control systems. *IEEE Transactions on Systems, Man, and Cybernetics, SMC-17*, 573-593.

Moll van Charante, E., Cook, R. I., Woods, D. D., Yue L., & Howie, M. B. (1993). Human-computer interaction in context: Physician interaction with automated intravenous controllers in the heart room. In H.G. Stassen (Ed.), *Analysis, design and evaluation of man-machine systems 1992*. New York: Pergamon Press.

Murray, C., & Cox, C. B. (1989). *Apollo: The race to the moon*. New York: Simon & Schuster.

Neisser, U. (1976). *Cognition and reality*. San Francisco: W. H. Freeman.

Norman, D. A. (1981). Categorization of action slips. *Psychological Review, 88,* 1-15

Norman, D. A. (1988). *The psychology of everyday things.* New York: Basic Books.

Norman, D. A. (1993). *Things that make us smart.* Reading, MA: Addison-Wesley.

Payne, J. W., Bettman, J. R., & Johnson, E. J. (1988). Adaptive strategy selection in decision making. *Journal of Experimental Psychology: Learning, Memory, and Cognition, 14,* 534-552.

Peirce, C. S. (1955). Abduction and induction. In J. Buchler (Ed.), *Philosophical writings of Peirce.* London: Dover. (Original work published, 1903).

Pittenger, J. B. (1991). Cognitive physics and event perception: Two approaches to the assessment of people's knowledge of physics. In R. R. Hoffman and D. S. Palmero (Eds.), *Cognition and the symbolic processes: Applied and ecological approaches.* Hillsdale, NJ: Lawerence Erlbaum Associates.

Rasmussen, J. (1986). *Information processing and human-machine interaction: An approach to cognitive engineering.* New York: North-Holland.

Rasmussen, J., & Lind, M. (1981). Coping with complexity. In H. G. Stassen (Ed.), *First European annual conference on human decision making and manual control.* New York: Plenum Press.

Rasmussen, J., & Rouse, W. (1981). (Eds.), *Human Detection and Diagnosis of System Failures.* New York: Plenum Press.

Reason, J. & Mycielska, K. (1982). *Absent minded? The psychology of mental lapses and everyday errors.* Englewood Cliffs, NJ: Prentice-Hall.

Reed, E. S. (1988). *James J. Gibson and the psychology of perception.* New Haven, CT: Yale University Press.

Roth, E. M., Bennett, K., & Woods, D. D. (1987). Human interaction with an intelligent machine. *International Journal of Man-Machine Studies, 27,* 479-525.

Sarter, N., & Woods, D. D. (1994). Pilot interaction with cockpit automation II: An experimental study of pilots' models and awareness of the flight management system. *International Journal of Aviation Psychology, 4,* 1-28.

Simon, H. A. (1969). *The sciences of the artificial.* Cambridge, MA: MIT Press.

Smith, P. J., Smith, J. W., Svirbely, J., Krawczak, D., Fraser, J., Rudman, S., Miller, T., & Blazina, J. (1991). Coping with the complexities of multiple-solution problems. *International Journal of Man-Machine Studies, 35,* 429–453.

Suchman, L. A. (1987). *Plans and situated actions: The problem of human-machine communication.* Cambridge University Press, Cambridge, England.

Tenney, Y. J., Jager Adams, M., Pew, R. W., Huggins, A. W. F., & Rogers, W. H. (1992). *A principled approach to the measurement of situation awareness in commercial aviation.* (NASA Contractor Report No. NAS1-18788). Hampton, VA: NASA Langley Research Center.

von Uexkull, J. (1957). A stroll through the worlds of animals and men. In C. Schiller, Ed. and Trans., *Instinctive behavior.* New York: International Universities Press. (Original work published, 1934).

Wiener, E. L. (1989). *Human factors of advanced technology (glass cockpit) transport aircraft.* (Technical Report 117528), Washington, D. C.: NASA.

Winograd, T. (1987). *Three responses to situation theory.* Technical Report CSLI-87-106, Center for the Study of Language and Information, Stanford University.

Woods, D. D. (1984). Visual momentum: A concept to improve the cognitive coupling of person and computer. *International Journal of Man-Machine Studies, 21,* 229–244.

Woods, D. D. (1986). Paradigms for intelligent decision support. In E. Hollnagel, G. Mancini, and D. D. Woods (Eds.), *Intelligent decision support in process environments.* (pp. 153–174). New York: Springer-Verlag.

Woods, D. D. (1988). Coping with complexity: The psychology of human behavior in complex systems. In L. P. Goodstein, H. B. Andersen, and S. E. Olsen (Eds.), *Mental models, tasks and errors.* London: Taylor & Francis.

Woods, D. D. (1991). The cognitive engineering of problem representations. In G. R. S. Weir and J. L. Alty (Eds.), *Human-computer interaction and complex systems.* London: Academic Press.

Woods, D. D. (1992). *The alarm problem and directed attention.* Cognitive Systems Engineering Laboratory Report 92-TR-06, Department of Industrial and Systems Engineering, The Ohio State University, Columbus, Ohio.

Woods, D. D. (1993a). Process tracing methods for the study of cognition outside of the experimental psychology laboratory. In G. A. Klein, J. Orasanu, R. Calderwood, and C. E. Zsambok

(Eds.), *Decision making in action: Models and methods.* Norwood, NJ: Ablex.

Woods, D. D. (1993b). The price of flexibility in intelligent interfaces. *Knowledge-Based Systems, 6,* 1-8.

Woods, D. D. (in preparation). *Visualizing function: The theory and practice of representation design in the computer medium.* Columbus, OH: Cognitive Systems Engineering Laboratory, Ohio State University.

Woods, D. D., & Elias, G. (1988). Significance messages: An integral display concept. *Proceedings of the Human Factors Society, 32nd Annual Meeting.* Santa Monica, CA.

Woods, D. D., Johanssen, L., Cook, R. I., & Sarter, N. (1994). *Behind human error: Cognitive systems, computers and hindsight.* Crew Systems Ergonomic Information and Analysis Center, Wright-Patterson AFB, OH (State of the Art Report).

Woods, D. D., O'Brien, J., & Hanes, L. F. (1987). Human factors challenges in process control: The case of nuclear power plants. In G. Salvendy (Ed.), *Handbook of Human Factors/Ergonomics.* New York: Wiley.

Woods, D. D., Potter, S. S., Johannesen, L., & Holloway, M. (1991). *Human interaction with intelligent systems: Trends, problems, new directions.* Cognitive Systems Engineering Laboratory Report, prepared for NASA Johnson Space Center, Washington, D. C.

Woods, D. D., & Sarter, N. (1993). Evaluating the impact of new technology on human-machine cooperation. In J. Wise, V. D. Hopkin, and P. Stager (Eds.), *Verification and validation of complex and integrated human-machine systems.* New York: Springer-Verlag.

Woods, D. D., Watts, J., & Potter, S. D. (1993). *How not to have to nagivate through way too many displays.* Cognitive Systems Engineering Laboratory Report 93-TR-02, Department of Industrial and Systems Engineering, The Ohio State University, Columbus, OH.

Woods, D. D., Wise, J. A., & Hanes, L. F. (1981). An evaluation of nuclear power plant safety parameter display systems. *Proceedings of the Human Factors Society, 25th Annual Meeting.* Human Factors Society, Santa Monica, CA.

Chapter 7

Active Psychophysics: The Relation Between Mind and What Matters

John M. Flach
Wright State University
Armstrong Laboratory
Wright-Patterson Air Force Base

Rik Warren
Armstrong Laboratory
Wright-Patterson Air Force Base

7.0   Introduction

"Psychophysics should be understood here as an exact theory of the functionally dependent relations . . . of the physical and psychological worlds. We count as . . . psychological . . . all that can be grasped by introspective observation or that can be abstracted from it; as . . . physical . . . all that can be grasped by observation from the outside or abstracted from it. Insofar as a functional relationship linking body and mind exists, there is actually nothing to keep us from looking at it and pursuing it from the one direction rather than from the other. . . . There is a reason, however, why psychophysics prefers to make the approach from the side of the dependence of the mind on the body rather than the contrary, for it is only the physical that is immediately open to measurement, whereas the measurement of the psychical can be obtained only as dependent on the physical . . . . This reason is decisive; it determines the direction of approach in what follows. . . . lawful relationships, which may be ascertained in the area of outer psychophysics, have their own importance. Based on them . . . physical

measurement yields a psychic measurement, on which we can base arguments that in their turn are of importance and interest. " Fechner (1860/1966, p. 7–11)

Our goal as psychologists and particularly as engineering psychologists concerned with the design of human-machine systems, is to *understand the functional relations between the physical and the psychological*:

$$\text{Psych} = f(\text{Physical}).$$

For Fechner and the tradition of psychophysics that has grown from his work, psychological measurement has predominantly meant sensation (e.g., detecting a light), and physical measurement has focused on standard descriptions of physical magnitudes in terms of number, mass, space, and time (e.g., the number of photons hitting the retina):

$$\text{Sensation} = f(\text{Physical Magnitude}).$$

In this chapter we attempt to expand both sides of this equation. On the psychological side, we explore questions of perception and cognition[1]. On the physical side, we will explore the possibilities for intrinsic measures that preserve functional (i.e., meaningful or task relevant) properties of the observable world:

$$\text{Perception}/\text{Cognition} = f(\text{Functional Properties}).$$

Like Fechner, we hope that this expansion will lead to further arguments that in their turn are of importance and interest.

---

[1] The distinctions between sensation, perception, and cognition are amorphous at best. Some make no distinction between sensation and perception; others include perception within the domain of cognition distinct from sensation. We are also inclined to group perception and cognition together, distinct from sensation. Thus, we often refer to perception/cognition to emphasize this point. However, whereas constructivists include perception with cognition because perception is seen as a kind of problem solving in which meaning must be solved for, we are more inclined to see problem solving as a subset of perception in which correct choices are recognized in terms of what they mean, rather than computed from among alternatives. For an example, see Klein, Calderwood, and MacGregor's (1989) model for Recognition Primed Decision Making.

7.1   Matter Versus What Matters

The first step to expanding our psychophysics into the domains of perception and cognition is to examine the relation between sensation and perception/cognition. This relation has traditionally been viewed as one of processing or elaboration as illustrated in Figure 7.1. Figure 7.1 shows a stimulus (i.e., matter) that is defined in terms of intensity, quality, extension, or duration. The stimulus is input into the sensory processor, and the result of this processing is a sensation. The sensation then undergoes further processing or elaboration to produce a percept (i.e., what matters). Now we are in the domain of cognition;

$$\text{Perception} = f(\text{Sensation})$$

or

$$\text{Perception} = f(g(\text{Physical})).$$

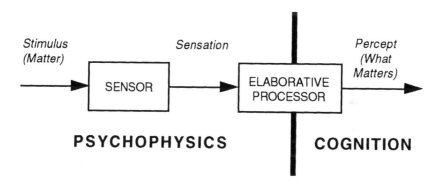

Figure. 7.1. *A constructivist view mapping matter to sensation and sensation to perception (i.e., meaning).*

The difference between the sensation and the percept is that sensations reflect properties of the sensory system (e.g., Law of Specific Nerve Energies) and intensive, qualitative, extensive, or temporal properties of matter; whereas perceptions reflect meaning or what matters. For example, the extent (size) or mass (weight) of an object can be sensed. The sensation, however, has no meaning in and of itself. Its meaning depends on the task context. An extent of 2 inches means one thing to an adult looking for an object to throw and something completely different to an adult looking for a gap to pass through. The *perception* of a throwable object, then is seen as a product of higher processing that is influenced by a person's *intentions* (throwing an object to ward off threat or escaping from threat), a person's *learning history* (the past discovery that thrown objects can cause threats to retreat), as well as the sensations resulting from the object. Thus, in this view meaning is constructed as a product of information processing.

In this constructivist scheme, the physical world of matter is dimensioned using external, observer independent units (meters, grams, seconds). These are the units of standard physics. These metrics were appropriately chosen by physicists because they allow observer (i.e., experimenter) independent descriptions of the physical world of matter. These dimensions are extrinsic to the measurement event. Meaning, from the constructivist's point of view, is considered to be an attribute of the psychological, not of the physical. For psychophysics, meaning (i.e., higher mental processing) is a confounding factor to be controlled or eliminated. For example, signal detection theory is a methodology for unconfounding the sensory dimension (d') from the cognitive effects of decision biases (beta).

Perception and cognition, on the other hand, are about the observer (i.e., subject) dependent (or relevant) properties of the physical world (i.e., what matters). For perception and cognition, meaning is not confound, but rather it is the phenomena of interest. Traditional approaches to perception/cognition therefore have given little attention to the physical world and have focused instead on the processing mechanisms that transform sensations into percepts and on the internal representations produced as a result of this processing. Thus, the psychophysical tradition of Fechner has had little influence on cognitive science.

Psychophysics has no theory of representation. What matters is defined in objective, physical terms inherited largely from Newtonian physics. Cognitive psychology starts with representations, but has no theory of environments. What matters is defined in terms of mental representations. Meaning, for most of cognitive psychology, is a purely mental construct. Is there some intermediate ground where meaning (i.e., what matters) can be defined objectively, but along dimensions that map more readily into the properties of representations (i.e., into the distinctions that are important for how people behave and think)? Active psychophysics is an attempt to find this intermediate ground.

7.2   The Physical Side

Perhaps the metrics that are best for the goals of physics are not best for the study of perception and cognition. That is, perhaps the problem in applying Fechner's psychophysical program to the phenomena of perception and cognition is in our physics. A first step for developing a psychophysics of perception and cognition may require a new physics. Is it possible to describe the physical world in terms of meaning without entering into the psychological world of introspective observation? We would like an externally observable description of the physical world, which is lawfully related to the psychological dimensions of perception and cognition. What is needed is what Gibson (1960/1982) called "ecological physics."

> "The sensory psychologists, of course, have read their physics and chemistry. But physical science portrays a sterile world. The variables of physics make uninteresting stimuli. Why is this true? I think it is because psychologists take for stimuli only the variables of physics as they stand in textbooks. We have simply picked the wrong variables. It is our own fault. . . . Consider, for example, the physics (that is to say acoustics) of speech sounds. As recently as 1951, in the <u>Handbook of Experimental Psychology</u> (Stevens, p. 869), the fact that a word is perceptually the same when whispered as it is when shouted was taken to prove that the physical characteristics of sound waves, frequency, intensity, and so on, cannot tell us about speech. Speech perception would require a psychological theory, not physical measurement. But the invention of the sound spectrograph seems to have shown that certain higher order variables of acoustic energy are the critical constituents of speech and the stimuli for hearing it. These newly discovered invariant patterns of sound are completely physical, even if they had not previously been studied in physics. What was needed to understand the psychophysics of hearing words was not more psychology but more physics" (Gibson, 1960, p. 345)

Traditionally, psychophysics has imported the metrics of physics into psychology without change. Ecological physics challenges psychology to discover its own metrics for describing the physical world. One basis for these new metrics is an intrinsic measurement system that "takes the action system as a natural standard for measuring the properties of the environment" (Warren, 1985, p. 6). Thus, instead of inches or meters as a standard for measuring extent, we take hand width, leg length, or eye height. In human-vehicle systems we might think in terms of safe following distance in car lengths or glide slope might be "measured" relative to a position on the wind screen. Taking the action system as a standard now makes our description of the environment observer-dependent or better observer-relevant[2]. However, it does not make the description any less observable or objective. It requires no mentalism or no introspection. It remains within the domain of the physical as conceived by Fechner.

It is important to note that the use of intrinsic measures is very important in physics. One example is the Reynolds number that indexes characteristics of fluid flow. As the Reynolds number increases, qualitative changes in flow arise. For low Reynolds numbers, flow is steady. At higher numbers vortices appear. At still higher numbers flow becomes chaotic and irregular. Thus, we are actually advocating a very traditional approach to physics. The key in doing our "psycho" physics is to discover the natural metrics of behavior, not to uncritically import metrics from other domains.

The advantage of using the action system as the standard for measuring the environment is that we can rescue "meaning" from the domain of the "psychological" and bring it into the domain of the "physical." Thus, we can talk in terms of graspability instead of extent. We can begin to specify critical boundaries—too large to grasp, too small to grasp. We can demarcate "comfort" or "optimal" regions for graspability. Just as the Reynolds number indexes qualitative boundaries in fluid flow, our metrics should index qualitative boundaries in behavior. Historically, action relevant metrics predated many of today's metrics: the inch was preceded by the thumb as a standard, pacing distances preceded the foot, and an acre was defined in terms of a day's plowing with a team of oxen. For example, a house painter might "see" or "measure" surface areas in terms of time (e.g., a 4-hour surface area) or paint (e.g., a 2-gallon surface area). In terms of human-vehicle systems, consider an approach to landing. Meaningful aspects of an approach would be safety margins— too low, too high, too steep, too shallow, too fast, too slow. These functional boundaries reflect the dynamic capabilities of the action system (i.e., the vehicle). These are not mentalistic constructions. They are real, objective physical

---

[2]Sometimes it seems as though psychophysicists go to extremes to make the stimuli observer irrelevant so that "cognitive noise" does not contribute unwanted variability. Ebbinghaus's (1885/1964) invention of the nonsense syllable is, of course, the classic example of this approach.

features that can be measured and specified based on the laws of aerodynamics. Thus, it would be possible, in the tradition of psychophysics, to study whether there are lawful relations between these "physical meanings" in the world and "psychological" and behavioral events:

$$Perception/Cognition = f(Meaning).$$

This is the challenge of active psychophysics (Warren, 1988a; Warren & McMillan, 1984). As Fechner notes, a functional relation such as the one described earlier can be pursued from one side or the other. As illustrated in Figure 7.1, constructivist approaches to perception and cognition have generally viewed meaning as a product of processing and have treated it experimentally as a dependent variable. In active psychophysics we are suggesting that meaning is the raw material for processing and should be treated experimentally as an independent variable. This choice is consistent with Fechner's choice if it is possible to define meaning in physical terms without reference to the introspective, psychological world.

7.3   The Psychological Side

In this section, we consider the psychological side of the equation. To this point, we contrasted active psychophysics to traditional psychophysical approaches and constructivist approaches to cognition and perception. In this section, we begin by considering the relation between active psychophysics and Gibson's (1966, 1979/1986) ecological approach to perception. Then we consider the nature of information couplings between actors and environments and the role of action in perception and cognition.

7.3.1   Affordances

The concept of intrinsic measurement in which the environment is described in terms of functional, action-relevant metrics (i.e., in terms of meaning) is the foundation for understanding Gibson's (1979/1986) concepts of affordance and direct perception. Gibson wrote that "affordances of the environment are what it offers the animal, what it provides or furnishes, either for good or ill" (p. 127). Affordances are properties of the environment, but properties that are scaled to the organism. Affordances are objective, physical properties; they are not mentalistic constructions. They do not depend on an actor's intentions, attention, or learning history. Although Gibson was undoubtedly influenced by Lewin's (1936) concept of valence (Gibson & Crooks, 1938; Lombardo, 1987), there are significant differences between the concepts of valence and affordance. In contrasting affordance with the more mentalistic term of valence, Warren (1984) wrote "a rock of a certain size, shape, and weight affords both throwing and pounding, the fact that a person in need of a hammer attends to certain aspects of the rock does not alter its properties or its utility for both activities" (p. 684). Intention, attention, and learning history may all be important factors in determining how a person behaves toward the rock, but the affordance properties are independent of these "psychological" dimensions. Affordance properties can be described on a purely objective basis by an external observer who can measure physical properties of the actor and the environment. This observer does not need to know anything about the introspective psychological world of the actor. Thus, affordances are relativistic. Affordances are not attributes of the environment or of the actor. They are relations defined over actor environments.

7.3.2   Information Coupling

"The whole of nature is a single continuous system of component parts acting on one another, within which various partial systems create, use, and transmit to each other kinetic energy of different forms, while obeying general laws through which the connections are ruled and conserved. Since in exact natural science all physical happenings, activities, and processes, whatever they may be called (not excluding the chemical, the imponderable, and the organic) may be reduced to movements, be they of large masses or of the smallest particles, we can also find for all of them a yardstick of their activity or strength in their kinetic energy, which can always be measured, if not always directly, then at least by its effects, and in any case in principle.

The uncertainty which we have from the start about the nature of the physical happenings on whose occurrence our sensations depend and which accompany our thoughts —- in short, of psychophysical processes —- is at any rate not accompanied by any doubt about the measure we have to apply to them. If they still find a place in physics, there is also a place for energy as their measure; if they do not, they are of no concern to us." (Fechner, 1860/1966, p. 23)

"For an information-based theory of perception that purports to replace sensation-based theories of perception the distinction between stimulation and information is crucial." (Gibson 1960, p. 348)

In the study of sensations, interest generally focuses on energy couplings between the physical world and the organism (e.g., how many photons of light strike the retina, how large a displacement of the tympanic membrane). However, the relation between too high, too low, too steep, too shallow, and a pilot landing an aircraft can not be understood at the level of photons of light. The nature of the coupling is fundamentally different. It is not a coupling of energy, but one of information. Certainly energy is necessary to carry information, however the behavior of the pilot cannot be understood at the level of energy exchanges. What is information? In this context we are not referring to information in the statistical sense of reduction of uncertainty, but in a more colloquial sense of specificity. Are there properties of light that are specific to (in one-to-one correspondence with) too high, too low, too steep, or too shallow?

The importance of this distinction between stimulus and energy was acknowledged by Garner (1974):

"In this manner I am agreeing with James Gibson (1966) in differentiating the informational properties of a stimulus from its energetic properties. While this distinction between stimulus as energy and the stimulus as information or structure is very general, in some of the research I shall report it assumes a very specific role, since it differentiates alternative ways in which stimulus redundancy can affect discrimination performance. Stimulus energy provides activation of the sense organ, but it is stimulus information or structure that provides meaning and is pertinent to what I call perception." (p.2)

Thus, active psychophysics is about the information coupling between the environment and the actor, rather than about the energy coupling. In fact, the pilot hopes to avoid a high energy coupling (crash) with the environment. Beyond the issue of affordances (i.e., what the environment means to an actor), two questions remain that must be addressed by active psychophysics. First, are there constraints (i.e., invariants) within the communication medium (e.g., optic array or computer-generated display) that correspond to meaningful properties of the environment (i.e., affordances)? Second, is the organism attuned to (able to pick up) these constraints? These two dimensions are illustrated in Table 7.1 adapted from Owen (1990, p. 289-326). One dimension is specificity within the perceptual array. Properties of this medium for perception are either specific to or nonspecific to meaningful properties (i.e., affordances) of the environment. The second dimension represents the perceptual skill of the actor. The actor is either attuned to the constraints within the medium or is unattuned. In this context, information is *functional* when the constraints in the medium are specific to environmental affordances and the actor is sensitive to the constraints. When constraints in the medium are specific, but the actor is not sensitive to them, then the information is *nonfunctional*. If the actor is tuned to nonspecific properties of the perceptual array, then this information is said to be *dysfunctional*. The final cell is for that information that is nonspecific and ignored, *afunctional* information. This matrix can provide an interesting basis for discussing training (Flach, Lintern, & Larish, 1990), interface design (Gaver, 1991), and human error. The similarity between Table 7.1 and the two-dimensional table (states of the world x responses) used in signal detection theory (e.g., hits, misses, false alarms, and correct rejections) reflects again the congruence between the ecological program for studying perception/cognition and the traditional psychophysical program for studying sensation (see Green & Swets, 1966; MacMillan & Creelman, 1991; Wickens, 1984, for discussion of signal detection theory).

Table 7.1. *A Two Dimensional Framework Illustrating the Functionality of Information. Adapted from Owen (1990).*

ACTOR

PERCEPTUAL ARRAY	ATTUNED	UNATTUNED
SPECIFIC	FUNCTIONAL (informative)	NON-FUNCTIONAL (noninformative)
NONSPECIFIC	DYS-FUNCTIONAL (misinformative)	AFUNCTIONAL (unimformative)

7.3.3 Direct Perception

Now to the curious notion of direct perception. If one accepts the notion that affordances are properties that have functional meaning for an actor, and if one accepts the fact that there is structure in a medium that is specific to these properties, then it seems possible that these properties can also be "seen" or "picked up" in the same sense that color can be "seen" or "picked up." That is, there is no need to postulate an elaborative or processing agent to construct meaning. Meaning is a physical attribute available to be picked up in the same way that color is. Graspable or safe glide slope are no more mental constructions than is red. Note that this does not mean that the mapping from the physical to the psychological is a simple one. The perception of "red" is not simply a function of the wavelength of light, but is affected by many contextual factors as well. In this sense, a claim for direct perception does not mean that there are no ambiguities, that there are no errors in perception, that our percepts are not context dependent. Rather, the claim is that percepts have their basis in physical reality in the same way as is usually accepted for sensations. Percepts are not elaborations or constructions from sensations, but percepts have their own direct connection to a physical reality.

For those who accept the concept of direct perception, the strategy of intrinsic measurement and the psychophysics that follow are essential. However, direct perception is still a difficult notion for most people. It is not necessary for one to accept direct perception in order to see the utility of the psychophysical approach we recommend here. This approach is still a viable strategy for cognitive psychology. Even if sensations and elaborative processing are intermediaries to perception, it still might be useful to look for the functional relations between the physical in terms of affordances and the psychological:

$$\text{Perception/Cognition} = f(\text{Affordances}).$$

This would simply be a top-down approach in which the direct functional relations provide a context for understanding the intermediate processing stages. Thus, of the two concepts, affordance and direct perception, only the concept of affordance is essential to the psychophysical program that we are recommending.

7.3.4   Role of Action

For sensory psychophysics, action, like meaning, is viewed primarily as a nuisance variable to be eliminated or controlled. Thus, researchers go to great lengths to eliminate action from the experimental context by constraining movements (e.g., bite boards) or by brief, tachistoscopic presentation of stimuli. Except for the method of adjustment, control over stimulation is completely in the hands of the experimenter. The subjects' responses generally have little or no effect on stimulation. However, for a psychophysics of perception/cognition, action capabilities are the basis for our measurement system. Meaning depends on action capability. Whenever action is constrained, meaning is altered. Thus, in active psychophysics action is no longer a nuisance variable, but a fundamental dimension of the phenomenon of interest. In this scheme perception and action cannot be treated as two independent processing modules. Perception and action are intimately coupled, and meaning is an emergent property of this coupling. Thus, the *psych-* in our psychophysics must include perception and action[3]. Warren and McMillan (1984) introduced the term *active psychophysics* to draw attention to the often ignored role of action in perception.

Gibson (1966) distinguished two kinds of activities: performatory and exploratory. The activities that are the basis for deriving intrinsic metrics for an actor-relevant description of the environment are performatory activities. These are activities that are directly linked to realizing the affordances offered to an organism. Reaching for an apple, steering a car away from an obstacle, piloting a plane toward a runway are all performatory activities. It is our capacity for performatory activity that is the metric for describing our environment in terms of affordances. Altering the capacity for performatory activity alters the functional significance of the environment. Thus, confining a person to a wheelchair has drastic consequences in terms of the meaning of environmental features (e.g., stairs that afford passage for a pedestrian now become a barrier). Similarly, constraining what a subject can do within an experiment can have important consequences for the "meaning" of the stimuli that are presented.

Exploratory activity refers to activities of perceptual systems. For Gibson, perceptual systems were not passive receptors of stimulation, but active seekers of information. Thus, the ability to pick up attributes

---

[3]This is another reason to include perception with cognition.

of the physical world depends on activity of the organism. For example, Gibson (1962) showed that a hand free to explore an object may be more discriminating than a hand that is constrained. Thus, exploratory activity does not affect meaning (in terms of affordances), but has significant implications for the quality of information. Smets (1994) illustrates this quite nicely with the design of a 3-D television system. The dimensionality of the screen image changed from 2- to 3-dimensional when the display image was slaved to the head position. In some sense it might be argued that dimensionality is not a property of space, but an emergent property of an interaction between an observer and a space. For a single, fixed eye all spaces may be two dimensional; whereas for a moving eye, spaces will be two dimensional; if relative positions of objects in the field of view are independent of eye position (e.g., normal television) and three dimensional if relative position of objects varies systematically with eye-position (see Smets, 1994).

The classifications *performatory* and *exploratory* reflect the consequences of actions. Performatory actions have "survival" or "goal" value for the actor. Exploratory actions have "informational" value for the actor. Any particular action (e.g., walking) may have both performatory (e.g., getting food) and exploratory (e.g., revealing previously occluded objects) consequences. Further, control of performatory activity may require informational support provided by exploratory activity. Thus, performatory and exploratory actions are intimately linked, and together they are intimately linked to perception. For this reason, a theory of perception must also be a theory of action and research on perception should be designed to incorporate opportunities for performatory and exploratory actions, similar to those available in natural environments.

7.4   Methodological Implications

"Sensitivity to action or behavior is clearly of a special kind, unlike the more familiar sensitivity to the prods and pushes of the external world or the pangs and pressures of the internal environment. The stimuli are self-produced, and the causal link is from response to stimulus as much as from stimulus to response. The classical stimulus-response formula, therefore, is no longer adequate; for there is a loop from response to stimulus to response again, and the result may be a continuous flow of activity rather than a chain of distinctive reflexes.

Action-produced stimulation is obtained, not imposed —that is, obtained by the individual not imposed on him. It is intrinsic to the flow of activity, not extrinsic to it; dependent on it, not independent of it. Instead of entering the nervous system through receptors it re-enters. The input is not merely afferent, in the terminology of the neurologist, but re-afferent — that is contingent upon efferent output. The favorite modern term for this action-produced input is one borrowed from electronic circuitry, namely, feedback." (Gibson, 1966, p. 31)

The problem of meaning, the significance of action, and the nature of information all have important methodological implications. Traditionally, experimental psychologists manipulate stimulation and measure responses. The intensive, extensive, qualitative, and temporal aspects of the stimulus were determined by the experimenter and were independent of a subject's actions. Control over stimulation has been viewed as critical to inferring causation between stimulus and response. However, to pursue an active psychophysics, it will be necessary for experimenters to release control of stimulation to subjects.

Traditionally, the link between perception and action has been viewed as a one-way street, with perception driving action. In the new active psychophysics, the link must be viewed as a dynamic coupling in which stimulation will be determined as a result of subject actions. It's not simply a two-way street, but a circle. We must give up the illusion of open-loop systems that has been the working assumption behind much of our experimental logic. We must recognize that we are dealing with closed-loop systems; not simply by adding feedback loops in our block diagrams, but by changing the fundamental design of our research programs.

Figure 7.2 provides a framework for thinking about the methodological implications of active psychophysics (Flach, 1990). Traditionally, experimenters have operated on only the forward loop (shaded boxes), manipulating stimulation as the independent variables and measuring responses as the dependent variables. In an active psychophysics, rather than stimulation, experimenters will constrain goals, dynamics, and information as independent variables. Experimenters will also provide disturbances. However, although disturbances have interesting properties of their own, disturbances will primarily have the role of catalysts in this paradigm.

Manipulation of goals will define functionality for a given task. Goals can be provided explicitly or implicitly. For example, the experimenter may ask the subject to maintain a particular track or speed

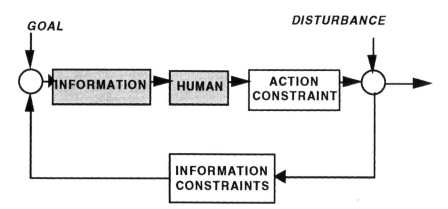

Figure 7.2. *A framework for addressing perception/action in the context of a closed-loop system.*

(explicit). On the other hand, the experimenter may implicitly create goals by introducing consequences for actions within the task environment, for example, introducing obstacles to define a safe path of travel. Any aspect of the task environment that has consequences for the subject are part of the goal structure.

Manipulation of dynamics will define the action coupling between the the actor and the environment, for example, specification of the control law for the subject's "vehicle." The world is a very different place, in terms of safe fields of travel and potential landing sites for a helicopter and a fixed wing aircraft. Flach, Hagen, and Larish (1992) discussed how the dynamics of hovering versus straight and level flight influence the pick up of information for the control of altitude. For laboratory tracking tasks, McRuer's Crossover Model is an excellent illustration of how performance adapts to changing constraints on dynamics (McRuer & Jex, 1967; McRuer & Weir, 1969; see Flach 1990, for a discussion of the Crossover Model in the context of active psychophysics).

Manipulation of information will define the perceptual coupling between the actor and the environment, for example, manipulating the optical texture available. Denton (1980) painted stripes on roadways so that they got progressively closer together as they neared traffic circles. This produced an increasing edge rate (specifying acceleration). The result was a reduction in approach speeds and a reduction in accidents. A number of studies have examined altitude perception and control as a function of optical texture (splay versus optical density) (Flach, et. al.,

1992; Johnson, Tsang, Bennett, & Phatak, 1989; Warren, 1988b; Wolpert, 1988; Wolpert & Owen, 1985, for an extended discussion of information for low altitude flight see Warren & Flach, 1994).

Dependent measures, for active psychophysics, will not be exclusively responses. But to the extent possible we want to measure the input and output for each of the boxes within Figure 7.2. We want to know not simply what the actor does, but we want to know the information consequences as well as the environmental (goal-relevant) consequences of those actions. As Power's (1978) claimed, "behavior is the control of perception." Gibson (1958) illustrated this nicely in the context of the visual control of locomotion. He wrote "an animal who is behaving in these ways is optically stimulated in the corresponding ways, or, equally, *an animal who so acts as to obtain these kinds of stimulation* is behaving in the corresponding way (p. 155, emphasis added). Thus, stimulation becomes a dependent variable. For example, we might ask a pilot to maintain an aircraft at a specified altitude and then measure how he makes the world look. Does he act so as to maintain global optical flow rate constant? Does he maintain a constant splay angle, a constant optical density? Again, see Warren and Flach (1994) for a more extended discussion of how things might look when traveling at a constant altitude.

We are also interested in the converse relations of the consequences of information or environmental contingencies on action. For example, if a pilot is instructed to maintain constant altitude, does he make adjustments when global optical flow rate is manipulated (e.g., by varying speed) or when the ground texture density is varied?

The relations across these measures (information, action, consequences) are critical to understanding human performance. Because of the closed-loop structure, no single measure is adequate, in the same way as knowing the number of hits without knowing the number of false alarms is inadequate to determine a sensory threshold, or knowing accuracy without speed is inadequate for characterizing performance in choice tasks. For example, behaviors may reflect tradeoffs between information generated (exploratory activity) and accomplishment of task goals (performatory activity).

Because of this closed-loop structure, psychology will have to depend more on control theory for the logic of experiments and inferences. This logic has been used quite successfully in the area of vehicular control (e.g., Sheridan & Ferrell, 1974; Wickens, 1984, 1986). However, we recommend it as a more general experimental paradigm for a wide range of perceptual and cognitive problems. It is important to keep in mind, however, that we do not recommend the servomechanism as a theoretical metaphor. Rather, we recommend the tools of control theory for designing experiments and for guiding the logic of measurement and inference. For example, analysis in the frequency domain using the Bode space provides an important tool for relating input and output to infer a transfer function for a particular process. For active psychophysics, control theory will have an importance equivalent to signal detection theory in traditional psychophysics. (See Flach, 1990, for a more extended discussion of the use of control theory within active psychophysics).

7.5   Summary

"Psychophysics, the oldest psychology, has continually adapted itself to the substantive concerns of experimentalists."
—MacMillan & Creelman (1991, p. 25)

The fundamental question addressed in this chapter is can psychophysics be adapted to the substantive concerns of perception and cognition. We believe that it can. However, adapting psychophysics to the substantive issues of perception and cognition will, first, require a new physics; an ecological physics that describes the world in actor-relevant or functional terms, that describes the world in terms of affordances. A second requirement will be a new understanding of information. Information must be understood in terms of mapping from structure in media to meaning in terms of affordances. Finally, a new and deeper appreciation for the intimacy of perception and action will be required. Experimental tasks must be designed to create opportunities for both performatory and exploratory actions. In doing this, experimenters will have to relinquish control of stimulation to their subjects and must learn to manipulate meaning (through specification of consequences and performatory capabilities) and information (through specification of the perceptual array and exploratory capabilities). With such a program of basic research, we are confident that the foundation will be set for direct generalizations to applied problems of human-machine systems.

In conclusion, we emphasize that we are not presenting active psychophysics as an alternative in opposition to traditional psychophysics, but as a legitimate and valuable complement. A complement that we feel is consistent in spirit with Fechner's intention to relate mind to matter. We try to relate mind to *what matters!* The complimentary relation between these alternative approaches to perception is illustrated quite effectively by Schiff and Arnone (1995) in the context of driving.

Acknowledgments

John Flach was supported by a research grant from the Air Force Office of Scientific Research, Directorate of Life and Environmental Sciences (AFOSR-91-0151), during preparation of this chapter. Support was also provided by the Armstrong Laboratory, Human Systems Division, Wright-Patterson AFB, OH, and the Psychology Department of Wright State University. Opinions expressed in this chapter are our own and do not represent an official United States Air Force position. Many people offered useful feedback on earlier versions of this chapter. We would like to particularly acknowledge Herb Colle, Peter Hancock, and Kim Vicente.

7.6   References

Denton, G. G. (1980). The influence of visual pattern on perceived speed. *Perception, 9,* 393–402.

Ebbinghaus, H. (1885). *Memory.* New York: Teacher's College, Columbia University. (Reprint ed. in 1964 by New York: Dover)

Fechner, G. (1966). *Elements of psychophysics.* (H. E. Adler, Trans., D. H. Howes & E. G. Boring, Eds.), New York: Holt, Rinehart, & Winston. (Originally published in 1868 as *Elemente der psychophysik*).

Flach, J. M. (1990). Control with an eye for perception: Precursors to an active psychophysics. *Ecological Psychology, 2,* 83–111.

Flach, J. M., Hagen, B. A., & Larish, J. F. (1992). Visual information for the active regulation of altitude. *Perception & Psychophysics, 51,* 557–568.

Flach, J. M., Lintern, G., & Larish, J. F. (1990). Perceptual motor skill: A theoretical framework. In R. Warren and A. H. Wertheim (Eds.), *Perception and control of self-motion.* (pp. 327-355). Hillsdale, NJ: Lawrence Erlbaum Associates.

Flach, J. M, & Warren, R.. (1995). Optic flow in low altitude flight. In

P. A. Hancock, J. M. Flach., J. K. Caird, and K. J. Vicente (Eds.), *Local applications of the ecological approach to human-machine systems*. Hillsdale, NJ: Lawrence Erlbaum Associates.

Gaver, W. W. (1991). Technology affordances. In S. P. Robertson, G. M. Olson, and J. S. Olson (Eds.), *Reaching through technology: CHI '91 Conference Proceedings*. (pp. 79–84). New York: ACM.

Garner, W. R. (1974). *The processing of information and structure*. New York: J. Wiley.

Gibson, J. J. (1958). Visually controlled locomotion and visual orientation in animals. *British Journal of Psychology, 49*, 182–194. Also in E. Reed and R. Jones (Eds.), (1982). *Reasons for realism* (pp. 148–163). Hillsdale, NJ: Lawrence Erlbaum Associates.

Gibson, J. J. (1960). The concept of the stimulus in psychology. *American Psychologist, 15*, 694–703. (Also in E. Reed and R. Jones (Eds.), (1982). *Reasons for realism*. (pp. 333–349). Hillsdale, NJ: Lawrence Erlbaum Associates.

Gibson, J. J. (1962). Observations on active touch. *Psychological Review, 69*, 477–491.

Gibson, J. J. (1966). *The senses considered as perceptual systems*. Boston: Houghton-Mifflin.

Gibson, J. J. (1986). *The ecological approach to visual perception*. Hillsdale, NJ: Lawrence Erlbaum Associates. (Original work published 1979).

Gibson, J. J., & Crooks, L. E. (1938). A theoretical field analysis of automobile driving. *American Journal of Psychology, 51*, 453–471. (Also in E. Reed and R. Jones (Eds.), (1982). *Reasons for realism*. (pp. 119–136). Hillsdale, NJ: Lawrence Erlbaum Associates.

Green, D. M., & Swets, J. A. (1966). *Signal detection theory and psychophysics*. New York: Wiley.

Johnson, W. W., Tsang, P. S., Bennett, C. T., & Phatak, A. V. (1989). The visually guided control of simulated altitude. *Aviation, Space, and Environmental Medicine, 60*, 152–156.

Klein, G. A., Calderwood, R., & MacGregor, D. (1989). Critical decision making for eliciting knowledge. *IEEE Transactions of Systems, Man, and Cybernetics, 19*, 462–472.

Lewin, K. (1936). *Principles of topological psychology*. New York: McGraw-Hill.

Lombardo, T. J. (1987). *The reciprocity of perceiver and environment: The evolution of James J. Gibson's Ecological Psychology*. Hillsdale, NJ: Lawrence Erlbaum Associates.

MacMillan, N. A., & Creelman, C. D. (1991). *Detection theory: A user's guide*. Cambridge: Cambridge University Press.

McRuer, D. T., & Jex, H. R. (1967). A review of quasi-linear pilot models. *IEEE Transactions on Human Factors in Electronics, HFE-8*, 231-249.

McRuer, D. T., & Weir, D. H. (1969). Theory of manual vehicular control. *Ergonomics, 12*, 599–633.

Owen, D. H. (1990). Perception and control of changes in self-motion: A functional approach to the study of information and skill. In R. Warren and A. H. Wertheim (Eds.), *Perception & control of self-motion*. (pp. 289-326). Hillsdale, NJ: Lawrence Erlbaum Associates.

Powers, W. T. (1978). Quantitative analysis of purposive systems. Some spadework at the foundations of scientific psychology. *Psychological Review, 85*, 417–135.

Sheridan, T. B., & Ferrell, W. R. (1974). *Man-machine systems*. Cambridge, MA: MIT Press.

Schiff, W., & Arnone, W. (1995). Perceiving and driving: Where parallel roads meet. In P. A. Hancock, J. M. Flach., J. K. Caird, and K. J. Vicente (Eds.), *Local applications of the ecological approach to human-machine systems*. Hillsdale, NJ: Lawrence Erlbaum Associates.

Smets, G. (1995). The theory of direct perception and telepresence. In P. A. Hancock, J. M. Flach., J. K. Caird, and K. J. Vicente (Eds.), *Local applications of the ecological approach to human-machine systems*. Hillsdale, NJ: Lawrence Erlbaum Associates.

Stevens, S. S. (1951). *Handbook of experimental psychology*. New York: Wiley.

Warren, R. (1988a). Active psychophysics: Theory and practice. In H. K. Ross (Ed.), *Fechner Day '88 (Proceedings of the 4th Annual Meeting of the International Society for Psychophysics)*. (pp. 47–52). Stirling, Scotland.

Warren, R. (1988b). Visual perception in high-speed low altitude flight. *Aviation, Space, and Environmental Medicine, 59* (11, Suppl.), A116-A124.

Warren, R., & McMillan, G. R. (1984). Altitude control using action-demanding interactive displays: Toward an active psychophysics. *Proceedings of the 1984 IMAGE III Conference*. (pp. 405–415). Phoenix, AZ: Air Force Human Resources Laboratory.

Warren, W. H. (1984). Perceiving affordances: Visual guidance of stair climbing. *Journal of Experimental Psychology: Human Perception &*

Performance, 10, 683–703.

Warren, W.H. (1985, June). *Environmental design as the design of affordances.* Paper presented at the Third International Conference on Event Perception and Action, Uppsala, Sweden.

Wickens, C. D. (1984). *Engineering psychology and human performance.* Columbus, OH: Merrill.

Wickens, C. D. (1986). The effects of control dynamics on performance. In K. R. Boff, L. Kaufman, and J. P. Thomas (Eds.), *Handbook of perception and human performance.* (Vol. II, pp. 39.1–39.60). New York: Wiley.

Wolpert, L. (1988). The active control of altitude over differing texture. *Proceedings of the Human Factors Society, 32,* 15–19.

Wolpert, L., & Owen, D. (1985). Sources of optical information and their metrics for detecting loss in altitude. *Proceedings of the Symposium on Aviation Psychology, 3,* 475-481.

Chapter 8

Constructing an Econiche

William H. Warren, Jr.
Brown University

8.0   Introduction

Ecological psychology is concerned with the relations that have evolved between organisms and their natural environments that support successful perceiving and acting. Yet, our species is currently unique in that it lives in an environment that is largely of its own construction, built in order to fine-tune or extend the job done by evolution. Most of the objects we grasp, surfaces we walk on, and shelters we inhabit are, for better or worse, artifacts. We are thus in the rather novel position of constructing our own econiche, or as the architect Lerup put it, "We design things and things design us." Yet, despite several decades of progress in ergonomics, there are still few general working principles for designing environments that fit the activities of human beings and anticipating the reciprocal effects on human activity.

Panero and Zelnick (1979) noted this when they described the assorted volumes of reference standards that are used by architects and designers: "Much of the available material is based almost exclusively on outdated trade practices or on the personal judgments of those preparing the standards. With few exceptions, most reference standards are simply not predicated on enough hard anthropometric data" (p. 12). Ecological psychology has something to contribute to such design problems, because ecological research is relevant to human action in the built environment as well as in the natural one.

8.1   Affordance Design

Several disciplines have emerged since the World War II that emphasize designing the built environment to a human scale. Ergonomics and human factors engineering have focused on the individual operator in the workplace, with the expressed purpose of "fitting the task to the man" (Grandjean, 1980; Shackel, 1976; Woodson, 1981). However, these fields are burdened with the often conflicting interests of increased productivity on the one hand and humanizing the workplace, including improved health and safety, on the other (Singleton, 1982). In the United States, much of this work was stimulated by the needs of the military and has tended to concentrate on rather specialized problems of the human-machine interface, such as cockpit design, process control, and instrumentation.

More recently, diverse research in environmental psychology and human ecology has examined larger scale interactions between environment and behavior (Canter & Lee, 1974; Ittelson, Proshansky, Rivlin, & Winkel, 1974; Stokols, 1977). This includes work on cognitive maps, personal space and crowding, the influence of behavior setting on social interaction, and preferences for environmental and architectural features. Also in the 1970s the field of environmental design emerged, emphasizing the human-scaled design of exterior and interior architectural spaces and, often, a participatory design process (Allsopp, 1974; Mikellides, 1980).

One of the contributions ecological psychology can make to this maze of problems is the concept of *affordances*. Gibson (1979) described an affordance as that which an environmental structure offers to a particular organism for activity, based on the relations between properties of the environment and that organism. Thus, a hard, flat, narrow surface may afford walking for me but not for a rhinoceros, and a horizontal surface at the height of my knees may afford sitting for me, but not for a small child. Now, it is the business of designers and architects to create places and objects such as these that afford walking, sitting, working, playing, and specific kinds of social interaction. In short, environmental design can be construed as the design of affordances.

As a nice example of this idea, Wise and Fey (1981) give their architecture students exercises in what they call Centaurian Design — designing things for use by four-legged, two-armed Centaurs. By radically altering the action capabilities of the organism, students are forced to confront the functional problems of designing new affordances to fit them. The entertaining results, which you may be able to imagine, include Centaurian umbrellas, elevators, and revolving doors.

The notion of an affordance offers both a unifying concept and a new perspective on problems of environmental design. It has the potential to unify previous work in the disparate domains of environmental psychology and ergonomics, because, in principle, an affordance analysis can be applied at a number of scales to the activities of an individual, group, or community. But the concept also provides a new perspective by emphasizing the material and informational bases for these activities in the fit between organism(s) and environment. Because most research on affordances has focused on the scale of individual actions, that is where I will concentrate. Some of these ideas have been recently articulated by Norman (1988) in his instructive and entertaining book, *The Psychology of Everyday Things*, even to the point of invoking the role of affordances in design. However, his interpretation of Gibson's concept is less materially based and more subjectively determined that the one I provide here.

I begin by offering four tentative criteria for the successful design of affordances:

1. The design must fit the action capabilities of the user, or what Turvey and Shaw (1979) have called, the user's *effectivities*. We all have our favorite examples of ill-fitting designs, such as the ungraspable doorknob (Figure 8.1) or the low doorway. The original F-111 military aircraft was plagued by a lever that was supposed to be pulled backward in order to close the wings for high-speed flight — consonant with the direction the wings moved, but quite contrary to the throttle, which had to be pushed forward in the opposite direction. Several crashes resulted before the lever was reversed. Occasionally things are intentionally designed so as to violate the user's effectivities, as in the old trick of the castle staircase with an irregular riser, to trip up the marauding hoards.

Successful affordance design requires a task-specific analysis of the organism-environment system that considers the relevant system variables and the biomechanics of the task. This is where ecological psychologists have made the most progress, and I'll return to it later.

2. An affordance must be perceptually specified to the user, which implies that the designer understand the informational basis for action. This concerns not only lighting and noise levels, the focus of much ergonomics research, but also the arrangement of surfaces and their optical properties. Well-designed affordances should, in Koffka's (1935)

Figure 8.1. *Ungraspable doorknob. Handle is too deep to grasp with hand and thumb, glass door is too heavy to move.*

words, "name themselves" and actually lead the actions of the user.

Violations of this principle are not uncommon. Gibson's favorite example of an unspecified affordance was the modern plate glass window, an invisible obstacle that fells birds and humans alike. This is often remedied by unsightly markers on the glass at eye level, thereby compromising the effect of open space sought by the architect. Another problem is surfaces with texture that is too fine-grained or lacking altogether, yielding edges that are poorly defined both statically and by dynamic occlusion of texture. Examples include the steps in Saarinen's Pan Am terminal at Kennedy Airport (see Figure 8.2; the black edging was a late addition and may only have made matters worse — on which side of the black strip is the edge?) and those in the carpeted lobbies of several conference hotels of my acquaintance. Due to the failure of underspecified affordances to "name themselves," the architect Acking (1980) noted, "As the architecture gets worse, the number of signs increases."

This is not to suggest that all affordances are immediately perceptible without what James Gibson (1966) called the education of

Figure 8.2. *Steps in the Pan Am terminal at Kennedy Airport. Fine, regular texture of floor tile makes edges difficult to see. The black vinyl edging was added later, but which side of it specifies the edges of the steps is still unclear.*

attention. Eleanor Gibson and her colleagues (Gibson et al., 1987) recently demonstrated that whereas young toddlers differentiate surfaces that afford walking and those that do not, crawlers of the same age do not. Thus, learning a new effectivity, walking, is accompanied by perceptual learning of the corresponding affordances. Such is the case with the discovery (or design) of new affordances by adults, as in the case of the "kneeling chair" with knee-pads and a seat but no back — learning the action of sitting in such a chair is accompanied by perceptual learning about its visual specification. Visual artists have played with violating the affordance properties of ordinary objects, producing such affordance puns as Surrealist Meet Oppenheim's fur-covered cup and saucer, Lucas Samaras' "Chair transformations" adorned with spikes and pins, Claes Oldenburg's "Soft toilet," and the left-handed coffee mug (with a hole below the lip if grasped by an unsuspecting right-hander).

3. Affordances must be designed to complement social patterns of use. This has been a major theme of research in environmental

psychology. But it is important to note that such social patterns are not immutable — not only do we design things to fit us, but, reciprocally, things design us back (Reed, 1985). Discovery of the affordances of sharp-edged rocks, fire, and other implements had a profound impact on hominid evolution. Today the introduction of new technologies such as microcomputer workstations or virtual reality systems, however ergonomically sound their design, has far-reaching implications for the structure of work and social interaction.

4. The designer should strive to create objects that are not only functional, but also aesthetically satisfying. There has long been a tension in architecture between the fascination with formal style and the concern with functional space. In the case of modernist architecture, critics have argued that formal considerations tended to dominate, paradoxically violating the Bauhaus dictum, "Form follows function" (Allsopp, 1974; Brolin, 1976; Newman, 1980). Other designers suggest that there may exist a natural aesthetics of design that has a basis in function, much as the forms of the paradigm of beauty, Nature, have a basis in the structural and functional demands of physics and biology (Ghyka, 1946/1977; Hale, 1993; Humphrey, 1980). Meanwhile, the art of designing affordances is to work some ineffable unity between function and form.

8.2   Intrinsic Metrics

The method developed by ecological psychologists for the analysis of affordances is applicable to these criteria for design, particularly the first two listed earlier. Underlying the method is the principle of *intrinsic metrics*, which takes the action system as a natural standard for measuring the properties of the environment, in contrast to the imposition of an *extrinsic* metrics such as feet or meters. Such measurement is of necessity task-specific, because the relevant dimensions of the environment and the action system vary from task to task. When applied to environmental design, this notion leads directly to the principle of body-scaled or, more generally, action-scaled design. Although, as implied by the term body-scaled, I emphasize geometric dimensions of the actor and environment, the approach is readily generalized to other dynamic dimensions as implied by *action-scaled*. The phenomenological power of body scaling is brought home by a story my father tells about walking into a rest room recently to find that the sinks were at chest height: He immediately felt like a little kid struggling to reach the faucets.

The notion of body-scaled design is, of course, not new. In fact, the first units of measurement were anatomical, such as the inch, the hand, the cubit, and the pace (Berriman, 1953). A number of systems have developed based on the proportions of the human figure, beginning with the Greeks' use of the Golden Section. This is the ratio of the distance between the head and the navel to that between the navel and the ground in a standing man, approximately 1:1.618, which is repeated in the ratio of navel height to total height. However, this mystical value was applied not to the body-scaled design of interior space, but rather to the overall proportions of the Greek temple. Similar geometrical observations were made by Vitruvius (1960) in the 1st Century B.C., who pointed out that a man in two canonical postures defines a square and an inscribed circle with its center at the navel, later captured in Leonardo's famous sketch of Vitruvian Man.

Le Corbusier (1954/1966), the supreme modernist, based his system of the Modulor on the Golden Section of a 6 ft. man with hand upraised. He extended two Fibonacci series down from these body ratios and decided that he had found "a harmonious measure to the human scale, universally applicable to architecture and mechanics," upon which he based many subsequent projects. Although some appropriate body-scaled standards may have emerged from this scheme, such as ceilings at the height of the upraised hand, the Modulor was more a case of Le Corbusier imposing his own formal ideal on human space.

An empirical approach to body scaling developed with the study of *anthropometry*, which originated in anthropology and was soon applied to ergonomics (Damon, Stroudt, & McFarland, 1971; Roebuck, Kroemer, & Thomson, 1975). Bodies of anthropometric data are typically available to designers in the form of Dreyfuss figures (Figure 8.3), which present the 5th, 50th, and 95th percentile values for the lengths of various body segments in certain populations (Diffrient, Tilley, & Bardogjy, 1974). Some of these anthropometric measurements have been translated into specific design recommendations for seating, work space, and so on, often without direct analysis of the tasks themselves.

There are three problems that I see in applying such anthropometric data to environmental design. First and foremost, anthropometric values are anatomical, not functional. Body dimensions are typically measured in standardized, static positions, whereas many environmental dimensions must be specific to the task and depend on the action performed. For example, the passage width of horizontal circulation spaces must take into account not only shoulder width, but also body sway, a comfortable safety margin, and cultural norms of "personal space." Some research on functional or dynamic anthropometry attempt to address these questions, as in the classic studies of the reach envelope of the seated operator (Dempster, Gabel, & Felts, 1959; Hertzberg, 1960). But even here, the envelope depends on what action is performed with the hand — pushing a button, flipping a switch, or grasping a knob. As the ergonomist Kroemer (1982) argued, "More research is needed to establish better founded procedures to translate static body position data into functional design recommendations."

The second problem is an extension of the first. Although design standards can be based on anatomical measurements, they typically provide no more than a range of recommended values. The notion of a *comfort mode* or *optimal* values scaled to the user is undefined, and no evaluation procedures for optimal action have been established.

The third problem is a practical one. Even when anthropometric data are available, they are seldom used by designers. In Ramsey and Sleeper's (1970) *Architectural Graphic Standards*, the bible of the profession, the Dreyfuss figures appear on the first few pages — separate from the actual guidelines in the rest of the volume. Architects I have talked to admit they seldom refer to the standards anyway, typically relying on their own rules of thumb and constrained by the components that are available from manufacturers. In some cases these rules of thumb may reflect ecological folk wisdom derived from common constraints provided by the human body and the task demands (Drillis, 1963), but in other cases they can be misleading historical anomalies (Templer, 1992; Warren, 1984).

An affordance analysis could contribute to these problems by providing a dynamic, task-specific approach to human action. Using intrinsic measurements of the fit between the organism and the built environment, such an analysis attempts to identify critical points or limits at which a particular action breaks down and optimal points at which the action is most efficient, comfortable, or safe, depending on the task goals. Subsequently, the informational parameters specifying the affordance can be manipulated to determine what information is necessary for the perceptual guidance of action. Such an approach has been successfully applied by Hallford (1984) to the activity of grasping objects of varying size, by Carello, Grosofsky, Reichel, Solomon, and Turvey, (1989) to reaching objects at varying distances, and by Mark (1987; Mark & Vogele, 1987) to the problem of sitting height. As an

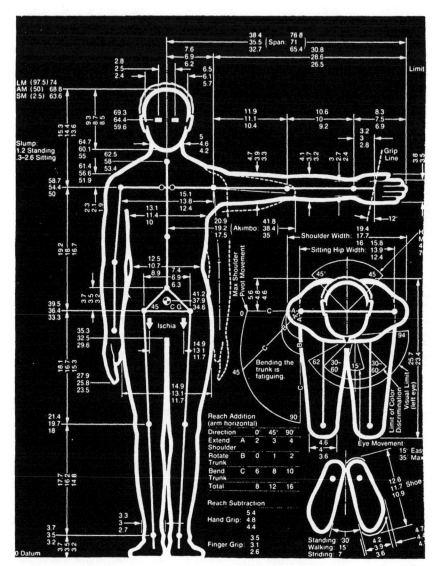

Figure 8.3. *A Dreyfuss figure with anthropometric means for the 5th, 50th, and 95th population percentiles.*

illustration, I outline two case studies from my own work on stair climbing (Warren, 1983, 1984) and walking through apertures (Warren & Whang, 1987).

8.3   An Affordance Analysis of Stair Climbing

Many of us have experienced the frustration of climbing "monument steps," standard equipment on many institutional buildings to provide an aggrandizing visual pedestal. The 5" to 6" risers on monument steps are too low to take them comfortably one at a time, but the 16" to 20" treads are too deep to take them two at a time. This seems to be a prime example of functional design sacrificed on the altar of aesthetics, but many architects also believe that a gentler slope is easier and safer to climb.

Why do monument steps feel so uncomfortable? How can we give the designer some guidance in creating environments that are better scaled to actors? For an ecological psychologist, these questions boil down to the problem of measuring an affordance, using methods of intrinsic measurement.

The relevant variables of this particular animal-environment system are complex, but by utilizing the techniques of dimensional analysis and similarity theory (Rosen, 1978; Schuring, 1977; Stahl, 1961, 1963) such a system of many variables can be reduced to a few dimensionless ratios, called π numbers. As a simple example, if we measure riser height (R) with respect to the climber's leg length (L), we obtain a dimensionless, or unitless, ratio:

$$\pi = R/L \qquad (8.1)$$

Thus, we are using a dimension of the organism as a natural standard for intrinsically measuring a reciprocal dimension of the environment. A specific value of this π number expresses a particular fit between the climber and the stairway. Other π numbers can also be derived using this method, but my experiments have focused on variations in riser height while holding the diagonal distance between stairs, corresponding to the climber's step length, constant.

Intuitively, as riser height is varied with respect to leg length, we would expect to find an optimal point (πo) at which the energy expenditure required to climb through a given vertical distance is at a minimum. Second, as riser height increases with respect to leg length, we will reach a critical point (π_{max}) at which the stair becomes impossible to climb bipedally, and the actor must shift to a quadrupedal hands-and-knees gait — a phase transition in behavior. This critical point can be estimated by a simple biomechanical model ($L+L1-L2 = R_{max}$), which yields a value of $\pi_{max} = .88$ based on empirical measures of leg segments (see Warren, 1984, for derivation of this model). It follows from similarity theory that these optimal and critical points are constant over scale changes in the system, that is, critical riser height should be a constant proportion of leg length, regardless of the absolute size of the climber.

Beyond merely describing an affordance in this way, we must also determine whether it is perceptually specified. If the affordance is perceived, then its critical point should predict the perceptual category boundary between "climbable" and "unclimbable" stairs, and its optimal point should predict visual preferences for stairways. Ultimately, we wish to understand the optical information that specifies action-scaled affordances and how it is detected and used by the perception-action system.

Thus, an affordance such as the "climbability" of a stairway can be characterized by its critical and optimal points. It is important to note that these qualitative properties emerge from the dynamics of the ecosystem, that is, they are condensed out of variation in the relationship between the organism and its environment. Hence, functional perceptual categories and preferences can be shown to have a natural basis in ecosystem dynamics.

To test whether the perceptual category boundary could be predicted from the biomechanical model of critical riser height, I presented slides of stairs with risers varying from 20" to 40" to two groups of subjects; a short group and a tall group. The subjects were asked to categorize each as "climbable" or "unclimbable" and to give a confidence rating of their judgment. Categorization judgments are plotted as a function of absolute riser height R in Figure 8.4a, which shows a difference in critical riser height between the two groups; the confidence ratings also dropped to a minimum at these category boundaries. When the same data are replotted as a function of the π number R/L, on what I will call *intrinsic axes* (Figure 8.4b), the curves are nearly congruent. This indicates that the critical point is a constant regardless of the size of the climber, as predicted from similarity theory. Further, the category boundary for both groups falls at $\pi_{max} = .88$, precisely as predicted by the biomechanical model. This result has been replicated and extended by Mark (1987) and Mark and Vogele (1987). Thus, it appears that the critical point of this affordance is accurately perceived.

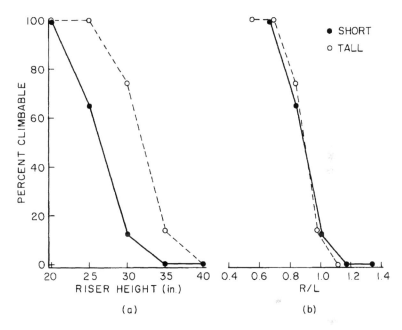

Figure 8.4. *Judgments of critical riser height: (a) Mean percentage of "climbable" judgments for each riser height, and (b) Intrinsic plot of "climbable" judgments as a function of the riser height to leg length ratio (R/L).*

The same should be true of the optimal point. Optimal riser height, defined in terms of energetic efficiency, was determined empirically by measuring oxygen consumption during climbing by short and tall climbers. Riser height varied from 5″ to 10″ with the diagonal distance between stairs held constant at 14″ and step frequency constant at 50 steps/min. The results appear in Figure 8.5a, which shows a minimum energy expenditure per vertical meter of travel at riser heights of 7.7″ for the short subjects and 9.5″ for the tall subjects. These values are somewhat higher than common indoor risers of 6″ to 7″ and help to explain the trouble many people have with even lower monument steps. Far from making long flights of stairs easier to climb, low risers increase total energy expenditure by as much as 15%, and deep treads may make an efficient step length impossible as well. When the same data are replotted on intrinsic axes in Figure 8.5b, the curves become parallel with an optimal $\pi_0 = .26$ for both groups. Thus, the optimal point also appears to be a constant over scale changes.

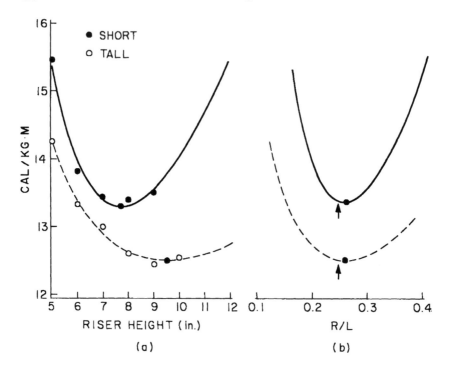

Figure. 8.5. *Optimal riser height. Energy expenditure as a function of (a) riser height, and (b) the R/L ratio.*

To see whether the optimal point predicts perceptual preferences, slides of pairs of stairways were presented to short and tall observers, who were asked to judge which stairway looked more comfortable to climb to the top. As shown in Figure 8.6, there is a group difference in preferred riser height, but the curves again collapse when plotted on intrinsic axes. The preferred riser falls at $\pi = .25$ for both groups, very close to the optimal point of .26.

Thus, it appears that critical and optimal points provide a useful characterization of an affordance, and also predict perceptual performance. When given functional tasks, subjects seem to perceive affordances, that is, they perceive the environment in functional body-scaled terms. This illustrates how affordance relationships can be objectively measured, and such information is essential to guide the successful design of affordances.

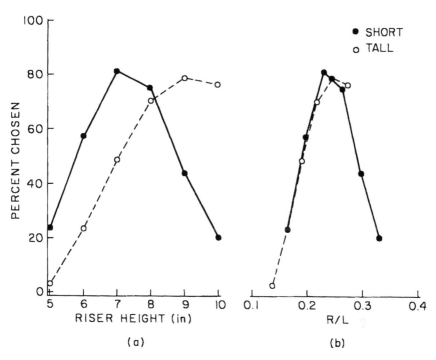

Figure 8.6. *Preferred riser height. Mean percentage chosen as a function of (a) riser height and (b) the R/L ratio.*

8.4   Walking Through Apertures

This type of analysis can be generalized to other affordances, and we subsequently applied it to the problem of walking through apertures or passageways of varying widths. What makes an aperture passable? The relevant intrinsic relation here is that between aperture width *(A)* and the widest body dimension, which in the case of typical men and women is shoulder width *(S)*:

$$\pi = A/S \qquad\qquad (8.2)$$

As this dimensionless ratio approaches 1, the actor must introduce shoulder rotation to pass through the aperture — another phase transition in behavior. However, the design of doorways, corridors, hatchways, and other horizontal circulation spaces cannot simply be based on this anatomical dimension, for the action of walking involves the additional factors of body sway and a comfortable safety margin. Thus, a task-specific affordance analysis is required to determine the critical point of minimum aperture width.

We evaluated critical aperture width empirically by videotaping small and large subjects walking through openings of different widths and measuring the amount of shoulder rotation. Figure 8.7a plots the angle of shoulder rotation as function of aperture width for each group. Critical aperture width was taken to be the point at which shoulder rotation increased above baseline levels of body sway, yielding values of 53 cm for the small group and 62 cm for the large group. When the data are replotted on intrinsic axes in Figure 8.7b, the curves become nearly congruent, with $\pi_{min} = 1.3$ for both groups. Thus, the critical point is again a constant over scale changes.

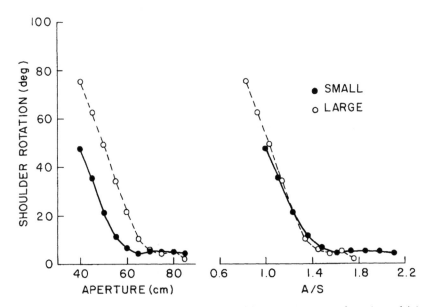

Figure 8.7. *Critical aperture width. Shoulder rotation as a function of (a) aperture width, and (b) the ratio of aperture width to shoulder width (A/S).*

When subjects viewed the apertures through a reduction screen at a distance of 5 m, their perceptual judgments of openings that are "passable" or "impassable" without turning their shoulders were similar. (The data appear in Figure 8.8.) The category boundary falls at $\pi = 1.15$, which indicates that subjects are slightly more conservative when actually walking through the aperture than when viewing it from 5 m away. Identical results were obtained when observers walked freely toward the aperture and stopped at a distance of 5 m, making more information about the layout of the room available from optical flow. Similarly, Carello et al. (1989) reported accurate judgments of the reachability of objects under static binocular viewing conditions. This raises the question of the visual information used to perceive affordances.

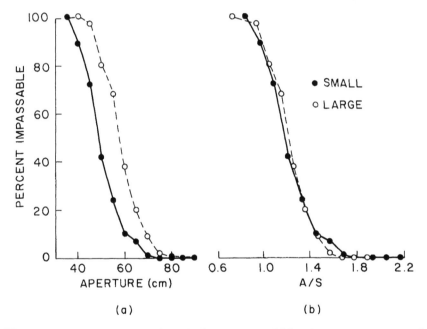

Figure 8.8. *Judgments of critical aperture width. Mean percentage of "impassable" judgments as a function of (a) aperture width, and (b) the A/S ratio.*

8.5   Intrinsic Information for Affordances

The last result suggests that there must be salient static information
available to specify the body-scaled size of objects in the environment.
A distinction can be drawn between metric and nonmetric information
for size, distance, and shape: Nonmetric information would specify only
the relative sizes of objects (e.g., on an ordinal or interval scale), whereas
metric information requires the introduction of a scalar or standard to
fix quantitative scale values (e.g., a ratio scale). Although there is
growing evidence that the perception of shape from motion, shading,
and texture is nonmetric (Koenderink & Van Doorn, 1991; Todd &
Bressan, 1990; Todd & Reichel, 1989), the successful control of actions
such as grasping, sitting, climbing, and walking through openings
seems to require that some metric properties of affordances be
perceived. Although it is often casually supposed that perceiving size
and distance implies an absolute extrinsic metric (for example, distance
perception tasks that ask for estimates in feet), it seems more likely that
the perceptual system makes use of action-relevant or intrinsic metrics,
for example, size and distance perceived in body-scaled units such as
eye height or step length. In a sense, this turns a metric-size task into a
relative size task in which one perceives the size of an object relative to
one's own body.

A compelling case can be made that the perceived sizes of objects in
a scene are scaled to the eye height of the observer. This was first
discovered during the Renaissance by the practitioners of linear
perspective, but was formalized only recently. A simple demonstration
is to look in a doll house window — the tiny furniture and accessories
appear life size because they are scaled appropriately to the observer's
eye height in the doll's room, that is, as ratios of the distance between
the eye and the doll house floor. This effect was exploited in 17th
Century Dutch peepboxes, parlor diversions that had an interior scene
illusionistically painted on the inside and a peephole at the station point,
creating the appearance of viewing a life-size room. The Dutch genre
painter Pieter de Hoogh was obviously experimenting with eye height
scaling when he painted an interior from two different station points
(cf., e.g., "The linen cupboard" (1663) from the eye height of the adults,
with "A mother beside a cradle" (c. 1659) from the eye height of the
child). I'm told that architects often similarly scale their perspective
drawings to the eyeheight of the client, so that structures appear
properly scaled for the viewer.

There is a plethora of static and kinematic eye height information.

The first formulation was in terms of an invariant horizon ratio (Figure 8.9a, b; Gibson, 1979; Sedgwick, 1980). In an open environment, the horizon line intersects all objects at eye level, thereby specifying the observer's eye height on the object. The height of the object (y) is specified relative to eye height (e) by the ratio of the visual angle of the whole object (β) to the visual angle of that portion that is below eye level (γ):

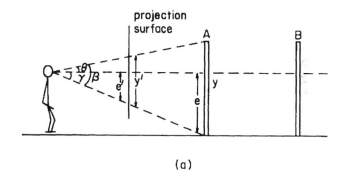

(a)

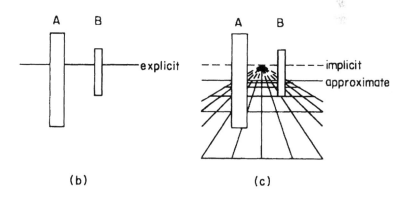

(b)   (c)

Figure 8.9. *Eye height information for object size: (a) Geometry of the eye height ratio for object A; (b) Explicit horizon: Eye level on an object is given by the height at which the visible horizon intersects the object; (c) Implicit horizon: Location of the horizon is given by the limit of optical texture convergence (vanishing point) and the limit of optical texture compression.*

$$y/e \approx \beta/\gamma \qquad\qquad (8.3)$$

This simple relation holds if the height of the object is small relative to its distance from the observer; the more general equation is:

$$y/e = (\tan\gamma + \tan\theta)/\tan\gamma \qquad\qquad (8.4)$$

where θ is the visual angle of that portion of the object above eye level (which can be negative for small objects). This provides a scale not only for the height of the object, but for any object dimension in the frontal plane. Contrary to the generally accepted size-distance invariance hypothesis, this sort of information raises the possibility that perceived size does not depend on perceived distance, a speculation that has recently received empirical support (Haber & Levin, 1992). However, there are two obvious problems with the idea: The horizon is often not visible, and an observer's eye height may change from task to task.

Although the explicit horizon cannot be seen indoors, in a cityscape, or in a forest, there is still abundant information for eye level. As Gibson (1979) pointed out, the "implicit horizon" is specified by texture gradients on visible surfaces: Convergence and compression of ground texture can be extrapolated to a "vanishing limit" at the horizon (Figure 8.9c), and the convergence of wall texture specifies the horizon at its vanishing point. When the observer walks, the focus of the optical outflow pattern is at eye level on surfaces ahead. And, finally, eye level is perpendicular to the gravity vector, for which there is proprioceptive information from the feet, ankles, and vestibular system. In recent experiments, when visual information is placed in conflict with other proprioceptive information in a "pitchroom" apparatus which tips the visual scene fore and aft, even minimal optical structure can bias the perception of eye level by as much as 60% (Matin & Fox, 1989; Stoper & Cohen, 1989). Nemire and Ellis (1991) found that converging lines on the walls and floor of a pitchroom display (parallel to the depth axis) were sufficient to bias the perception of eye level, whereas lateral lines (orthogonal to the depth axis) had a much weaker influence. This is consistent with convergence information specifying an implicit horizon at eye level. These results demonstrate that there is usable visual information for eye height in enclosed environments.

Related evidence shows that such eye height information is used not only to perceive eye level, but also to scale the sizes of objects in the scene. Stoper (1990) reported systematic effects of the pitchroom on judgments of object size, such that when the specified eye level was higher, objects resting on the floor appeared smaller — exactly what would be predicted by the eye height ratio. In our line of research on affordances, we showed that judgments of passable apertures were systematically affected by raising a false floor behind a peephole by 21.5 cm (Warren & Whang, 1987). This artificially reduced the specified eye height, making the aperture appear larger than it actually was and leading to a decrease of 4.5 cm in perceived critical aperture width.

The eye height hypothesis even offers a possible explanation of a famous size illusion often used to flog ecological accounts of perception. The Ames room is a trapezoidal room with a slanted floor that appears rectangular when viewed through a peephole at the appropriate station point. A friend standing in the far corner looks like a midget and expands into a giant upon walking to the near corner, an illusion that persists somewhat even if the observer is allowed to move (Gehringer & Engel, 1986). The standard explanation is based on perceptual expectancies: The observer selects a rectangular interpretation of the room from the family of all projectively equivalent rooms because of past experience in a carpentered environment. This distorts perceived distance and hence perceived size, following size-distance invariance. But why do we not rely instead on past experience with our friend, or on the radical violation of object constancy that occurs when she walks from one corner to the other?

In fact, the textures in the room are distorted so as to specify a rectangular layout of surfaces with a level floor (Runeson, 1988). The size "illusion" could thus be determined by the eye height ratio, independent of perceived distance: The far corner has a low floor, so that the friend's head is below eye level and thus she appears smaller than the observer; conversely, the near corner has a raised floor, so that the friend's head is above eye level and she appears larger. We see what is specified by body-scaled size information, contrary to any expectancies about our friend, and such size-scaling effects would apply to any object in the room. This is precisely what is reported by Stoper (1990) when he altered eye level in a different manner, by pitching the room.

This points to the fact that eye height information is only valid under certain conditions: when the ground surface is approximately level and when objects are resting on that surface. With sloping or uneven ground or objects suspended in midair, we would expect and often find systematic errors and illusions (e.g., Gibson, 1950, pp.

174–180) — evidence that eye height information is actually used by the visual system. Stoper has tried to account for the "magnetic hill" illusion, in which objects appear to roll uphill, in just these terms.

A second problem arises because eye height changes as we stand, sit, step up, and so on, and yet we do not experience a corresponding expansion and contraction of our surroundings. Or do we? When Mark (1987) strapped 10-cm blocks to his observers' feet, he found that judgments of maximum stair height shifted systematically upward, as a constant ratio of the new eye height. This occurred despite the fact that the maximum stair that could be climbed was not altered by wearing blocks, because thigh length was unaffected. However, after walking around briefly on the blocks, judgments rapidly shifted back to the original maximum height. This suggests that the visual system can rescale perceived size for changes in eye height with a bit of perceptual exploration. In effect, under some conditions we can see that our own eye height has changed. How this process works is a fascinating but unanswered question.

Thus, research from different traditions supports the idea that eye height information can be used to scale the sizes of objects and their affordances. The implication for designers is to provide surfaces with sufficient optical texture and to locate objects on visible surfaces. Sloping floors, multileveled terraces, and objects suspended in midair may induce the kinds of body orientation and size illusions observed in the pitchroom and the Ames room. However, the eye height ratio is surely not the only body-scaled information for affordances. It would appear to be useful primarily for medium-to-large objects at medium-to-large distances over a ground surface. Other information is presumably available at different ranges and scales, such as binocular information for objects within the reach envelope. Given how little we really know about perceiving the layout of surfaces, designers should play it safe by providing sufficiently rich optical structure so that the locations and orientations of surfaces are adequately specified.

8.6   Conclusion

Such an affordance analysis has three major advantages over traditional anthropometry. First, it is dynamic and task-specific, analyzing actions as they are actually performed rather than relying on static anatomical measurements. Second, the analysis yields π numbers, that is, body-scaled values for actors of any absolute size. The designer can then use the ratios to translate known anthropometric data into affordance values for a specific population — men, women, children, the elderly, and so on. Finally, the method considers not just the physical dimensions of the environment, but also the perceptual basis for successfully guiding action within it.

Although I have emphasized body-scaling and size, the affordances of the environment are clearly multidimensional, encompassing not only the geometric dimensions of objects and limbs, but dynamic properties such as object mass, rigidity and elasticity, surface friction and compliance, muscle strength and force production, joint stiffness, bone and joint stress, rates of energy expenditure, and so on. For example, the upper limit on climbable stairs in older populations may not be the geometric one due to leg length, but a dynamic one due to joint flexibility or muscle strength (Konczak, Meeuwsen, & Cress 1992; Warren, 1983). The present approach can be generalized to incorporate these variables in more complex π ratios, as is typically done in similarity theory. The more difficult challenge lies in determining the informational basis for perceiving such affordances, given that the information is not merely body-scaled, but action-scaled. There is potentially both static and kinematic information available to specify such dynamic properties across modalities, but we are only beginning to explore it (Bingham, Schmidt, & Rosenblum, 1989; Warren, Kim, & Husney, 1987).

To demonstrate that such a method really does have a contribution to make to environmental design, Table 8.1 compares current architectural standards from Ramsey and Sleeper (1970) with the results of the ecological studies on stair height, seating height, passage width, and grasping size cited earlier. The so-called "ecological standards" are derived from π numbers by applying them to the appropriate anthropometric percentile value for the general population and, of course, are highly tentative. How these percentile values are selected is itself an issue. The well-known fallacy of the "average man" (Daniels, 1954; Panero & Zelnick, 1979) implies that one cannot best accommodate the majority of a population by designing for the 50th percentile; rather, different design problems require different criteria. In Table 8.1, the "ecological standard" for riser height is based on the 5th percentile to insure safe risers for the shorter end of the population, that for passage width is based on the 95th percentile to insure that the largest individuals can pass easily, and that for seating height is based on a range from the 5th to the 95th percentile, on the assumption that chairs should be adjusted to the individual user. The differences between architectural and ecological standards are more marked in some cases than in others, but there is clearly room for improvement based on an affordance analysis.

In sum, this reconstrual of environmental design as the design of affordances could bear fruit for the architecture of public and private spaces, the design of furniture, equipment, and other implements, and for health and safety in the workplace. And the message for ecological psychologists is that when we study real problems of perception and action in real environments, the gap between basic and applied research is not so wide.

Table 8.1. *Architectural and ecological standards for several affordances*

Architectural and Ecological Standards for Several Affordances				
		Architectural Standard	Ecological Standard	π
Riser height	Opt	5-7 in	7.5in	.26L
Seat height	Opt	17.5 in	14.5 - 17.5in	.47L
	Max	2 ft 6 in	2ft 3in - 2ft 8in	.87L
Passage width	Min	21 in	25in	1.3S
Graspable object diameter	Max	-	6in	.9G

Acknowledgments

This chapter was originally presented at the Third International Conference on Event Perception and Action, Uppsala, Sweden, June 1985. The present version remains largely faithful to the original, but adds a section on intrinsic information.

8.7   References

Acking, C. A. (1980). Humanity in the built environment. In B. Mikellides (Ed.), *Architecture for people.* (pp. 101-104). London: Studio Vista.

Allsopp, B. (1974). *Towards a humane architecture.* London: Frederick Muller.

Berriman, A. E. (1953). *Historical metrology.* New York: Dutton.

Bingham, G. P., Schmidt, R.C., & Rosenblum, L.D. (1989). Hefting for maximum distance throw: A smart perceptual mechanism. *Journal of Experimental Psychology: Human Perception and Performance, 15,* 507–528.

Brolin, B. C. (1976). *The failure of modern architecture.* London: Studio Vista.

Canter, D., & Lee, T. (Eds.), (1974). *Psychology and the built environment.* New York: Holstead.

Carello, C., Grosofsky, A., Reichel, F. D., Solomon, Y., & Turvey, M. T. (1989). Visually perceiving what is reachable. *Ecological Psychology, 1,* 27–54.

Le Corbusier (Charles Edouard Jeanneret Gris). (1966). *The modulor,* (P. de Francia and A. Bostock, Trans.) Cambridge, MA: Harvard University Press. (Original work published 1954).

Damon, A., Stoudt, H. W., & Mc Farland, R. A. (1971). *The human body in equipment design.* Cambridge, MA: Harvard University Press.

Daniels, G. S. (1954). *The "Average Man"?* (Technical Note Number WCRD 53-7). Wright-Patterson AFB, OH: Wright Air Development Center.

Dempster, W. T., Gabel, W. C., & Felts, W. J. L. (1959). The anthropometry of manual work space for the seated subject. *American Journal of Physical Anthropology, 17,* 289–317.

Diffrient, N., Tilley, A. R., & Bardogjy, J. C. (1974). *Humanscale 1/2/3.* Cambridge, MA: MIT Press.

Drillis, R. J. (1963). Folk norms and biomechanics. *Human Factors, 5,*

427–441.

Gehringer, W. L., & Engel, E. (1986). Effect of ecological viewing conditions on the Ames' distorted room illusion. *Journal of Experimental Psychology: Human Perception and Performance, 12,* 181–185.

Ghyka, M. (1977). *The geometry of art and life.* New York: Dover. (Original work published 1946).

Gibson, E. J., Riccio, G., Schmuckler, M. A., Stoffregen, T. A., Rosenberg, D., & Taormina, J. (1987). Detection of the traversability of surfaces by crawling and walking infants. *Journal of Experimental Psychology: Human Perception and Performance, 13,* 533–544.

Gibson, J. J. (1950). *Perception of the visual world.* Boston: Houghton-Mifflin.

Gibson, J. J. (1966). *The senses considered as perceptual systems.* Boston: Houghton-Mifflin.

Gibson, J. J. (1979). *The ecological approach to visual perception.* Boston: Houghton-Mifflin.

Grandjean, E. (1980). *Fitting the task to the man.* London: Taylor & Francis.

Haber, R. N., & Levin, C. A. (1992). *The perception of object size is independent of the perception of object distance.* Paper presented at the 33rd Annual Meeting of the Psychonomic Society, St. Louis, MO.

Hale. N. C. (1993). *Absraction in art and nature.* New York: Dover.

Hallford, E. W. (1984). *Sizing up the world: The body as referent in a size-judgment task.* Unpublished doctoral dissertation, Ohio State University, Columbus, OH.

Hertzberg, H. T. E. (1960). Dynamic anthropometry of working positions. *Human Factors, 2,* 147–155.

Ittelson, W., Proshansky, H., Rivlin, L., & Winkel, G. (1974). *An introduction to environmental psychology.* New York: Holt, Rinehart, & Winston.

Koenderink, J., & van Doorn, A. (1991). Affine structure from motion. *Journal of the Optical Society of America A, 8,* 377–385.

Konczak, J., Meeuwsen, H. J., & Cress, E. M. (1992). Changing affordances in stair climbing: The perception of maximum climbability in young and old adults. *Journal of Experimental Psychology: Human Perception and Performance,18,* 691-697.

Koffka, K. (1935). *Principles of gestalt psychology.* New York: Harcourt, Brace.

Kroemer, K. H. E. (1983). Engineering anthropometry: Work space and equipment to fit the user. In D. J. Oborne and M. M. Gruneberg (Eds.), *The physical environment at work*. New York: Wiley.

Mark, L. S. (1987). Eye height-scaled information about affordances: A study of sitting and stair climbing. *Journal of Experimental Psychology: Human Perception and Performance, 13,* 361-370.

Mark, L. S., & Vogele, D. (1987). A biodynamic basis for perceived categories of action: A study of sitting and stair climbing. *Journal of Motor Behavior, 19,* 367–384.

Matin, L. & Fox, C. R. (1989). Visually perceived eye level and perceived elevation of objects: Linearly additive influences from visual field pitch and gravity. *Vision Research, 29,* 315–324.

Mikellides, B. (Ed.) (1980). *Architecture for people*. London: Studio Vista.

Nemire, K., & Ellis, S. R. (1991). *Optic basis of perceived eye level depends on structure of the optic array*. Paper presented at the 32nd Annual Meeting of the Psychonomic Society, San Francisco, CA.

Newman, O. (1980). Whose failure is modern architecture? In B. Mikellides (Ed.), *Architecture for people*. (pp. 45-58) London: Studio Vista.

Norman, D. (1988). *The psychology of everyday things*. New York: Basic Books.

Panero, J., & Zelnik, M. (1979). *Human dimension and interior space*. New York: Whitney Library of Design.

Ramsey, C. G., & Sleeper, H. R. (1970). *Architectural graphic standards* (6th ed.). New York: Wiley.

Reed, E. S. (1985). *Technological change and the psychology of human action*. Unpublished manuscript.

Roebuck, J. A., Kroemer, K. H. E., & Thomson, W. G. (1975). *Engineering anthropometry methods*. New York: Wiley.

Rosen, R. (1978). *Fundamentals of measurement and representation of natural systems*. New York: North Holland.

Runeson, S. (1988). The distorted room illusion, equivalent configurations, and the specificity of static optic arrays. *Journal of Experimental Psychology: Human Perception and Performance, 14,* 295–304.

Schuring, D. J. (1977). *Scale models in engineering*. New York: Pergamon Press.

Sedgwick, H. A. (1980). The geometry of spatial layout in pictorial representation. In M. Hagen (Ed.), *The perception of picture,VI,* 33-90. New York: Academic Press.

Shackel, B. (Ed.), (1976). *Applied ergonomics handbook*. Guildford, Surrey:

IPC Business Press.

Singleton, W. T. (Ed.), (1982). *The body at work: Biological ergonomics.* Cambridge, England: Cambridge University Press.

Stahl, W. R. (1961). Dimensional analysis in mathematical biology. I. General discussion. *Bulletin of Mathematical Biophysics, 23,* 355–376.

Stahl, W. R. (1963). Similarity analysis of physiological systems. *Perspectives in Biology and Medicine, 6,* 291–321.

Stokols, D. (Ed.), (1977). *Perspectives on environment and behavior.* New York: Plenum Press.

Stoper, A. E., & Cohen, M. M. (1989). Effect of structured visual environments on apparent eye level. *Perception and Psychophysics, 46,* 469–475.

Stoper, A. E. (1990). *Pitched environments and apparent height.* Paper presented at the 31st Annual Meeting of the Psychonomic Society, New Orleans, LA.

Templer, J. (1992). *The staircase.* Cambridge, MA: MIT Press.

Todd, J. T. & Bressan, P. B. (1990). The perception of 3-dimensional affine structure from minimal apparent motion sequences. *Perception and Psychophysics, 48,* 419–430.

Todd, J. T., & Reichel, F.D. (1989). Ordinal structure in the visual perception and cognition of smoothly curved surfaces. *Psychological Review, 96,* 643–657.

Turvey, M. T., & Shaw, R. E. (1979). The primacy of perceiving: An ecological reformulation of perception for understanding memory. In Nilsson, L. G. (Ed.), *Perspectives on memory research.* (pp. 167-222). Hillsdale, NJ: Lawrence Erlbaum Associates.

Vitruvius (Marcus Vitruvius Pollio). (1960). *The ten books on architecture* (M.H. Morgan, Trans.). New York: Dover.

Warren, W. H. (1983). A biodynamic basis for perception and action in bipedal climbing. *Dissertation Abstracts International, 43,* 4183B. (University Microfilms No. 83-09263).

Warren, W. H. (1984). Perceiving affordances: Visual guidance of stair climbing. *Journal of Experimental Psychology: Human Perception and Performance, 10,* 683–703.

Warren, W. H., Kim, E. E., & Husney, R. (1987). Visual perception of elasticity and control of the bounce pass. *Perception, 16,* 309–336.

Warren, W. H. & Whang, S. (1987). Visual guidance of walking through apertures: Body-scaled information for affordances. *Journal of Experimental Psychology: Human Perception and Performance, 13,* 371–383.

Wise, J. A., & Fey, D. (1981). Principles of Centaurian design. *Proceedings of the Human Factors Society, 25th Annual Meeting,* (pp. 245–249). Santa Monica, CA: Human Factors Society.

Woodson, W. E. (1981). *Human factors design handbook.* New York: McGraw-Hill.

Chapter 9

Designing with Affordances in Mind

Brian S. Zaff
Logicon Technical Services, Inc.

9.0   Introduction: The Ecological Approach

The ecological approach owes much to the thinking of Gibson (1950, 1966, 1979). From the ecological perspective, the activities of an individual are assumed to always be constrained by an interaction between the individual's capabilities and the properties of the environment that envelops that individual. In order to understand the individual's activities and to design artifacts that meet the individual's needs, it is necessary to understand the ways in which that individual's capabilities interact with the surrounding environment. From an ecological perspective, the individual and his or her environment is conceived of as an inseparable pair, in the sense that one cannot be defined, let alone understood, without reference to the other. It is easy to appreciate the fact that the individual simply could not exist without an environment, but it is equally true, although perhaps not as obvious, that the environment, insofar as it exists as a niche for a specific individual, could not be defined without the individual that it environs or surrounds (Gibson, 1979).

The concept of organism-environment mutuality is one of the fundamental tenets of Gibson's ecological approach to perception and action and has significant implications for the practice of design. This principle serves to identify the *ecosystem* as the primary unit of analysis. Thus, instead of units that are physically defined in a fashion independent of the individual, from the ecological perspective, the units are defined in terms of the physical relationship between the individual and the environment. With regard to the activities of an individual, the ecosystem defines the sum total of all things that are possible for the individual within the environment, or as Gibson (1979) described it, the ecosystem is, from the perspective of an individual, a set of affordances.

9.1 The Concept of Affordances

The concept of affordances is used to capture the reciprocal relationship that exists between the individual and his or her environment and follows directly from the concept of organism-environment mutuality. Affordances describe what the objects, substances, and event of the environment are good for or what harm they can do with regard to a particular individual (Gibson, 1977, 1979). Gibson explains, however, that unlike values and meanings, which are typically considered to be subjective or phenomenal, affordances are material properties of one's environment. That is, they are measurable aspects of the individual's environment that support that individual's actions and intentions, and they are measurable aspects of the environment that can only be measured in terms of the individual.

Affordances define the sum total of measurable actions that are possible by a given individual within his or her environment, but as Turvey, Shaw, Reed, and Mace (1981) point out, simply because affordances are possibilities for action does not imply that they are merely epistemological categories. In other words, simply because affordances define the potential for action does not reduce them to a phenomenological status. Affordances have a physical reality. Nevertheless, affordances can also be distinguished from the physical properties of an environment which can be considered independently of the individual.

Affordances are not simply attributes of the environment, nor are they solely attributes of the individual. They have no existence independent of a joint consideration of the individual and the environment. They cannot be specified independently of the individual, nor do they represent subjective properties of the individual's phenomenal experiences. Like a line that is created by the intersection between two planes that cannot be thought of at the same time without considering both planes, affordances cannot be conceived of without considering both the individual and his or her environment.

Affordances are the real physical properties of the individual–environment interaction which can only be defined at the level of an ecosystem (Gibson, 1977, 1979). The affordance of any thing according to Gibson, is a specific combination of its substances and surfaces taken in reference to an individual. Alternatively, the same relationship between the individual and his or her environment can be defined in terms of the individual's effectivities. The *effectivities* of an individual are the dynamic capabilities of that individual taken with reference to a set of action-relevant properties of the environment (Turvey & Shaw, 1979).

The individual-environment interaction can be described in either of two ways; either in terms of a set of affordances, or in terms of a set of effectivities. Both expressions can describe the same interaction, but they do so from different perspectives. Affordances focus on the set of environmental properties taken in reference to a set of action-relevant properties of the individual: the individual's effectivities. Effectivities, alternatively, focus on a set of properties of an individual taken with reference to a set of action-relevant properties of the environment: the environment's affordances. It is important to note that both terms are, first and foremost, referential.

From the ecological approach, the perception of whether or not an object is within reach, for example, is the perception of an affordance or the absence of an affordance. It involves perceiving the relationship between oneself and the to-be-reached-object. From this perspective, the distance of the object is not resolved in absolute terms with respect to some arbitrary unit of measurement, and then compared with or converted to an action appropriate scale, as suggested by Foley (1978, 1980). Instead, the ecological approach assumes that the affordance is perceived without mediating inferences or computations, that it is perceived directly (Carello, Grosofsky, Reichel, Solomon, & Turvey, 1989).

The individual's continued existence may depend on an ability to detect the available affordances, but the existence of those affordances cannot be said to depend on their felicitous detection. The perception or, more specifically, the visual perception of affordances requires the satisfaction of several conditions. First, a particular relationship between a set of properties of the environment taken in reference to a set of action-relevant properties of the individual must exist. Second, the relationship between the individual and his or her environment must be specified in the light in the form of an optical invariant, or more generally specified in the perceptual array. Third, the individual must be sensitive to the optical structure that is specifying his or her relationship to the environment. And, last, the individual must attend to a particular relationship from among a number of potential relationships that are consistent with the intentions of the individual.

9.2   Optical Specification of Affordances

Gibson (1977, 1979) suggested insofar as the individual is capable of
guiding his or her behavior, that it is necessary for the individual to
perceive what the objects and events of the environment offer or afford
for action. For an individual to guide his or her activities by perceiving
affordances implies that he or she must be capable of perceiving the
relationship between environmental properties and the properties of his
or her own action system. Thus, according to this description, an
individual is likely to make use of an intrinsic or body-scaled metric,
rather than an extrinsic or arbitrarily defined metric (e.g., meters, liters,
seconds, etc.) when gauging the environment in terms of his or her own
capabilities.

Gibson (1950, 1966, 1979) observed that the optical information
specifying the environment is accompanied by optical information
specifying the self, including the head, body, arms and hands in relation
to the environment. Whenever we perceive the environment, we quite
literally perceive ourselves in relation to the environment at the same
time. According to Gibson (1979):

> "When a man sees the world, he sees his nose at the same time; or
> rather, the world and his nose are both specified and his awareness can
> shift. Which of the two he notices depends on his attitude; what needs
> emphasis now is that information is available for both." (p. 116)

The two sources of information (i.e., about the world and oneself) are
coexistent. Gibson (1979) emphasizes that the one cannot exist without
the other. The ecological approach asserts that the logical separation of
the individual from the environment is an unnecessary, indeed artificial
abstraction. The two allegedly separate realms of the subjective and the
objective, as they are typically characterized, are actually only the two
poles of attention. Self perception and perception of the environment go
together; they are mutually constraining and allow for the possibility of
perceiving affordances, insofar as the individual always sees him or
herself in relationship to the environment.

Although several potential sources of visual information about the
relationship of the self to the environment have been mathematically
isolated (Gibson, Olum, & Rosenblatt, 1955; Lee 1976; Sedgwick, 1973,
1980, 1983), the existence of an invariant structure in the light does not
ensure that that potential source of information will have any particular
salience for a given individual performing a particular task. It is not at

all inappropriate from an ecological perspective to regard relevant information as being possibly distinct from the potential information. Any number of potential sources of information may prove irrelevant for the task at hand, but evidence of frequently successful control of one's activities in a changing environment indicates that one possesses the ability to acquire the relevant information about one's own affordances. The concept of affordances, as Castelfranchi and Miceli (1987) point out, is not perceptually bound. The existence of an invariant structure in the optic array specifying an affordance does not ensure that that potential source of information will have any particular relevance for a given individual performing a particular task, or that the individual will always be able to detect the affordances that are relevant.

A variety of errors can occur with regard to the individual's attempt to detect affordances. The individual can misperceive an affordance by identifying what appears to be an affordance when none exists. For example, the individual may see the horizontal, opaque, and rigid surface of a frozen lake as affording support for bipedal-locomotion when in fact the affordance is lacking. Assuming, however, that there is a reciprocity between a set of properties of the environment and the effectivities of the individual, and that the affordance exists, the individual may still fail to detect the affordance. Failure on the part of the individual to detect an affordance may result (a) because the relationship between the individual and his or her environment is not specified in the light, or (b) because the individual lacks the necessary attunement. Thus, the individual's capacity to detect the potential affordances is an empirical question.

9.3   The Acquisition of Information About Affordances

A necessary condition for the perceptual guidance of behavior is the acquisition of information by the individual about his or her relationship to the environment, in short, the individual must be able to acquire information about the available affordances. Gibson (1979) argued that affordances can be directly perceived by virtue of the fact that there is typically enough information in the light to specify the individual's relationship to the environment and to support the visual perception of affordances. In order for the individual to guide his or her activities by perceiving the available affordances, he or she must be capable of perceiving the relationship between a set of environmental properties and the properties of his or her own action system. The activities of the individual would then necessarily be adjusted to his or her own capabilities, and the relationship between those action properties and the properties of the environment would be perceived both in terms of the individual's body dimensions, such as those required for grasping (Hallford, 1984) or for passing through an aperture (Warren & Whang, 1987), and in terms of dynamic variables, such as mass, force, and work, as they are related to the metabolic expenditure of energy (Warren, 1984). The implication from the perspective of a theory of direct perception is this: When object-size information is an important factor in the perception of an object's function, then that information is likely to include a reference to the size of the individual's corresponding body part.

9.3.1 Mediational Approach to the Acquisition of Information About Affordances

Before going on to discuss the direct perception of affordances, it should be noted that a recognition of the importance of the concept of affordances to the control of behavior does not necessarily imply that the acquisition of information about one's affordances is direct. More traditional theories of perception and action, such as those of Hochberg (1974), Norman(1988), or Rock (1977, 1983), although denying the possibility of direct perception, recognize that the individual must at some level be capable of acquiring information about his or her relationship to the environment, and therefore these theories may find the concept of affordances useful. Although some of the more traditional approaches may find the concept useful, they nevertheless challenge the ecological claims that affordances can and are directly perceived. Norman (1988), for example, finds the concept of affordances particularly useful for the analysis of the relationship between the individual and his or her environment, but he has difficulty accepting the unconventional idea that affordances can be directly perceived. Norman claims that our perception of affordances arises from an interpretation of things in the environment which is made in light of our past knowledge and experiences.

The idea that the perception of affordances is influenced by our experiences, it must be noted, is not, however, inconsistent with the ecological perspective (see Gibson & Spelke, 1983). Nor does it preclude the possibility that the perception of affordances can occur without the aid of mediational processes. Nevertheless, traditional approaches to perception, such as those proposed by Hochberg, (1974), Norman (1988), or Rock (1977, 1983), are dominated by the assumption that the individual's epistemic contact with the surrounding environment must be mediated by internal knowledge and processes, and that the direct perception of higher order relationships (i.e., affordances) is finally impossible.

9.3.2   Ecological Approach to the Acquisition of Information About Affordances

From an ecological perspective, the direct perception of affordances is in fact considered necessary in order for perception to be considered at all direct. Mace (1977) suggests that insofar as direct perception is possible, it is necessary that the organism be capable of directly detecting higher-order invariants that specify the ecological significance of things and events in the environment. Without an acceptance of the idea that affordances are directly perceived, it would be necessary to conclude that the individual's perception of the world is indirect. For even if certain properties of the ambient light were directly detected, as Fodor and Pylyshyn (1981) are willing to grant, the individual would still be required to make a transition from the invariants in the light, to the properties of the environment that are ecologically relevant, by inferring the latter from the former on the basis of knowledge of the correlation that connects them. To use Mace's example, the individual might presumably have to connect the invariant structure of a perceived pattern with certain properties such as hardness and opacity, and then on the basis of these properties, infer what activities can be performed with objects containing these properties. The direct acquisition of information about one's affordances makes this process of inference unnecessary, as one is thought to be able to directly perceive what objects and events of the environment offer or afford given one's capabilities and intentions.

In order for these higher order invariants, that is, affordances, to be directly perceived, they must be specified in the structure of the ambient energy array. If the claims regarding the existence of higher order invariants in the structure of the ambient array can be substantiated, and if the perceiver can be shown to be sensitive to this structure, then the ecological account of the direct acquisition of information about one's affordances would appear to be more theoretically parsimonious than mediational accounts that rely, by way of explanation, on processes of inference.

From the ecological perspective, information about one's affordances refers to a physical state of affairs that is specific to the control and coordination required of activity (Gibson, 1958; Turvey & Carello, 1981). According to Turvey and Kugler (1984), the requirements for information that would be useful for the guidance of life-sustaining activities can be found in the properties of structured patterns of energy relating the individual to his or her environment. The layout of surfaces in the environment is specified by the pattern of structured light that is reflected from environmental surfaces and converges at every point in the medium. Gibson (1958) termed this converging pattern of differential reflectance that is projected to the place of observation the ambient optic array. Invariants in the pattern of the optic array specifying an event not only specify the relationship among the layout of surfaces in the environment, but also the relationship of the perceiver to those surfaces. That is, each observation point is somewhat different, so that an individual occupying one of these will have available not only information about the layout of surfaces, but also information about his or her relationship to those surfaces.

9.3.3   Distinctions Between the Ecological and Mediational Approaches

The central question concerning the theory of affordances as Gibson (1979) saw it "is not whether they exist and are real, but whether information is available in ambient light for perceiving them" (p. 140). People do perform relevant actions on their environment, meaning that they undoubtedly acquire information about their affordances. But how this information is acquired is one of the distinguishing features between the direct and mediational approaches to perception. A mediational account of affordance perception would assume that information about the individual's affordances is either not specified in the light, or if it is specified, that the individual inherently lacks the capacity to detect the specification. There are in fact two different ways to have a mediational approach to affordance perception. The first involves an assumption of insufficient information in the light to specify the affordance (e.g., Fodor & Pylyshyn, 1981; Norman, 1988; Rock, 1977, 1983). According to this account the individual is said to supplement the information by any of a number of extraretinal sources of information, such as, inference, prior experience, expectancies, and so on. The other mediational account assumes that the information is in the light, but that either the individual is incapable of perceiving the higher order invariants that would specify an affordance for the individual, or that such higher order invariants are simply not specified in the structure of the ambient optic array (e.g., Ullman, 1980). In this case the individual would be required to perform a variety of computations on the lower order variables in order to arrive at the higher order invariant which is information about the affordance.

In contrast to both the extra retinal and computational versions of the mediational approach, the direct theory of perception claims that there is not only information in the form of invariants in the ambient array that specify the layout of surfaces, but that invariant combinations of invariants are present that also specify the affordances of the environment for the perceiver (Gibson, 1979). Direct perception of affordances is thus possible to the extent that the individual is capable of detecting the invariant structure of the ambient array that specifies the relationship between the individual's action-relevant properties and the environmental properties needed to support the intended activity.

9.4   Body-scaled Information in the Perception of Affordance

As defined earlier, the affordance of anything "is a specific combination of the properties of substances and surfaces taken with reference to an animal" (Gibson, 1977, p. 67). Although having physical properties, affordances have not been described in terms of classical physics, but rather, in terms of ecological physics. Ecological physics implies the use of intrinsic or "body-scaled" metrics that take into account the relationship between environmental properties and the properties of the individual's action system. Therefore, in order to formally characterize an affordance, a method of intrinsic measurement can be employed, in which one part of the system is taken as the "natural standard" against which a reciprocal part of the system is measured (Warren, 1983). Following Warren (1983), an individual's property, I, may be taken as a standard for measuring an environmental property E. If I and E are measured in the same conventional unit and are expressed as a ratio, the result is a dimensionless Π (pi) number that uniquely expresses a particular individual-environment relationship.

As the individual-environment relationship varies, the affordances are altered. Affordances, as such, are likely to be both dichotomous at critical points that correspond to transitions for behavior (e.g., either a surface affords support, or it does not) and possess a preferred range corresponding to the "best fit" between the environment and the individual's action system.

Several studies in the literature have examined the perception of affordances as it pertains to the perceiver's ability to relate his or her activities to his or her own body dimensions. Hallford (1984), for example, conducted a series of experiments that examined the relations between object size and body size, in terms of a person's ability to make functional judgments regarding the size of the object with reference to an affordance or action standard. The study demonstrated that judgments based on the relevant body-scaled metric were consistently made with greater accuracy and confidence when compared to similar judgments based on an arbitrary physical standard.

Warren (1984), in his study of the affordance of stair climbing, found that the perceptual category boundaries between climbable and unclimbable stairs corresponded to a critical riser height that was a constant proportion of the perceivers' leg lengths. In addition, Warren demonstrated that the individual is able to identify the optimal relationship between his or her capabilities and the environmental support for those actions in the absence of any previous experience with the specific quantity of that attribute. Specifically, he found that perceivers were able to visually identify riser heights that closely matched an ideal riser height in terms of minimum energy expenditure, even though the ideal riser heights were considerably higher than those found in ordinary stairways (see Warren, this volume).

In another study on the perception of the affordances for sitting and stair climbing, Mark (1987) found that the perceiver's judgments of the maximum sittable chair height and the maximum climbable stair height were both highly accurate and a constant ratio of the perceiver's eye height. Warren and Whang (1987) studied the perception of passability through an aperture for visually guided walking and found that the perceiver was able to accurately identify the aperture that afforded passage. These judgments also appeared to be based on eye–height scaled information.

However, as Mark and Vogele (1987) have pointed out, the various studies pertaining to the perception of affordances, such as Warren's (1984) and their own, do not rule out the possibility that the individual performs several highly accurate computations which specify his or her relationship to the layout of environmental surfaces. They also note, that although these studies are not able to distinguish between the direct perception of affordances and perception that is mediated by the computation of several independent measures, the viability of the direct account would be supported by analyses of the optic array that demonstrate the existence of information specifying the relationship between the individual and the specific layout of environmental surfaces. Analyses have been conducted by Lee (1980), Sedgwick (1973, 1980, 1983), R. Warren (1982), and W. Warren (1983) which have demonstrated that from both static and changing points of observation, there exist potential sources of visual information specifying the perceiver's relationship to the environment. However, to date, only in the field of self-motion perception has there actually been a demonstration that the perceiver is indeed using the mathematically specified, potential information to control changes in his or her rate of self motion (e.g. Owen, Warren, Jensen, Mangold, & Hettinger, 1981; Wolpert, 1988; Zaff & Owen, 1987). These studies are unique insofar as they involved the manipulation of the available optical information in order to determine if any of the various potential sources of information have a functional utility.

In an attempt to distinguish between the direct perception of affordance and the acquisition of information about one's own affordances that is mediated by computational processes, Carello et al. (1989) studied the perception of the affordance of reaching. From the computational perspective, the detection of one's relationship to the environment for the task of reaching involves a series of computations, including the computation of the distance of the target object from the observer, the computation of the furthest possible extension of the limbs from a given postural position, and a comparison of the two resulting quantities. The results of the Carello et al. (1989) study demonstrated that on the basis of only optical information, the perceivers were very sensitive to the functional consequences that postural restrictions placed on the activity of reaching. More importantly, Carello et al. noted that the comparison of the results from various studies that have explicitly examined estimations of target distance (a necessary component of the computational approach) and the results from their own study revealed a number of inconsistencies between the pattern of performance for judgments of distance and the performance of perceivers attempting to detect their reaching affordances. Although the Carello et al. study did not completely rule out the possibility of a mediational account of the acquisition of information about what is reachable, it did illustrate the lack of parsimony inherent in a computational solution.

Mark (1987) similarly found that when he manipulated the perceivers' sitting and climbing affordances by placing them on 10-cm high blocks, the individuals quickly (without prompting or concerted practice) regauged their perceptual boundaries to reflect the changed affordances. Mark also recognized that the results of his study could not exclude the possibility that the individuals were readjusting their judged categorical action boundaries using either an algorithm or heuristic to recalculate their maximum sittable or climbable heights. However, he did observe that their ability to estimate the actual block height did not improve in conjunction with the noted improvements in their ability to detect the critical action boundary between the surfaces that afforded the intended action and those that did not. Because a computational account of the acquisition of information about one's affordances would presumably require an accurate estimation of block size as one of the computational components necessary for the rescaling of the critical action boundary, the results suggest that the perceiver is not using a computational framework to acquire information about his or her own affordances. Instead of mediating the acquisition of information about one's affordances by summing various lower order variables, it appears that the subjects were detecting the higher order invariant that directly specified the critical boundary between climbable and unclimbable stair heights and sittable and unsittable chair heights.

From the results of previous research, it seems clear that the visual perception of affordances is based on body-scaled, optical information which specifies the relationship between environmental properties and the perceiver's own action system. It is also evident that the perceiver is typically quite proficient at detecting his or her own affordances[1].

9.5   Detecting Affordances for Another Individual

From the research to date, it is apparent that the perception of one's affordances is based on task-intrinsic, body-scaled information about the relevant relationship between environmental properties and the

---

[1] A distinction is made between the person whose affordances are being judged, that is, the person carrying out the action (i.e., attempting to reach an object located on a shelf overhead) and the person who is doing the judging. If a person is attempting to judge his or her own affordances then that person is both the judge and the judged person. However, if one person is judging the affordances that exist for another person than a distinction is made between the

perceiver's own action system. A number of these studies have also identified potential sources of optical information, specifying the relationship of the individual to his or her environment, thereby allowing for the possibility that the affordances may be perceived directly. It is evident on the basis of these previous studies that the perceiver is typically quite proficient at judging his or her own affordances. Evidently, the individual is capable of acquiring information about his or her own affordances, and the evidence that there is information specifying the affordance available in the structure of the ambient light implies that the perception of affordances is very likely to be direct.

Gibson (1979) argued that specification of an affordance takes the form of "an invariant variable that is commensurate with the body of the observer himself and is more easily picked up than one not commensurate with his body" (p. 143). Thus, the perception of affordances for another person whose effectivities are different from those of the perceiver constitutes a situation in which a set of affordances would not be commensurate with or scaled to the body of the perceiver and would therefore presumably be more difficult for the perceiver to detect. In spite of the fact that this claim is consistent with the concept of affordances as it has been defined, the accuracy with which an individual judges the affordances for another person has received only limited consideration.

It should be noted at this point, that a distinction is being maintained between the affordances of another individual and the affordances for another individual. The first case concerns the affordances that are provided by other people. The affordances arising from the relationship between the individual and another person in the environment are likely to be extremely elaborate, because of the complex nature of that part of the environment that is affording a particular activity, namely, the other person. "Sexual behavior, nurturing behavior, fighting behavior, cooperative behavior, economic behavior, political behavior — all depend on perceiving what another person or other persons afford, or sometimes on the misperceiving of it" (Gibson, 1979, p. 135). The perception of what another person affords may be enormously complex, but as Gibson points out, its optical specification will nevertheless be lawful. "It is just as much based on stimulus information as is the simpler perception of the support that is offered by the ground under one's feet" (p. 135).

In the latter case, the case of perceiving affordances for another person, the issue is no longer what that other person affords the individual doing the perceiving, but what that other person's environment affords him or her, given that person's particular set of effectivities. Thus, rather than the more typical situation in which an individual is attempting to perceive what the environment affords for his or her own actions, the individual is now attempting to perceive what another individual's environment affords that other individual. The person doing the perceiving is attempting to acquire information about the relationship between a set of properties in the environment that are taken in reference to the action-relevant properties of another individual.

Although it is true that we are most often concerned with judging and acting on what the environment affords us given our set of effectivities, there are several notable instances when the individual becomes concerned with the affordances for another individual. A common instance arises in the context of parental super-vision, a particularly apt term for describing what parents must have when caring for a young child insofar as its definition can be extended to include being able to see what the environment affords for the young child.

A predator–prey situation also involves an instance in which one individual may attempt to perceive not only what another individual affords, but the affordances for another individual as well. In addition to what the prey affords the predator, and vice versa (i.e., eating versus avoiding being eaten), it is likely to be important for both the potential prey trying to elude its predator, and for the predator trying to capture its prey, to be able to judge the affordances for the other as a way of anticipating the pursuit or avoidance strategies that the other is likely to employ. This anticipatory capability is based on the idea that affordances serve to constrain the activities of an individual by determining, in conjunction with that individual's effectivities, what effectuations are and are not possible.

The task of judging what the environment affords for another individual also constitutes one of the primary tasks facing designers and architects, insofar as they are responsible for the shape of the built environment. If we begin by asking the question, "Why have humans changed the shape and substances of their environment?," the importance of the concept of affordances to designers should at once become apparent.

When taking into consideration the concept of affordances as it applies to the detection of affordances for another individual, it is important to note that the concept embodies the claim that the environmental properties are related to a particular action and to the specific requirements of the individual who is performing that action. If the anthropometric differences between individuals are minimal, then the available affordances would undoubtedly be similar. However, even within members of the same species, there are many significant anthropometric variations that have important consequences for each individual's action capabilities. These differences in action capabilities, by the very nature of what is embodied in the concept of affordances, would indicate that what the environment affords different individuals would also be different. Thus, the activity of acquiring information about the affordances for another individual may mean that the perceiver is attempting to identify affordances that may be very different from his or her own.

If the anthropometric differences between individuals are typically small, then the consequences of being relatively inaccurate at judging another's affordances would, in that case, be minimal because the research to date has shown that perceivers are extremely accurate at detecting their own affordance. However, the anthropometric differences between individuals are often significant, especially when making comparisons across species or between individuals of the same species that vary greatly in age or physical health.

9.6 Empirical Investigation: Detection of Affordances for Another Individual

On the basis of a series of studies examining an individual's ability to judge the affordances for another person (Zaff, 1989), it was shown that a person is, on average, consistently more accurate at detecting his or her own affordances than at detecting the affordances that exist for another individual. It was also apparent that increasing differences between the action-relevant properties of the person doing the judging and the person whose affordances are being judged resulted in a decline in the judge's ability to acquire information about the affordances that exist for that other individual.

In this series of experiments the subjects were asked to judge the reachability of a overhead shelf for themselves and for other individuals whose height, arm length, and hence reaching ability differed from their own. In one experiment each subject was required to judge the reaching affordances for four different individuals whose height and reaching ability differed from their own. The persons whose affordances were being judged ranged in height from 0.91 to 1.83 m. In turn, each of the four individuals as positioned in front of an adjustable shelf, and stood with arms at their sides in full view of the subject. The subjects were presented with a forced-choice task and simply asked to state whether the shelf at its current adjusted height could be reached by the person standing in front of it. Twenty different shelf heights were used in each condition, which ranged from 22.86 cm below the maximum reachable shelf height for that person to 25.40 cm above the maximum in 2.54 cm steps.

The results clearly indicate (see Figure 9.1.) that as the difference between the reaching abilities of the person doing the judging and the person whose affordances were being judged increased, the accuracy of judgments concerning the affordances available for that other person declined. It was also apparent that the systematic pattern of distortion resulted from the relative differences in maximum reach. The judges tended on average to overestimate the maximum reachable shelf height of a judged person whose reaching abilities exceeded their own and underestimate the maximum reachable shelf height of the judged persons whose reaching abilities were less than their own.

9.7 Effect of Experience on the Detection of Affordances for Another Individual

The results from Zaff (1989) also indicated that experience in judging the affordances for another person affected the accuracy of those judgments[2]. The experienced judges were still more accurate and sensitive to their own affordances than to the affordances for another individual, and when judging their own affordances, the level of accuracy was no different than the level of accuracy for the inexperienced judges. However, experience clearly affected the performance of the judges and did so most notably in terms of the pattern of under- and overestimation errors. Unlike their inexperienced counterparts, the experienced judges tended to overestimate the

---

[2]In the case of detection the reaching affordances, the experienced judges were those individuals that had extensive prior experience in judging the affordances for individuals with significantly different effectivities (i.e., daycare workers with five or more years of practical experience).

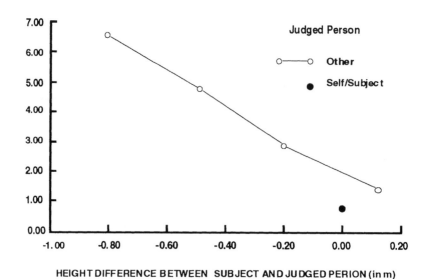

Figure 9.1. *The mean underestimation errors for judgments of maximum reachable shelf height as a function of the difference in height between the subject and the person whose reaching affordances the subject is judging (From Perceiving Affordances for Oneself and Others: Studies in Reaching and Grasping (p.123), by B.S. Zaff, 1989, Unpublished Doctoral Dissertation. Copyright © 1989 Reprinted by permission.)*

reaching affordances of the persons whose actual maximum reach was substantially less than their own.

It should be noted that the difference between the experienced and inexperienced judges was limited to estimations of the reaching affordances of the shorter judged persons. On the basis of these results, it is impossible to determine whether experience simply curtails the tendency of the judges to underestimate the maximum reachable shelf height of the judged persons, or whether the apparent reversal in the pattern of judgments was limited to judged persons whose relative maximum reach was less than the judges, because judges only had an appreciable amount of experience judging the affordances for individuals whose relative maximum reach was less than their own.

Given these two possibilities, it seems more plausible to attribute the limitation in the reversal of directional errors to the limitations in the experience of the judges. Having been in situations in which they were called on to judge the affordances for a young child, these judges are likely to have learned of their own inherent tendency to underestimate the affordances of that child and taken steps to correct it. In fact, one experienced judge, while attempting to evaluate the reaching affordances of the 0.92-m judged person, expressed the idea that, "Little children can always reach higher than it looks like they can." This judge then proceeded to overestimate the reaching affordances of that judged person.

The experienced judges, rather than simply reducing the magnitude of underestimation and thereby becoming more accurate in their judgments, however, showed a tendency to overestimate the reaching affordances of the judged persons whose actual capabilities were less than their own. The fact that the experienced judges now overestimate the reaching affordances of the shorter judged persons has several practical considerations.

In terms of "child-proofing" the child's environment by making all potentially hazardous and breakable material beyond the reach of the young child, directional errors of the sort produced by the experienced judges would clearly be more acceptable. If the caretaker as to error in one direction, overestimating the maximum reachable shelf height would provide for a margin of safety. If the experienced judges were responsible for designing the built environment of a young child, ideally, one would hope that the fit between the environment and the child's action-relevant properties would be as close as possible. However, if some errors are unavoidable, the tendency to overestimate the child's affordances is likely to be more desirable. Although errors in either direction can be hazardous for the child who is strongly motivated to use an artifact that is inappropriately scaled to his or her action-relevant properties, the simple fact that the child will grow to fit an environment that has been scaled to an overestimation of his or her affordances may render that type of directional error somewhat more desirable than underestimations errors.

9.8   Systematic Distortions in the Pattern of Under- and Overestimation Errors

A possible explanation for the pattern of under- and overestimation errors can be derived from a careful consideration of the relationships that the judges are presumably perceiving while attempting to acquire information about the affordances for another individual. Initially, the inexperienced judges appear to be, in a sense, attending more to their relationship to the judged person, rather than to the judged person's relationship to his or her environment. The judges' behavior appears to be influenced by the difference in their own height and the height of the judged person, rather than simply reflecting an evaluation of the judged person's affordances. For example, the inexperienced judges perceive that the judged person is shorter than themselves, and perceive that that judged person's reach would be less than their own reach. These judges then proceed to underestimate the maximum reachable shelf height of the shorter judged person and do so in a manner that reflects the fact that they are being influenced by the relative difference in their own and the other person's effectivities. The greater the difference, the greater the magnitude of the error. In general, the inexperienced judges' behavior could be characterized by an apparent inability to take the perspective of the other individual. They are apparently unable to see what the environment would afford a person with a different set of action-relevant properties. Thus, it is possible to conclude that the inexperienced judges lack super-vision.

With experience, however, the judges appear to take the judged person's perspective. The success of their attempt appears to be tempered by the fact that they continue to be influenced by their own relationship to their own environment. The behavior of the experienced judges, when attempting to detect the affordances for a judged person whose actual maximum reachable shelf height is less than their own, also seems to be impaired by a tendency to attend to their own relationship to their own environment. The consequence of this attention to their own relationship to their own environment may be the overestimation of the reaching affordances of the shorter judged person. In other words, the judges are judging their own affordances in addition to the affordances of the judged person whose affordances they are attempting to evaluate.

9.9   The Importance of the Individual's Relationship to the Environment

In an attempt to distinguish between direct and mediational accounts of the detection of one's affordances, Zaff (1989) simultaneously manipulated both judge's familiarity with the task and the judge's relationship to his or her environment. The argument could be made, in concordance with the mediational account of affordance detection, that the judges were more accurate at judging their own affordance because they were more accustomed to judging their own relationship to the environment and had accumulated knowledge about the specific height that they could reach. From this perspective, one could claim that the judges were not directly perceiving the relationship between their own action-relevant properties and the set of environmental properties that were needed to support the intended action, but that they were mediating the perception with the addition of information acquired from prior experience with the task of reaching an object while standing. The judges, one might argue, were more accurate at judging what they could reach overhead, than at judging what another person could reach, because they were simply more familiar with the task. According to this mediational line of reasoning, the judges had acquired an absolute sense of what they were capable of reaching, because of repeated experience with the task, and were using that additional information when evaluating their reaching affordances.

The ecological claim is that the judges were able to acquire information about their affordances because they were able to directly perceive the relationship between their own action-relevant properties and the environmental properties needed to support the intended action. The judges, according to the ecological perspective, could literally see the relationship between themselves and the shelf, and thus, directly see whether a particular shelf height, within a very narrow margin of error, afforded reaching. From this perspective, the judges' familiarity with the task need not have any influence on their performance of the task. The assumption from the ecological perspective is that the judges were able to identify the critical boundary between reachable and unreachable shelf heights because they could literally see the relationship between themselves and the shelf, and not because they had any prior knowledge of what they could reach.

In order to distinguish between these two perspectives with regard to their claims about the detection of affordances, it was necessary to manipulate both the familiarity with the height being reached and the judge's relationship to the environment when attempting to detect a particular affordance. The simultaneous manipulation was accomplished by having judges from a kneeling and standing perspective judge what they could reach from a kneeling and standing posture. The set of predictions from each perspective as fairly straightforward. In order for a mediational perspective that assumes the addition of extraoptical information to be consistent with its explanation of the judge's performance for detecting his or her own affordance and the affordances for another person, the judges would be expected to be generally more accurate at judging what they could reach from a standing posture, regardless of whether they were standing at the time or not. The assumption guiding this prediction is that the judges are more familiar at judging what they can reach from a standing posture than at judging what they can reach from a kneeling posture, and since the judges are not directly using the information that is in the light, their perspective should have little or no effect.

Alternatively, from the ecological perspective, the judges would be expected to be more accurate at detecting their own affordances when they could easily see the relationship between their own action-relevant properties and the environmental properties necessary to support the intended action. Thus, familiarity with the task of reaching from a standing posture is not considered to be sufficient to account for the judges' ability to evaluate their affordances. Rather, it is expected that the judges should be highly accurate because the relationship they are judging is the relationship that is currently being specified in the optic array. In other words, the judges should be highly accurate when attempting to detect their reaching affordance from a standing posture when perceiving the relationship from a standing perspective and when attempting to detect their reaching affordances from a kneeling posture when perceiving their relationship from that perspective. The evaluation of the affordances in these two situations are expected to be accurate. The judges would also be expected to be considerably less accurate when the relationship between themselves and the environment, was not the relationship that they were being asked to evaluate. That is, when kneeling and asked to judge what is reachable from a standing posture, the judges would be unable to see the relationship between the standing self and the maximum reachable shelf height from that standing position. The judges, in short, would be required to mediate the detection of affordance in this situation, insofar as they were being asked to evaluate the affordance characteristic of a relationship that is not specified in the light.

The results from Zaff (1989) clearly support the ecological claim that the detection of affordances is facilitated by the direct perception of the relationship between one's own action-relevant properties and the environmental properties necessary to support the intended action. When the judges were able to see their relationship to the environment, their ability to identify the critical boundary between reachable and unreachable shelf heights was characterized by a consistently high degree of accuracy. However, when required to evaluate affordances for a relationship that was not present in the light, and hence not capable of being directly perceived, the judges' performance revealed errors of considerable magnitude.

In summary, the results from Zaff (1989) have suggested that the perceiver's relationship to the environment is important to his or her ability to accurately identify the critical boundary region between reachable and unreachable shelf heights. More specifically, the results have shown that the individual is more accurate at evaluating his or her affordances when he or she is able to position his or her eye where it typically would be during the performance of the intended activity.

In addition, the results of this series of studies have shown that an individual is typically more accurate at judging his or her own affordances than he or she is at judging the affordances for another individual. With experience the judges improved in their ability to evaluate the affordances for another individual; however, they still remained better at judging their own affordances.

The judge's ability to evaluate the affordances for another individual was shown to be influenced by differences in body scale. Differences in body scale were also found to be related to a systematic pattern of under- and overestimation errors.

The results of these studies are likely to have profound implications for designers. Although the specific type of errors encountered (i.e., over versus underestimation of reachable shelf heights) are likely to be task relative, the simple facts that a) the individual is better able to judge his or her own affordances than the affordances for another person, and b) the individual's relationship to his/her environment influences the ability to detect one's own affordances, are likely to have profound implications for designers. From the perspective of designers, the problem may be that people are generally so proficient at judging their own affordances that they may take such capabilities for granted, whereas at the same time fail to recognize the inherent difficulties associated with judging another's affordances. A significant number of design errors may be directly attributable to a failure on the part of the designer to accurately assess the affordances for another person.

The discussion now turns to the concept of affordances and how it has been and can be applied to design and the implications that the concept has for the design process.

9.10   Application of the Concept of Affordances for Design

The concept of affordances characterizes the environment in organism-relevant terms, and affordances are, in essence, synonymous with what the environment means for the individual in terms of his or her capacity for action. Insofar as the distribution of energy in the medium which surrounds the individual is structured by an affordance, sensitivity to that structure will permit the direct perception of that affordance.

In applying the affordance concept to the design procedure, the designer is compelled to take into account what a user can do to, in or with a product, as well as what harm the product can cause the user, or what harm the user can cause the product. It is likely to be intuitively obvious that a particular object affords certain behaviors and disallows others, but in order for the concept of affordances to prove useful, the gap between theory and practice must be filled. It is one thing to comprehend the concept of affordances and recognize the fact that affordances can be directly perceived, it is but quite another to apply that information to specific design problems.

Application of the affordance concept involves, as Owen (1985) stated, simultaneously taking into account both the structural and dynamic properties of the individual who will be using the product and the structural and dynamic properties of the product. A specific combination of physical properties of an environment taken in reference to a specific set of action capabilities of the individual defines the affordance structure of the environment for that particular individual. Because of this reciprocal relationship between the individual and the environment, the affordance structure under consideration will be as general or as specific as the combination of the action capabilities of the individual and the physical properties of the environment when taken together. This condition of reciprocity makes application of the affordance concept likely to yield design recommendations that pertain to a specific kind of encounter between the individual and the environment, rather than a general set of anthropometric standards that can be applied across a variety of behavioral settings.

There are several good examples of analyses of the affordance structure of the environment and its relationship to design that can be pointed to in the literature. For example, Whyte, in his book *The Social Life of Small Urban Spaces* (1980), conducts an extensive affordance analysis of public plazas in an effort to identify the physical structures that result in an imbalance between the settings that are used and those that are not. Vast differences between the plazas in New York City existed in terms of the frequency with which they were used. But with few exceptions the plazas appeared (at least on the surface) to be fairly comparable in design. The aesthetics, location, shape, overall size, and amount of sunlight were not found to be critical factors influencing the frequency of use. What Whyte (1980) found to be the single most important factor was the availability of sittable space. The most popular plazas tended to have considerably more sitting space; in essence, "people tend to sit most where there are places to sit" (p. 28). The design recommendation that can be gleaned from Whyte's (1980) affordance analysis of urban plazas is unambiguous — the amount and variety of available sitting space is directly related to the ultimate success of the place in terms of the frequency with which it is used.

Newman, in his book *Defensible Space: Crime Prevention Through Urban Design* (1973), obviously took very seriously the idea that human behavior is affected by the built environment and that the affordance structure of the environment needs to be explicitly considered during the design process. He wrote:

> "Architecture is not just a matter of style, image and comfort. Architecture can create an encounter and prevent it. Certain kinds of space and spatial layouts favor the clandestine activities of criminals. An architect, armed with some understanding of the structure of criminal encounter, can simply avoid providing the space which supports it." (p. 12)

On the basis of his affordance analysis of the built environment, Newman was able to identify several design recommendations which are intended to alter the affordance structure of the environment. He suggested that establishing territorial definitions of space serve to extend the occupants' realm of responsibility beyond their immediate dwelling by creating a nested structure of private, semiprivate, semipublic, and public spaces. This feature plus (a) the improvements in "social surveillance" capabilities which allow residents to naturally survey the exterior, and (b) the interior of public places and the avoidance of the opportunity for concealment will reduce the vulnerability of a city's inhabitants.

On a somewhat smaller scale, Owen (1985) offered an affordance analysis of tripping, in which he stated that it is not sufficient to simply conclude that a person falls because they have tripped, it is necessary to know what they have tripped over. Most things have multiple affordances, and some, in addition to their intended benefits, also have injurious affordances. The floor-mounted doorstop, to use Owen's example, prevents a doorknob from striking a wall. But because it is approximately 1.25-in. high and mounted anywhere from 2.5 to 12 in. from the wall, encounters with these artifacts have potentially injurious consequences for the pedestrian. Frequently, the injurious situations are not obvious, and as Owen (1985) suggested, an analysis of the affordance structure can be conducted by examining accident survey data.

Zohar (1978) developed an experimental method and data–summary technique in his analysis of objects that afford bumping into, which could be widely applied as a method of analysis to yield specific design guidelines pertaining to the affordances of an environment. By studying the individual's sensitivity to objects that afford being bumped into, Zohar was able to construct an "anthropometric bumping likelihood profile" which described the likelihood that an object would be bumped into as a function of its position relative to the individual. Zohar also discovered that the shape of bumping likelihood profile could be dramatically altered by manipulating the pedestrian's line of sight. When the pedestrian's line of sight was manipulated downward, for example, the frequency of bumping at ankle height decreased by 29% whereas the incidence of neck height bumping increased 14%. By similarly manipulating an individual's line of sight, the frequency with which individual's stumble or fall over objects that afford tripping could be dramatically reduced (Owen, 1985).

9.11   Application of the Concept of Organism-Environment Mutuality: Designing with Affordances in Mind

The notion that the environment provides constraints on action, and that the performance of intentional behavior requires information about the environment taken in reference to the action capabilities of the individual, is a concept that is widely accepted. In other words, the fact that there is a need for the individual to perceive what the environment affords is not a point generally in contention.

It is evident that both natural constraints (laws at an ecological scale) and inferences influence the individual's behavior. The issue, as Turvey et al. (1981) saw it, is over where to draw the line. The ecological approach, they claimed, seeks to extend the application of natural law as far as possible, a strategy that offers the hope of discovering the natural constraints that influence behavior and, it must be added, natural constraints that need to be considered during the design process. On the other hand, the information processing theorists with an emphasis on the mediational processes are seen as attempting to limit the application of natural law, whereby choosing to extend cognition and intelligence as far as possible. In limiting the application of natural laws to the level of conventional physical properties, meaningful relations between the individual and the environment are restricted to the domain of mental operations and representations, so that an element of arbitrariness is introduced.

By rejecting or severely limiting the domain of ecological laws, any awareness on the part of the individual of the reciprocal relationship between his or her capabilities and the environment become confined to the domain of representations. In other words, from this perspective the individual's relation to his or her environment is ultimately reducible to a correspondence between the manipulation of formal tokens and the attribution of a meaning to those tokens as representing properties in the world of some kind (see Winograd & Flores, 1986, for a more detailed discussion of this issue). Within this domain, there is inevitably an element of arbitrariness, for although the manner in which the tokens are manipulated may follow rules of formal logic, as Winograd and Flores (1986) pointed out, the representations are blind in the sense that arbitrary relations are constructed between the token and the property being represented. Thus, the processing of representations places the burden on the individual to bring meaning to an inherently meaning-free situation.

This element of arbitrariness, in the relationship between the individual and the behaviorally relevant aspects of the environment, is frequently viewed as one of the advantages of a mediational account of perception and action. It is employed as an explanation of nonveridical perception and changes in behavior which involve detecting properties of the external world that had gone previously undetected. Although it is possible to explain these occurrences without recourse to mediating representations, illusions are often accounted for by assuming that (a) the "mental model" formed is a never-quite-faithful (i.e., schematic, heuristic) representation of reality, and (b) changes in the individual's ability to respond to environmental properties involves a restructuring of the various mental structures.

What is likely to be of particular importance to designers is how seriously (a) to treat the principle of organism-environment mutuality and the existence of natural constraints that are derived from this principle, and (b) to what degree an element of arbitrariness can be properly described as being inherent to the relationship between the individual and his or her environment.

The degree to which this element of arbitrariness in the description of the individual's relationship to his or her environment is accepted by the designer as a guiding principle will influence the form of the design solution. The implications of this perspective for the designer are likely to be far-reaching and include, for instance, the assumption that any system of constraints would be as effective in facilitating the individual's ability to comprehend the workings of a particular product as any other system of constraints given similar levels in complexity of the representations. There would, for example, be no general way of determining why certain action-consequence mappings work better than other action-consequence mappings because each would presumably involve the formulation of equally arbitrary representations.

Passini (1987) pointed out that because the designers of large public buildings, such as airports, hospitals, and state office buildings, are likely to experience the buildings as wholes in their plans, rather than sequentially, and at a scale that is inconsistent with the ecological scale at which the users experience the buildings, significant way-finding problems are common. The typical solution to the way finding problem involves a reliance on signage, a solution that is consistent with the belief that arbitrary mappings ought to, in principle, function as well as the so called natural mappings. In a separate study, Seibel (1983) suggested that using signage as a solution to way finding problems is generally ineffective. He went on to suggest that a vastly superior solution involves the use of natural constraints, specifically, a visible destination. When it comes to way finding, no sign system can substitute for being able to directly see where one needs to go.

The idea that the individual both shapes and is shaped by the environment is certainly not a new one. Darwin (1859/1975) introduced the concept of organism-environment mutuality in reference to his theory of natural selection, and Marx (1844/1959) independently made explicit use of this reciprocal relationship in his formulation of a theory of human nature. Despite the force of the argument made on behalf of the concept of organism-environment mutuality, psychologists and designers alike, with a few notable exceptions, have paid little attention to the implication that such a position has for their respective disciplines.

For many individuals, the designed environment is no more conspicuous than the surrounding air—only attracting attention when it causes particular harm, discomfort, pleasure, or awe (Zimring, 1982). Thus, the claim that architecture is simply a matter of style, image, and comfort, although apparently mistaken, is nonetheless understandable. Despite a general lack of awareness and frequent claims to the contrary, there is growing evidence to suggest that the designed environment does indeed influence behavior, both indirectly as it shapes our preferences, and directly as it serves to constrain our behavior.

A number of studies have pointed to the fact that the design of the built environment influences the amount and type of social interaction. Hall, in his influential book, *The Hidden Dimension* (1966), noted that the distance between individuals both influences the type and frequency of the interactions that occur and is itself influenced by the intended interaction. Sommer (1969) conducted a series of studies that demonstrated that something as simple as rearranging the chairs in a room could influence the frequency with which casual conversations occur between strangers. Most people believe that hotel lobbies, railway stations, and airport terminals are full of people sitting and talking, but this Sommer (1969) suggested is a common misconception. In actuality, as his observations have lead him to conclude, most people sit alone reading or casually watching others. With very few exceptions, the only people talking were people that arrived together; whereas those who came alone invariably sat alone and did not interact with others sitting near them. A significant contributing factor to the lack of social interactions, which stands apart from the fact that people may be shy or that cultural norms may frown on such contact, is as Sommer pointed out the fact that the seating arrangements employed in most lobbies and waiting rooms are intended to insure an ease and economy of maintenance and are simply not designed to encourage social interactions. The built environment typically does not afford social interactions. Modular seating which is anchored to the floor, whether lined up in rows in the center of the room or against one wall, place people with their backs to each other and at angles that are awkward for conversation.

Sommer (1969) observed that when the arrangement of chairs in the dayroom of a state psychiatric hospital was changed from a pattern closely resembling a typical waiting room, with long straight rows of chairs shoulder to shoulder against the wall, to an arrangement in which the chairs were moved away form the wall and placed around the tables in various parts of the room, a remarkable increase in the amount of both transitory and sustained social interactions was observed. There was also a substantial increase in the amount of reading engaged in by the patients and a reduction in the tendency for patients to hoard magazines in their rooms, apparently due to the fact that there was now a place for the magazines to be stored which did not result in their swift removal by maintenance personnel (Sommer, 1969).

The study by Sommer (1969) serves to demonstrate the profound influence that the environment has on behavior, insofar as it illustrates the fact that initial advent of changes in the layout of the environment can affect behavior in ways that produce secondary changes in the environment which then also serve to affect behavior. In addition to illustrating the influence that the designed environment has on behavior and the reciprocal relationship that exists between behavior and the environment, Sommer's research paradigm also served to show the utility of field research for identifying the environmental influences on behavior.

In another experimental situation designed to examine the influence of seating arrangements on behavior, students were asked to choose from among a number of available seating arrangements in order to engage in a variety of different tasks. The students overwhelmingly selected a corner-to-corner or face-to-face arrangement for casual conversations, side-by-side arrangements for cooperative activities, and a catty-corner arrangements for students engaged in separate activities (Russo, 1967; cited in Sommer, 1969). In an attempt to verify this study, Norum, Russo, and Sommer (1967; cited in Sommer, 1969) observed cooperative children using a side-by-side arrangement, competing pairs of children occupying the corner arrangements, and children working on separate tasks almost invariably using a catty-corner arrangement.

It is worth noting, especially with respect to the influence that the particular arrangement of chairs has on social interactions, that the existence of the chair itself and the influence that it has apart from its location in the environment has yet to be seriously considered. Nadin (personal communication, November 1986) suggested that the chair has had a profound impact on the individuals that occupy them. The chair, in addition to affording a given individual a place to sit, also enables that individual to conduct a myriad of activities over extended lengths of time which could not be carried out as easily and as comfortably were it not for the existence of the chair. Nadin suggests that perhaps there is more than a spurious correlation between the existence of chairs in a culture and the development of a written language.

A considerable amount of the research over the last two decades on the topic of environmental influence on behavior has focused on the designed environment as a source of stress. If we return for a moment to Gibson's (1979) concept of affordances, and the related idea that people have altered the shape of their surroundings so as to change what the environment affords, it should become apparent why such an emphasis has been placed on the designed environment as a source of stress. The power of the environment to influence behavior is perhaps at no time more apparent than when it fails to fit the activities, needs, and preferences of the occupants.

Although Norman (1988) apparently rejected the idea that direct perception is possible, he went a long way toward accepting the gist of the ecological position. In recognizing that natural constraints restrict the possibilities for action, and that sensitivity to this naturally occurring structure constitutes information for the guidance of action, which is as Norman put it, "in the world," he demonstrated an appreciation of the principle of animal-environment mutuality (Gibson, 1979).

The stated goal of Norman's book, *The Psychology of Everyday Things* (1988) was to explain how it is possible to design understandable and easy-to-use things. He offered a number of design principles that were intended to facilitate the achievement of this goal. Several of these design principles reflected an appreciation of the principle of organism-environment mutuality and the existence of ecologically scaled laws which were derived from this principle. For example, Norman suggested that designers should attempt to keep "knowledge in the world." By this he meant that designers should design items so that a potential user could literally see what he or she must do to operate the device. Keeping knowledge in the world, Norman also noted, involves making use of a system of natural constraints or "natural mappings."

The concept of natural mappings and the similar notion of stimulus-response compatibility imply that laws of an ecological scale operate in the built environment as well as the so-called natural environment. The point that Gibson (1979) made in his discussion of affordances in the built environment is that there is not an artificial environment, operating in accordance with a set of arbitrary laws, and a natural environment, operating by a different set of natural laws. Rather, there is just one environment, which although modified by human design, still constrains the behavior of the individual by a system of laws that operate on an ecological scale (p. 130).

The very existence of anything that could constitute natural mappings between a set of actions and the consequences of those actions necessitates the existence of a system of ecological laws which, according to Turvey et al. (1981), epistemologically bind an individual to his or her environment. The existence of nomological or lawful relations between the individual and his or her environment allows for the possibility that the individual can directly detect the behaviorally significant properties of the environment, and the research conducted for this thesis has demonstrated that the direct perception of one's affordances is more than a mere possibility. There is ample evidence to warrant the conclusion that affordances are directly perceived when the perceiver acquires information available in the light about the relationship between his or her own action-relevant properties and the environmental properties that are necessary to support the intended action.

9.12 References

Carello, C., Grosofsky, A., Reichel, F. D., Solomon, H. Y., & Turvey, M. T. (1989). Visually perceiving what it reachable. *Ecological Psychology, 1*, 27–54.

Castelfranchi, C., & Miceli, M. (1987, August). *Perception of affordances versus evaluation.* Paper presented at the Fourth International Conference of Event Perception and Action, Trieste, Italy.

Darwin, C. (1975). *The origin of species.* New York: W. W. Norton. (Original work published 1859).

Fodor, J. A., & Pylyshyn, Z. W. (1981). How direct is visual perception? Some reflections on Gibson's "ecological approach." *Cognition, 9*, 139–196.

Foley, J. M. (1978). Primary distance perception. In H. Leibowitz and H. L. Teuber (Eds.), *Handbook of sensory physiology: Vol. 8. Perception.* (pp. 181–213). Berlin: Springer-Verlag.

Foley, J. M. (1980). Binocular depth perception. *Psychological Review, 87*, 411–434.

Gibson, E. J. (1969). *Principles of perceptual learning and development.* New York: Appleton-Century-Crofts.

Gibson, E. J., & Spelke, E. S. (1983). Development of perception. In P. Mussen (Ed.), *Handbook of child psychology.* (Vol. 3, pp. 201–310). New York: Wiley.

Gibson, J. J. (1950). *The perception of the visual world.* Boston: Houghton-Mifflin.

Gibson, J. J. (1958). Visually controlled locomotion and visual orientation in animals. *British Journal of Psychology, 49*, 182–194.

Gibson, J. J. (1966). *The senses considered as perceptual systems.* Boston: Houghton-Mifflin.

Gibson, J. J. (1977). The theory of affordances. In R. E. Shaw and J. Bransford (Eds.), *Perceiving, acting and knowing.* (pp. 67-82). Hillsdale, NJ: Lawrence Erlbaum Associates.

Gibson, J. J. (1979). *The ecological approach to visual perception.* Boston: Houghton-Mifflin.

Gibson, J. J., Olum, P., & Rosenblatt, F. (1955). Parallax and perspective during aircraft landings. *American Journal of Psychology, 68*, 372–385.

Hall, E. T. (1966). *The hidden dimension.* Garden City, NY: Doubleday.

Hallford, E. W. (1984). *Sizing up the world: The body as referent in a size-judgment task.* Unpublished Doctoral Dissertation, The Ohio State University, Columbus, OH.

Hochberg, J. (1974). Organization and the Gestalt tradition. In E. C. Carterette and M. P. Friedman (Eds.), *Handbook of perception* (Vol. 1, pp. 179–210). New York: Academic Press.

Lee, D. N. (1976). A theory of visual control of braking on information about time-to-collision. *Perception, 5,* 437–459.

Lee, D. N. (1980). The optic flow field: The foundation of vision. *Philosophical Transactions of the Royal Society of London, 290,* 169–179.

Mace, W. M. (1977). James J. Gibson's strategy for perceiving: Ask not what's inside your head but what your head's inside of. In R. E. Shaw and J. Bransford (Eds.), *Perceiving, acting and knowing.* (pp. 43-65). Hillsdale, NJ: Lawrence Erlbaum Associates.

Mark, L. S. (1987). Eyeheight-scaled information about affordances: A study of sitting and stair climbing. *Journal of Experimental Psychology: Human Perception and Performance., 13,* 361–370.

Mark, L. S., & Vogele, D. (1987). A biodynamic basis for perceived categories of action: A study of sitting and stair climbing. *Journal of Motor Behavior, 19,* 367-384.

Marx, K. (1959). *Economic and Philosophical Manuscripts of 1844.* (M. Milligan, Trans.) Moscow. (Original work published 1844).

Newell, K. M. (1974). Decision processes of baseball batters. *Human Factors, 16,* 520–527.

Newman, O. (1973). *Defensible space: Crime prevention through urban design.* New York: MacMillan.

Norman, D. A. (1988). *The psychology of everyday things.* New York: Basic Books.

Norum, G., Russo, N. & Sommer, R. (1967). Seating patterns and group task. *Psychology in the Schools, 4,* 276–280.

Owen, D. H. (1985). Maintaining posture and avoiding tripping: Optical information for detecting and controlling orientation and locomotion. *Clinics in Geriatric Medicine, 1,* 581–599.

Owen, D. H., Warren, R. , Jensen, R. S., Mangold, S. J., & Hettinger, L. J. (1981). Optical information for detecting loss in one's own forward speed. *Acta Psychogica, 48,* 203–213.

Passini, R. (1987, May). *Architectural and graphic communication in wayfinding.* Paper presented at the Eighteenth International Conference of the Environmental Design Research Association, Ottawa, Canada.

Rock, I. (1977). In defense of unconscious inference. In W. Epstein (Ed.), *Stability and constancy in visual perception.* (pp. 321–373). New York: Wiley.

Rock, I. (1983). *The logic of perception.* Cambridge, MA: MIT Press.

Russo, N. (1967). Connotation of seating arrangements. *Cornell Journal of Social Relations, 2,* 37–44.

Seibel, A. (1983). Way-finding in public spaces: The Dallas/Fort Worth, USA Airport. In D. Amedeo, J. B. Griffin, and J. J. Potter (Eds.), *Proceedings of the Fourteenth International Conference of the Environmental Design Research Association.* (pp. 129–138). Lincoln: University of Nebraska.

Sedgwick, H. A. (1973). The visible horizon: A potential source of visual information for the perception of size and distance. *Dissertation Abstracts International, 34,* 1301B-1302B. (University Microfilms No. 73-22530).

Sedgwick, H. A. (1980). The geometry of spatial layout in pictorial representation. In M. A. Hagen (Ed.), *The perception of pictures.* (Vol. 1, pp. 33–90). New York: Academic Press.

Sedgwick, H. A. (1983). Environment-centered representation of spatial layout: Available information from texture and perspective. In E. C. Carterette & M. P. Friedman (Eds.), *Handbook of perception* (Vol. 10, pp. 425–458). New York: Academic Press.

Sommer, R. (1969). *Personal space.* Englewood Cliffs, NJ: Prentice-Hall.

Turvey, M. T., & Carello, C. (1981). Cognition: The view from ecological realism. *Cognition, 10,* 313–321.

Turvey, M. T., & Kugler, P. N. (1984). An ecological approach to perception and action. In H. T. A. Whiting (Ed.), *Human motor actions: Bernstein reassessed.* (pp. 373–412). Amsterdam: North Holland.

Turvey, M. T., & Shaw, R. E. (1979). The primacy of perceiving: An ecological reformulation of perception for understanding memory. In L. G. Nilsson (Ed.), *Studies of memory: In honor of Uppsala University's 500th anniversary.* (pp. 167–222). Hillsdale, NJ: Lawrence Erlbaum Associates.

Turvey, M. T., Shaw, R. E., Reed, E. S., & Mace W. M. (1981). Ecological laws of perceiving and acting: In reply to Fodor and Pylyshyn (1981). *Cognition, 9,* 237–304.

Ullman, S. (1980). Against direct perception. *Behavioral and Brain Sciences, 3,* 373–415.

Warren, R. (1982). *Optical transformation during movement: Review of the optical concomitants of egomotion.* (Final Technical Report for Grant No. AFOSR-81-0108). Columbus, OH: The Ohio State University, Department of Psychology, Aviation Psychology Laboratory.

Warren, W. H. (1983). A biodynamic basis for perception and action in bipedal climbing. *Dissertation Abstracts International, 43,* 4183-B. (University Microfilm No. 83-09263).

Warren, W. H. (1984). Perceiving affordances: Visual guidance of stair climbing. *Journal of Experimental Psychology: Human Perception and Performance, 10,* 683–703.

Warren, W. H., & Whang, S. (1987). Visual guidance of walking through apertures: Body-scaled information for affordance. *Journal of Experimental Psychology: Human Perception and Performance, 13,* 371–383.

Whyte, W. H. (1980). *The social life of small urban spaces.* Washington, D. C.: The Conservation Foundation.

Winograd, T., & Flores, C. F. (1986). *Understanding computers and cognition: A new foundation for design.* Norwood, NJ: Ablex.

Wolpert, L. (1988). *Altitude control over different planar environments.* Paper presented at the 13th Annual Society Meeting of the International Society for Ecological Psychology, Yellow Springs, OH.

Zaff, B. S. (1989). *Perceiving affordances for oneself and others: Studies in reaching and grasping.* Unpublished Doctoral Dissertation, The Ohio State University, Columbus, OH.

Zaff, B. S., Owen, D. H. (1987). Perceiving and controlling changes in the speed of self motion. In D. H. Owen (Ed.), *Optical and event-duration variables affecting the perception and control of self motion* (Final Technical Report for AFHRL Contract No. F33615-83-K-0038). Columbus, OH: The Ohio State University, Department of Psychology, Aviation Psychology Laboratory.

Zimring, C. (1982). The built environment as a source of stress. In G. W. Evans (Ed.), *Environmental stress.* New York: Cambridge University Press.

Zohar, D. (1978). Why do we bump into things while walking? *Human Factors, 20,* 671–679.

Chapter 10

Use of a Means-End Abstraction Hierarchy to Conceptualize the Ergonomic Design of Workplaces

Marvin J. Dainoff and Leonard S. Mark
Miami University

10.0   The Problem: Ergonomic Design of Electronic Workplaces

There is now widespread international awareness of the potential problems associated with prolonged work in static postures. The United States Bureau of Labor Statistics recently estimated that so-called cumulative trauma disorders (CTD), now constitute almost half of all occupational illnesses. The term *cumulative trauma* refers to the presumed nature of these disorders, a kind of overuse of the musculoskeletal system resulting from prolonged work in awkward postures. Other terms used in the popular press are *repetitive strain injury* or *repetitive motion injury*. In the past, the term, *CTD*, has been restricted to disorders of the upper extremities (neck, shoulder, arm, wrist, hands); recently, there has been a tendency to include certain lower back problems.

The large increase in CTD rates refers, of course, to the working population as a whole. It is generally supposed that the particular workers who are most at risk for CTD are those in industrial settings in which lifting and vibration is a regular part of their work. However, a substantial fraction of these disorders are associated with static postures, in which workers are sitting for prolonged periods at video display terminals (VDTs).

A number of studies have linked VDT work with musculoskeletal disorders (Dainoff & Dainoff, 1986). Hettinger (1985) examined the incidence of musculoskeletal disorders across 12 occupational groups, including two groups of office workers. Not surprisingly, workers in the construction trades, which entail traditional risk factors of heavy lifting and vibration, showed the highest incidence of musculoskeletal disease. However, the third highest incidence was found for the group of office workers whose jobs required them to maintain fixed work postures throughout the workday. In contrast, office workers whose jobs permitted variations in posture had the lowest incidence of musculoskeletal disorders of all occupational groups. These results led Hettinger to conclude that fixed (static) work postures should be added to lifting and vibration as risk factors for musculoskeletal disorders.

Although CTD-related problems are likely to have multiple determinants, proper ergonomic design is almost always seen as an important, if not essential, preventative measure (Feuerstein, 1991). The assumption is that levels of static load will be reduced if there is a better postural fit between the worker and the workstation. This entails providing a degree of adjustability in chair configuration and work surface height. Thus, normal variability in human body dimensions and proportions is matched by adjustability in furniture, thereby allowing the operator to attain work postures which, from a biomechanical perspective, yield reduced static loads.

Consequently, the specification of ergonomic attributes of workplace equipment, VDTs, chairs, and work desks is now being reflected in a variety of standards and regulations at both national and international levels (Dainoff, 1990). In the United States, the Human Factors Society (HFS) has published a standard for VDT workstations (ANSI/HFS 100-1988). Within this standard, specific dimensions and ranges of adjustability are provided for work surfaces and chairs. For example, seat pan height is to be adjustable between 40.6 to 52 cms; if the seat back is adjustable, it must include a range from 90 to 105 degrees.

Although the HFS standard does have some anthropometric rationale for the particular dimensions specified, there is little discussion of functional purpose, that is, job or task, within the workplace. That is, whereas the framework of the standard accommodates various postural orientations, including forward and backward leaning postures as well as upright seated posture, there is no discussion of the context within which such alternative postures might be used. In our judgment, failure to specify the function of work and the worker's goal or intention is a serious problem with most ergonomic guidelines, not only the HFS standard.

The problem manifests itself in the implementation of the ergonomic standard. All too often we have observed that busy decision makers (e.g., facility managers, purchasing officers) believe that merely purchasing workplace equipment which matches published physical specifications is all that is required for successful implementation of an ergonomic intervention. This approach is likely to be counterproductive. To achieve the ergonomic goal of fitting the environment to the person, the focus must be on the VDT operator as an active, goal-directed (intentional) explorer of his or her environment, using the tools at his or her disposal to search for an efficient working posture. Efficiency, in this context, can be defined in terms of minimizing the energy expenditure associated with the muscular effort required to achieve some postural goal or work within some postural constraints (cf. the minimum principle of Nubar & Contini, 1961; see also Zacharkow, 1988).

Successful implementation of an ergonomic intervention requires the solution to (at least) two problems: The physical compatibility problem entails fitting the work environment to the user's body scale and biomechanical capabilities. Traditionally, this problem has been addressed using anthropometric and biomechanical methods. The psychological compatibility problem is to ensure that the operator is both properly informed about the purpose and capabilities of the ergonomic workstation as well as motivated to use those capabilities. If both the and psychological compatibility problems are not addressed, the operator is likely to assume poor working postures, despite the availability of expensive equipment. In short, the ergonomic intervention will not be successful.

In the case of seated work, the traditional approach to the physical compatibility problem is to provide sufficient ranges of adjustability and flexibility through chair and workstation controls so that the operator can assume efficient work postures and orientations. However, this solution to the physical compatibility problem creates a significant psychological problem for the user, namely, how to manage (control) multiple degrees of freedom. Each independent degree of freedom added to a system has a (cognitive) cost associated with the processing required for its control. At some point, the cognitive overload will be such that the user may simply stop adjusting the furniture.

The decision maker, who is responsible for implementing an ergonomic program, faces a similar problem in managing the degrees of freedom. Different vendors offer a variety of different chairs, workstations, and other accessories, all with various adjustability features. In a sense, the ergonomic problem space of the operator becomes a subset of the decision maker's problem space.

The control and coordination of systems with multiple degrees of freedom has come to be known as the *Bernstein problem*, after the Russian physiologist who characterized it (Bernstein, 1967; see Turvey, 1990, for a recent discussion). The general solution to this problem is to specify sets of constraints that will reduce the number of degrees of freedom, thereby decreasing the dimensionality of the problem space. To take a simple example, imagine an unpowered wheeled vehicle (a child's wagon) in which each of the four wheels must be steered independently (cf., Mark, Dainoff, Moritz, & Vogele, 1991). The poor child trying to ride the wagon down a hill would most likely be quickly overwhelmed by the complexity of this 4-degree-of-freedom system. We can help the child control this system by providing a pair of constraints in the form of front and rear axles. Now each pair of wheels must move in the same direction. Finally, the wagon can be steered, albeit crudely, by a third constraint in the form of a mechanism that allows the front axle to be rotated. Thus, four degrees of freedom are reduced to one. In the remainder of this chapter, we will examine how a means-end abstraction hierarchy (Rasmussen, 1988; Rasmussen & Pejtersen, this volume; Vicente, 1990, this volume; Vicente & Rasmussen, 1990) can be used to identify constraints for managing the multiple degrees of freedom inherent to the problem space of the ergonomic decision maker.

10.1   Means-End Abstraction Hierarchy

Rasmussen (1988) presented the means-end abstraction hierarchy as a conceptual tool for representing functional properties of complex real–world environments within which human information processing, control, and decision making must occur. Vicente and Rasmussen (1990) established direct links between the means-end abstraction hierarchy and the ecological approach to perceiving and acting through the concept of an affordance.

For Gibson (1979), an affordance represented, in physical terms, the potential for action which the environment will support for a given organism (actor). For example, a child's chair (describable in terms of specific ecologically relevant physical dimensions, such as seat pan height and width) affords sitting on for only a certain class of actors: those individuals—young children—who possess a specific range of physical attributes (e.g., knee height, buttock width). The same environmental object will not afford sitting (or at least the same sitting act) for most adults.

Affordance theory can, in principle, be useful in conceptualizing the ergonomic compatibility problem described earlier (e.g., Mark & Dainoff, 1988; Mark, Balliett, Craver, Douglas, & Fox, 1990) However, although most current ecologically-based experimental and analytic work tends to focus on individual affordances in isolation (e.g., Mark, 1987; Mark & Vogele, 1987; Warren, 1984; Warren & Whang, 1987), real-world work environments necessarily involve complex collections of affordances, some of which may be nested within others.

To meet the challenge of characterizing such environments, Vicente and Rasmussen (1990) argued that groups of affordances can be hierarchically organized by function. At any point in the hierarchy, the affordances at one level act as the *ends* (goals) with respect to the affordances at the next level down, the *means*. For example, the small chair (a means) discussed earlier affords sitting (an end) for a young child, say in nursery school. Sitting (a means), in turn, affords a posture in which the child's boots can be put on easily (an end). Furthermore, boots-on (now a means) affords playing in the snow (an end), and so on. This simple analysis can be placed in an operational context: As part of an overall functional analysis of a nursery school operation, it might alert the school operator to the need for having a sufficient number of chairs available in order not to delay the departure of children during bad weather.

The value of a means-end hierarchy as a information tool for decision making lies in its representation of actions throughout a complex system with respect to both their immediate goals and their contribution to the overall functioning of the system. Decision makers need this information in order to design the physical, psychological and organizational structure of a working environment that affords[1] the intended actions.

---

[1]The challenge in describing a specific affordance is to produce a description of the environmental properties that reveals the lawful constraints on the action that are invariant across actors of different body scales. For instance, the maximum height of a surface that affords climbing on bipedally is a function of the actor's leg length; not surprisingly, people with longer legs can climb on higher surfaces than people with shorter legs. A description of the affordance for bipedal climbing has to capture what is invariant across people with different leg lengths, thereby establishing a lawful constraint on this action capability. By rescaling environmental properties in terms of the actor's leg length, W. Warren (1984) was able to show that this critical boundary for the affordance of bipedal climbing was a constant proportion (0.88) of each actor's leg length.

10.2   Ergonomic Decision Making: An Illustrative Example

In our own ergonomic interventions we have found the conceptual frameworks provided by Rasmussen (1988) and Vicente and Rasmussen (1990) to be a useful tool for guiding decision-making processes. To illustrate the practical value of a means-end hierarchy, we examine the case of a hypothetical manufacturing company. The focus will be on those aspects of the organizational environment that are of concern to the decision maker, say a facility manager, responsible for implementation of ergonomic programs.

At each level of the hierarchy, there will be particular components that fall within the operational span of attention for decision making for the facility manager. Some of these components will be nested within others. A definition of these relationships is described by Rasmussen (1988) as a *part-whole decomposition*.[2] Thus, the overall problem space for the decision maker is two-dimensional, with one dimension referring to levels of the means-end hierarchy, and a second dimension comprising levels of part-whole decomposition. In most cases these two dimensions will be tightly coupled; components at higher levels of the hierarchy will be relatively more global, components at lower levels will be local. Following Rasmussen (1988), we construct a means-end hierarchy consisting of five major descriptive levels. Sublevels exist within each of the major levels.

10.2.1 Level 1: Goals, Values, and Constraints

The means-end relationships at level 1 constitute a statement of organizational philosophy and corporate values. The overall organizational goal (end) of this company is *productivity*. Two broad categories of means are used to achieve this goal: (a) operational

---

[2]Part-whole decomposition recognizes that any event, behavior, object, etc. can be viewed at different scales of analysis, ranging from the molecular level to more molar scales. There is no single right or wrong scale; rather, the appropriate scale of analysis is a consequence of the aim of the person who is performing the analysis. One of James Gibson's key points was to appreciate the significance of the chosen scale of analysis for examining perceiving and acting. His insistence on defining the "ecological" scale of analysis and describing the environment to be perceived at that ecological scale constitutes a cornerstone of his approach to perceiving and acting.

effectiveness, and (b) taking care of people. These means, in turn, serve as ends for the next sublevel of the hierarchy. *Operational effectiveness* is specified by both efficiency and the production of a valued service (or product). Likewise, people are taken care of to the extent that their health, safety, and job satisfaction are provided for.

These brief goal statements are meant to represent a distinct management philosophy held by top management of our hypothetical company. This philosophy is now generally referred to as *total quality management* (cf. Deming, 1986; Peters, 1987; Senge, 1990). In a quality-oriented organization, efficiency entails a focus on reduction in waste—either through unnecessary work procedures or in excess inventory. In Japan, this approach is called *just in time* (JIT). Likewise, creation of a service or product, which is perceived as valuable by customers, requires continual focus on improvement and service.

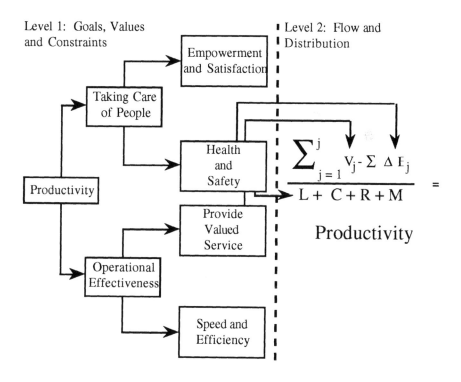

Figure 10.1. *Relationship between level 1 (goals, values, and constraints) and level 2 (flow and distribution) of the means-end hierarchy.*

Attainment of these goals is unlikely without an educated and motivated work force that is empowered to make decisions in service of higher level ends. Thus, level 1 goal statements are taken as serious principles used to inform daily decision making at all levels of the organization. The system's functions propagate down the hierarchy, thereby acting as global constraints at subsequent levels. We contend that the presence of a TQM orientation will have a great deal to do with effectiveness of implementation of an ergonomic program.

10.2.2 Level 2: Flow and Distribution

Rasmussen (1988), in his specific description of the problem space of an office environment, characterizes level 2 in terms of quantities of flow, distribution, and accumulation of material, energy, monetary values, and manpower. At this point, quantitative indices, of the type familiar to cost accountants, may be used to evaluate the extent to which level 1 ends are being achieved. Muckler (1982) provided an excellent discussion of how productivity indices can be used to evaluate human performance; for illustration, we employ here his description of the Hanes-Kriebel index:

Where:
V_j = total value to user of each j product or service
ΔE_j = cost penalty for each factor causing inefficiencies
C = capital cost
L = labor cost
R = materials cost
M = cost of miscellaneous supplies and services.

The usefulness of the Hanes–Kriebel index to the present discussion lies in its ability (assuming accurate measures of V and E) to reflect both of the productivity subcomponents described earlier. In particular, the taking care of people end can be expressed in terms of the E variable, in the sense that high-cost penalties may arise through inefficiencies caused by workers: who are absent from work for medical or other reasons; who quit, thereby requiring that replacements be trained; or who are less willing or able to work at peak efficiency because of occupationally related illnesses or poor morale. At the same time, explicit costs for these "overhead' factors of health, safety, and job satisfaction appear in the denominator as factor *M* (or possibly *L*), in which they reflect medical costs, sick leave, and expenses in recruiting and training new workers, and/or hiring substitutes. One consequence is that overhead costs, like occupational illnesses, have a dual reduction on productivity, subtractively in the numerator to reflect increased inefficiency of operation, and additively in the denominator reflecting a direct outflow of money for medical expenses.

These overhead costs are not trivial. Cyert and Mowery (1989) argue that one of the organizational barriers to technological innovation in the United States lies in the paradox that initial financial benefits from such efforts are typically found in reduced overhead, but conventional accounting procedures of American corporations are not set up to track overhead. These general conclusions can be explicitly related to the ergonomic arena. Aarås and his colleagues (Spilling, Eitrheim, & Aarås, 1986) examined the case of a Norwegian telephone cable assembly factory with very high turnover rates and medical costs. By introducing adjustable chairs and workstations, Åaras and his colleagues achieved dramatic and long term decreases in both turnover and medical costs. An economic analysis indicated that over an 11 year period the savings to the company was over 850% the original investment in ergonomic equipment. As predicted by Cyert and Mowrey (1989), the introduction of new technology (ergonomic design of the workplace) can result in financial benefits through reduction of overhead costs (turnover and absenteeism).

10.2.3 Level 3: Generalized Function

At this point, we narrow our focus of attention to two particular components derived from higher levels. The first component is order entry, which is derived from the operational effectiveness goal of level 1. Most purchase orders for goods manufactured by the company come by mail. Therefore, to meet the combined ends of efficiency and valued service to customers, a computerized order-entry system has been developed in which all purchase orders are entered into the system on the day they are received. Accordingly, a relatively large staff of order–entry operators is required, and they are under considerable time pressure to keep up with the flow of orders. (It is possible, in principle, to describe detailed mappings from level 3 through level 2, up to levels 1, in terms of specific operational and financial benefits of rapid response to customer orders. However, this level of detail is not relevant to the facility manager's domain.)

Within a functional level of description, order entry can be further broken down into smaller components (part-whole decomposition). During the input stage, data on each purchase order are entered into the computerized database. The processing stage requires the operator to review the input record for accuracy and exceptions. This work is aided by an online expert system. Finally, the output stage consists of a final approval by the operator after which the order is electronically shipped to the warehouse.

The second major component to be focused on at level 3 is prevention and control of cumulative trauma disorders (CTD) among data-entry operators. This function can be derived from a level 1 concern with overall health of employees, linked through a level 2 concern with a dramatic overall rise in worker compensation claims for CTD. That is, CTD prevention—as a means—can be directly mapped

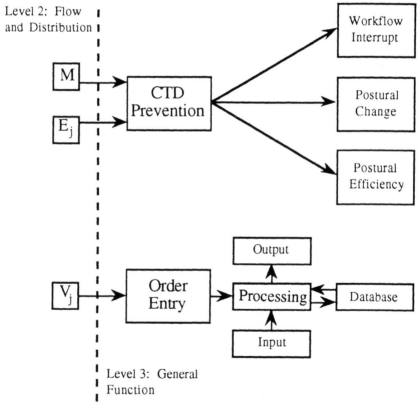

Figure 10.2. *Relationship between level 2 (flow and distribution) and level 3 (general function) of the means-end hierarchy.*

onto factor M in the denominator of the Hanes–Kriebel equation, which includes medical costs. In addition, efficiency is reduced to the extent that skilled data entry operators who suffer from CTDs either resign or go on medical leave and need to be replaced. Thus, CTD prevention can also be mapped onto factor ΔE of the Hanes–Kriebel equation.

Three subcomponents of CTD prevention remain within the general level of functional description at level 3: *postural efficiency, postural change,* and *workflow interruption.* Each represents a functional component of the organizational plan for prevention and control of CTD by reduction of known risk factors. These risk factors, as outlined by Putz-Anderson (1988), include excess force, awkward postures, high repetition rates, and inadequate recovery time. At this point in the

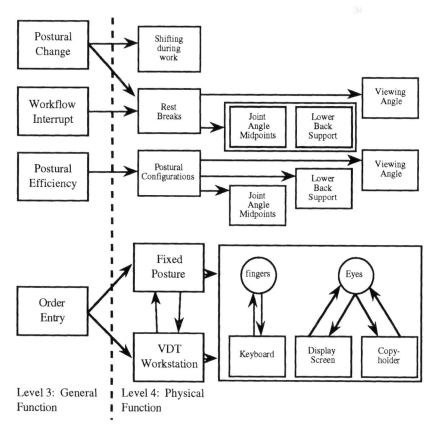

Figure 10.3. *Relationship between level 3 (general function) and level 4 (physical function) of the means-end hierarchy.*

hierarchy, the CTD prevention and control components serve as broad functional ends for specific implementation at lower levels. Postural efficiency specifies a biomechanical requirement that force and posture be constrained so that a given muscular output is attained with a minimum of input energy. Postural change delays the onset of fatigue by unloading compressed blood vessels, thereby increasing blood flow to muscles. Finally, workflow interruption allows fatigue to be dissipated.

10.2.4 Level 4: Physical Function

The generic level 3 functional descriptions for the order entry task are now translated into a more specific set of physical functions. Because the description at this level must incorporate both equipment functionality and operator posture, this portion of the hierarchy necessarily becomes somewhat more complicated. As will be discussed, the equipment configuration or VDT workplace entails a set of specific postural requirements. These can be expressed in the phrase *fixed posture*.

Most order-entry work involves transactions with the VDT. The essential functional elements of the VDT workstation consist of a keyboard, display screen, and copy holder. Given the higher level goal (end) of rapid customer response, the requirement of entering all orders immediately demands intense, high speed data-entry performance. This requirement, in turn, imposes certain postural constraints—there should be tight couplings between fingers and keyboard, eyes and copy holder, and eyes and display screen. Restating these relationships in more ecological terms, these couplings afford high rates of data input by the operator.

A parallel set of physical functions provides the level 4 means for the level 3 ends of CTD prevention and control. These functions represent ergonomic recommendations from the existing literature as applied to this particular work setting (Dainoff & Dainoff, 1986; Putz-Anderson, 1988). At level 3, postural efficiency is achieved through postural configurations, which minimize force exertions and awkward orientations. The list of desired postural configurations, which are the means of achieving this end, includes joint angles of trunk and limbs at or near the mid points of their range of motion, support for the lumbar region of the lower back, and viewing angles of the lines of sight to screen and copy between 0 and 15 degrees below the horizontal.

Recovery from operator fatigue and strain can be achieved by periodically interrupting the flow of work (level 3). In the context of order entry work, the means for accomplishing this end is by rest breaks. We can distinguish two types of rest breaks: regularly scheduled periods of time away from the workstation, and informal frequent pauses. (The latter are sometimes called minibreaks, or microbreaks, and refer to brief self-initiated periods of work interruption without leaving the work station.)

Postural changes, at level 3, can be achieved in two ways: by allowing for shifts in posture during work, and through stretching, standing, or simple exercises during both kinds of rest breaks. (Note that simply taking a rest break does not necessarily entail a change in posture.) Here, a level 3 end is mapped onto two different level 4 functional means.

10.2.5   Level 5: Physical Layout of the Workstation and Posture

Level 5 is literally the bottom line. We have, thus far, moved down the hierarchy from broad goals to increasingly more specific functional specification. In ecological terms, we have been identifying actions that the physical arrangement of objects and surfaces must afford. By implementing the functional specifications developed at higher levels of the hierarchy, we are, in effect, building affordances (Mark & Dainoff, 1988; Warren, this volume). This means that the two parallel, but independent, mappings under Order Entry and CTD prevention must now be combined. We must find physical means which afford both ends simultaneously.

The order-entry operator's task is characterized by repetitive, stereotyped actions performed under time stress. It is necessary, therefore, to provide workplace equipment which will afford efficient working postures as well as the capability for postural changes during work. Adjustable chairs and worksurfaces offer these affordances. Recent work (Winkle & Oxenburgh, 1990) has suggested that intense work of this type benefits from adjustable worksurfaces that afford periodic alternation between sitting and standing postures. Employees can be encouraged (or directed) to alternate postures on a regular basis. Attaining these postures requires that the ranges of adjustability of the workstation match the appropriate anthropometric dimensions of the target operator population.

Efficient seated posture can be achieved when the operator's forearm is parallel to the floor, the upper arm hangs vertically

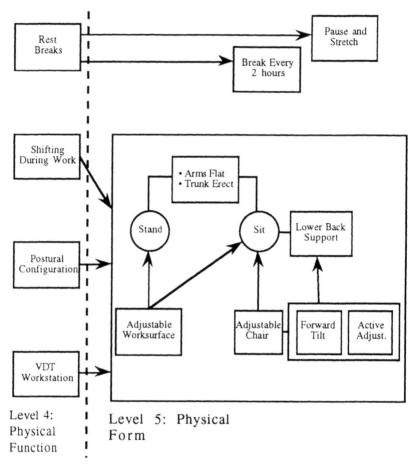

Figure 10.4. *Relationship between level 4 (physical function) and level 5 (physical form) of the means-end hierarchy.*

(unflexed), the wrists are flat, and the line of sight to the center of the display screen is 15–30 degrees below horizontal. Given the tight couplings (fingers–keyboard, and eye–screen) required by the order-entry task, efficient seated posture requires the trunk to be nearly vertical. Other upper limb requirements are similar to those just described for standing posture.

People in upright postures tend to drift into a forward slump, whereby the pelvis rotates backward transforming the optimum (standing) lordotic curve of the lumbar spine into a kyphosis, with attendant pressure on spinal disks, ligaments, and internal organs (Zacharkow, 1988). Lumbar lordosis can be restored by tilting the seat pan of the chair forward, with the backrest moved to vertical; the pelvis now rotates forward, easing pressure on disks, ligaments, and organs (Mandal, 1981). Forward-tilting chairs, however, place the operator in an unfamiliar orientation and seem to require a period of familiarization. In addition, the mechanism that controls the adjustability of both seat pan and backrest should be active in nature, that is, the operator is required to directly manipulate a button/lever in order to move into a desired posture. Once the control is released, the chair surfaces are locked into position, affording a stable platform of support (Mark & Dainoff, 1988).

Finally, the order-entry operator's work is, by necessity, externally paced, with little or no "dead time." Accordingly, interruptions of the workflow to dissipate fatigue have been explicitly designed into the workday. The operators are given formal work breaks every 2 hours and are encouraged to take frequent pauses during which they should stretch and move their bodies (cf. Genaidy, 1991).

10.2.6 Implications

The preceding exercise has tried to illustrate the type of information that a means-end hierarchy can provide decision makers as a tool for workstation design. In this example, the means-end hierarchy allows the facility manager to "see" the linkages by which the interrelations between broad corporate goals (level 1) and specific pieces of equipment to be purchased and work practices that could be changed (level 5) to achieve the broad goals. In doing so, the means-end hierarchy makes explicit the functional interconnections between components of a complex system and reveals the affordances that must be present in order to support the desired work activities.

Our field experience is that this approach has a marked advantage over conventional purchasing and management practices that often have an arbitrary component. Purchasing decisions typically involve lists of specifications which are arrived at in an ad hoc, nonsystematic fashion. On the other hand, workload decisions are mostly driven by some notion of maximum production rates. The means-end hierarchy provides a rational, systematic alternative, firmly rooted in the most basic goals of the organization and directly derivable from them.

Consider, for example, a different kind of job title, that of *personal benefits counselor*. Without working through the hierarchy in detail, it can be seen that the level 3 description of this function would be relatively more complex than that of the order-entry function. That is, the benefits counselor would be required to work with a variety of inputs—employees seeking advice, information and policy statements from management, medical plan information, and so on. These would be manifest in a variety of forms: computer files, phone calls, mail, or visits. The variety of inputs would be paralleled by a similar variety of output and processing requirements.

Accordingly, the physical requirements at levels 4 and 5 entailed by this functional specification would be rather different. The VDT workstation would require a more complex configuration to support the variety of postural requirements entailed in the job. Hence, adjustability needs for the furniture would be different; the focus would be on affording smooth movement among different workplace elements (e.g., telephone to computer to paper manuals). Presumably, the resulting increase in task component variety would produce a corresponding increase in postural variation (i.e., movement). As a result, there would be less need for highly adjustable workstations that afford alternating between sitting and standing postures. Likewise, we can assume that these employees might have a greater degree of personal control over the way in which their own time is allocated; consequently, the formal specification of work pauses becomes less important.

10.2.7 Participative Ergonomics and Total Quality Management

There is a broader issue rooted in management philosophy of the corporate organization. This approach to decision making is unlikely to be implemented unless the highest levels of the organization are willing to empower lower levels of the organization to base daily operating decisions on basic principles that are characteristic of Total Quality Management philosophies. From our experience, ergonomic interventions must be participative in nature and, as such, fall clearly within the scope of modern organizational approaches to Total Quality Management (TQM).

A set of ergonomic intervention techniques might well be based on a solid scientific basis; but without an effective implementation strategy, that intervention is likely to fail (see Pressman & Wildavsky, 1984 for an extended discussion in the realm of public policy). TQM characteristics, such as local empowerment and decentralization, emphasis on communication, the learning organization, and managers as teachers, have clear implications for the crucial details of how an ergonomic intervention should be implemented.

Participative ergonomics (Shipley, 1991) is a natural extension of TQM, in that the prime focus of decision making and responsibility lies with local teams of lower level managers, technical and design support people, and line workers. Ergonomic experts act as advisors and facilitators (coaches). The final goal of the process is to have the local team "own" and manage their ergonomic solutions. In this regard, such team organizations might well be regarded as quality circles.

Although the concept of participative ergonomics has been around for several years, it has not been widely implemented (Shipley, 1991). This is not surprising, because participative approaches are both conceptually and practically more difficult to implement. However, there are now case studies which indicate that the positive benefits are likely to be well worth the extra effort.

Lifshitz, Armstrong, Seagull, and Smith (1991) describe a successful participative ergonomic program in a Ford automotive plant. Primary problem-solving and decision-making responsibilities were vested in an autonomous ergonomic work team. Team members included area manager, industrial process engineer, power tool engineer, upgrader, union representative, hourly operator (at the worksite in question) plant physician, employee involvement coordinator, and outside ergonomic consultant. Over a 3 year period, team efforts resulted in a 50% reduction in CTDs.

A similar study was carried out at Nissan (Nagamachi, 1991). An ergonomic team, similar in membership to that at Ford, redesigned certain assembly operations. The outcome was not only a significant reduction in biomechanical loading on operator limbs, but a drastic reduction in frequency of improper assembly.

10.2.8 TQM and the Means-End Hierarchy

How does this all relate to the use of the means-end hierarchy as a decision-making tool? There are several points to be considered. First, the means-end hierarchy can be conceived as an alternative to traditional task analysis. Whereas traditional task analysis tends to structure problem spaces in terms of single "correct" pathways (Flach, 1990), a means-end hierarchy can be used to clarify real-world complexity by representing alternative solution possibilities. An abstraction hierarchy identifies both the goals for action (i.e., the actor's intention) and the possibilities for action (i.e., the affordances which are provided). Users are encouraged to explore the environment to find new pathways for realizing the goal.

This approach is crucial in an information-rich, rapidly changing industrial environment. Put simply, traditionally managed organizations with top-heavy staffs of technical experts cannot respond quickly enough to immediate, current pressures. Therefore, information processing and decision-making responsibilities should be widely distributed at all levels of the organization (Zuboff, 1988). This requires a TQM approach. For TQM to work, however, tools for effective decision making, such as the means-end hierarchy, must be widely available.

In the particular case of ergonomic decision making used in this chapter, the end result is a combination of flexible equipment and work procedures which will only be effective if the operator is properly informed and motivated. There are many degrees of freedom in the proposed operator workstation. Their effective coordination requires active participation by the operator. Such participation is highly unlikely to be achieved in a traditional management framework.

10.3   References

American National Standards Institute. (1988). *American National Standard for Human Factors Engineering of Visual Display Terminal Workstations* (ANSI/HFS 100-1988). Santa Monica, CA: Human Factors Society.

Bernstein, N. (1967). *The coordination and regulation of movements*. Oxford, England: Pergamon Press.

Cyert, R. M., & Mowery, D. C. (1989). Technology, employment and U.S. competitiveness. *Scientific American, 260,* 54–62.

Dainoff, M. J. (1990). Reducing health complaints in the computerized workplace: the role of ergonomic education. *Journal of Interior Design Education, 16,* 31–38.

Dainoff, M. J., & Dainoff, M. H. (1986). *People and productivity; A manager's guide to ergonomics in the electronic office.* Toronto: Carswell.

Deming, W. E. (1986). *Out of the crisis.* Cambridge, MA: Massachusetts Institute of Technology.

Feuerstein, M. (1991). A multidisciplinary approach to the prevention, evaluation, and management of work disability. *Journal of Occupational Rehabilitation, 1,* 5–12

Flach, J. (1990). The ecology of human-machine systems I: Introduction. *Ecological Psychology, 2*, 191–206.

Genaidy, A. M. (1991). A training program to improve human physical capability for manual handling jobs. *Ergonomics, 34*, 1–11.

Gibson, J. J. (1979). *The ecological approach to visual perception.* Boston: Houghton-Mifflin.

Hettinger, T. (1985). Statistics on diseases in the Federal Republic of Germany with particular reference to diseases of the skeletal system. *Ergonomics, 28*, 17–20.

Lifshitz, Y. R., Armstrong, T. J., Seagull, F. B. J., & Smith, T. D. (1991). The effectiveness of an ergonomics program in controlling work related disorders in an automotive plant - a case study. In Y. Queinnec & F. Daniellou (Eds.), *Designing for everyone.* London: Taylor & Francis.

Mandal, A. C. (1981). The seated man (homo sedens). The seated work position. Theory and practice. *Applied Ergonomics, 12*, 19–26.

Mark, L. S. (1987). Eyeheight-scaled information about affordances: A study of sitting and chair climbing. *Journal of Experimental Psychology: Human Perception and Performance, 13*, 361–370.

Mark, L. S., Balliett, J. A., Craver, K. D., Douglas, S. D., & Fox, T. (1990). What an actor must do in order to perceive the affordance for sitting. *Ecological Psychology, 2*, 325-366.

Mark, L. S., & Dainoff, M. (1988). An ecological framework for ergonomic research. *Innovation, 7*, 8–11.

Mark, L. S., & Vogele, D. (1987). A biodynamic basis for perceiving categories of action: A study of sitting and stair climbing. *Journal of Motor Behavior, 19*, 367–384.

Mark, L. S., Dainoff, M. J., Moritz, R., & Vogele, D. (1991). An ecological framework for ergonomic research and design. In R. R. Hoffman and D. A. Palermo (Eds.), *Cognition and the symbolic processes.* Hillsdale, NJ: Lawrence Erlbaum Associates.

Muckler, F. A. (1982). Evaluating productivity. In M. D. Dunnette and E.A. Fleishman (Eds.), *Human performance and productivity, Vol. 1: Human capacity assessment.* (pp. 13-49). Hillsdale, NJ: Lawrence Erlbaum Associates.

Nagamachi, M. (1991). An application of participatory ergonomics to automotive production processes. In Y. Queinnec and F. Daniellou (Eds.), *Designing for everyone.* London: Taylor & Francis.

Nubar, Y., & Contini, R. (1961). A minimal principle in biomechanics. *Bulletin of Mathematical Biophysics, 23*, 377–391.

Peters, T. (1987). *Thriving on chaos.* New York: Knopf.

Pressman, J. L., & Wildavsky, A. (1984). *Implementation,* (3rd ed.), Berkeley, CA.: University of California Press.

Putz-Anderson, V. (1988). *Cumulative trauma disorders: A manual for musculo-skeletal diseases of the upper limb.* London: Taylor & Francis.

Rasmussen, J. (1988). A cognitive engineering approach to the modeling of decision making and its organization in: process control, emergency management, cad/cam, office systems, and library systems. *Advances in Man-Machine Systems Research, 4,* 165–243.

Senge, P. (1990). *The fifth discipline: The art and practice of the learning organization.* New York: Doubleday.

Shipley, P. (1991). Participation ideology and methodology in ergonomics practice. In J. R. Wilson and E. N. Corlett (Eds.), *Evaluation of human work.,* London: Taylor & Francis.

Spilling, S., Eitrheim, J., & Aarås, A. (1986). Cost benefit analysis of work environment investment at STK's telephone plant at Kongsvinger. In N. Corlett, J. Wilson, and I. Manenica (Eds.), *The ergonomics of working postures.* (pp. 380-397). London: Taylor & Francis.

Turvey, M. T. (1990). Coordination. *American Psychologist, 45,* 938–953.

Vicente, K. (1990). A few implications of an ecological approach to human factors. *Human Factors Society Bulletin, 33,* 1–2.

Vicente, K., & Rasmussen, J. (1990). The ecology of human-machine systems II: Mediating "direct perception" in complex work domains. *Ecological Psychology, 2,* 207–249.

Warren, W. H. (1984). Perceiving affordances: Visual guidance of stair climbing. *Journal of Experimental Psychology: Human Perception and Performance, 10,* 683–703.

Warren, W. H., & Whang, S. (1987). Visual guidance of walking through apertures: Body-scaled information for affordances. *Journal of Experimental Psychology: Human Perception and Performance, 13,* 371–383.

Winkle, J., & Oxenburgh, M. (1990). Towards optimizing physical activity in VDT/office work. In S. Sauter, M. Dainoff, & M. Smith (Eds.), *Promoting health and productivity in the computerized office.* London: Taylor & Francis.

Zacharkow, D. (1988). *Posture: Sitting, standing, chair design and exercise.* Springfield, IL: Charles C. Thomas.

Zuboff, S. (1988). *In the age of the smart machine.* New York: Basic Books.

Chapter 11

Dimensionless Invariants for Intentional Systems: Measuring the Fit of Vehicular Activities to Environmental Layout

Robert E. Shaw, Oded M. Flascher, and Endre E. Kadar
Intentional Dynamics Laboratory
Center for the Ecological Study of Perception and Action
The University of Connecticut

"When in use, a tool is a sort of extension of the hand, almost an attachment to it or a part of the user's own body, and thus is no longer a part of the environment of the user. But when not in use, the tool is simply a detached object of the environment . . .the boundary between the animal and the environment is not fixed at the surface of the skin but can shift."

—Gibson (1979, p. 41)

"The field of safe travel, it should be noted, is a spatial field but it is not fixed in physical space. The car is moving and the field moves with the car *through* space. Its point of reference is not the stationary objects of the environment, but the driver himself. It is not, however, merely the subjective experience of the driver. It exists objectively as the actual field within which the car can safely operate, whether the driver is aware of it. It shifts and changes continually, bending and twisting with the road, and also elongating or contracting, widening or narrowing, according as obstacles encroach upon it and limits its boundaries. "

—Gibson & Crooks (1938, p. 454, emphasis in original)

11.0   Introduction

11.0.1 Aims and Motivations

Ecological psychology holds the belief that theory and basic research must ultimately aspire to practical applications. This follows directly from its primary aim to understand an actor's functional relationship to natural and manufactured environments. Although laboratory research is useful, it is no substitute for observations and measurements in the field. Even highly realistic simulations may be misleading. The safe flight of aircraft (Gibson, 1950) and the much earlier study of the field of safe travel for automobiles (Gibson & Crook, 1938; see above) presages the work we have undertaken here. The explicit aim of the evolving ecological approach is to develop methods of scientific investigation with ecological validity. Methods with ecological validity treat the organism and its environment as a system with variables defined on the econiche—environmental variables that make reference to and are scaled to the organism as a perceiver and actor in that environment. Methods are sought to explain, first, how actors achieve success on ecologically significant tasks performed in their natural environments and, second, to extend these methods to architectural and landscaped environments. This contrasts sharply with those methods that merely seek to understand how subjects attain statistical significance on arbitrary tasks performed in the laboratory. In this sense, the ecological approach aspires to treating the laboratory as an extension of the subjects' econiche.

More than any other trait, the explicit commitment to strive for both ecological validity and ecological significance in one's research is the hallmark of an ecological psychologist. This commitment qualifies many to contribute as ecological psychologists, even though they call themselves by other names. One needs only share the belief that success on practical problems involving the perceptual control of action is an important, perhaps the most important, way to validate the consistency and the significance of one's theory and research. Thus, the attempt to wed ecological psychology with human factors and human engineering, as this volume tries to do, seems quite natural and overdue.

Our specific goal in this chapter is to present a new approach to the wheelchair navigation problem. How are wheelchair users able to select and follow the best route though cluttered architectural spaces, such as office and factory workplaces, or residential and public spaces? This chapter is, in part, a progress report of an ongoing research and theory on this topic.

Regarding this problem we have three aims—one practical and the others theoretical. One aim is to offer a practical solution to the problem of measuring the dynamic fit of active wheelchairs to their functional spaces under the constraints of a given navigation goal. The second aim is to place this problem under *intentional dynamics*—a general method of ecological psychology for modeling goal-directed activities. Although we use the wheelchair navigation problem as our focus, these methods and principles should apply to other activities as well. To meet these aims, we explore the use of dimensionless analysis—a mathematical engineering technique for finding dimensionless measures (called π-numbers) of the similarity in the structure and functioning of systems which may appear quite different. We shall need to tailor this technique to certain prospective control problems. This entails our third aim, namely, to show how, in principle, dimensionless measures (π-numbers) may allow us to compare the invariants of the information detected about the layout of the environment relative to an intended goal path to the invariant aspects of the control law that must be applied to navigate the path so as to attain that goal.

A caveat and apology are in order. Our aim is to introduce the steps required to arrive at this conclusion without developing or explaining in detail the mathematics involved. We recognize, therefore, that our explanations are somewhat opaque to the reader, but our aim is to point the way to a solution to this problem rather than to provide such a solution, for such details would go well beyond the space permitted for this chapter. We hope the reader nevertheless is stimulated to help develop the approach sketched here.

11.1 Task Similarity, Intentional Dynamics, and the Prospective Control Problem

The general navigation problem is a prospective control problem; it asks how perceptual information about a future state of affairs—an intended goal—can be used to control a current state of affairs (that is, the current forces) in order to reach that goal. More specifically, it asks how a rule for the perceptual control of action allows an actor to find his or her way through a cluttered environment over a preferred path to an intended goal. We explore the prospective control problem by focusing on adult human actors who locomote through architectured environments by wheelchair. The task selected for them is a simple one—to pass without collision through passageways of various clearances that connect enclosed cubic chambers. We assume that the ceiling, wall, and floor surfaces of these chambers are smooth, flat, level, and with sufficient roughness to minimize wheel slippage. Although choosing this type of actor, environmental situation, and task narrows down the possible solutions to the prospective control problem considerably, it still leaves the major issues of the general navigation problem intact.

Consequently, any steps toward solving this general prospective control problem should help solve similar versions of the problem involving different actors, situations, and tasks. Of course, how general the method can be depends on how similar the tasks are. A formal measure of task similarity would therefore be desirable. Discovering such a similarity measure is one of the major goals of the current project.

What would a generic description of the prospective control problem in navigation tasks involve? One requirement is that it makes clear what variables of the task situation are most relevant to a solution. Should the variables be mechanical, biomechanical, physiological, psychological, or ecological? If the description depends on the first three kinds of variables, then the explanation will be causal. Causal explanations, however, are incomplete because they do not address the roles played by perceptual information, skillful control, and intention in goal-directed activities. On the other hand, psychological explanations are likewise incomplete because they typically ignore forces, masses, and friction. An understanding of the prospective control problem, therefore, involves both causal and intentional variables.

Clearly, an ideal explanation must address how goal-specific information specifies the forces by which actors intentionally control their goal-directed movements. In this regard, there are two problems: the *scaling problem* and the *transduction problem*. The *scaling problem* is a measurement problem at the ecological scale or, more briefly, a problem for *ecometrics*. Ecometrics asks how information at the scale of geometry and kinematics, the scale at which intended goals are specified, might be transformed into force units at the scale of the statics and kinetics of control, the scale at which goals might be accomplished. Taken together, these problems combine to form a more general problem of measuring the similarity of environmental information detected by the actor to the actor's control of the behavior required to accomplish an intended task in that environment. Put simply: We must answer the question of how the information detected is similar to the action produced whenever different actors succeed in solving the same task. This is a natural question for similarity theory, and its natural answer is to be found in discovery of the appropriate dimensionless numbers.

The *transduction problem*, which presupposes a solution to the ecometrics scaling problem, is a problem for *ecomechanics*. Ecomechanics asks how forceless information might be made efficacious in directing control processes so that the actor might reach an intended goal state. When classical mechanics is the study of laws governing *motions* of inanimate bodies and biomechanics the study of laws *governing movements* of biological systems, so ecomechanics is the study of laws that govern *actions* (goal-directed movements) of agents (Shaw, 1987; Shaw & Kinsella-Shaw, 1988). Unlike the first two forms of mechanics, the last one involves a special relationship holding between information and control which goes beyond mere force or energy flow descriptions. The details of this relationship have recently been spelled out as a theory that information detection and energy control must be *self-adjoint* in the sense of having mutual and reciprocal quantities (Shaw, Kugler, & Kinsella-Shaw, 1990; Shaw, Kadar, Sim, & Repperger, 1992).

Solving the scaling and transduction problems entails a new approach to prospective control problems, one that is a hybridization of physics, biology, and psychology. Attempts to develop such an approach are being made. A branch of ecological psychology called *intentional dynamics* subsumes ecometrics and ecomechanics. The transduction problem of ecomechanics and the scaling problem of ecometrics comprise dual aspects of the central problem of intentional dynamics. Hence, the prospective control problem falls naturally under this new discipline. (For an overview of intentional dynamics, see the earlier studies mentioned as well as Kugler & Shaw, 1990, and Kugler, Shaw, Vicente, & Kinsella-Shaw, 1990).

Before addressing the questions raised, we provide a statement of the social motivation for this project and explain the practical significance of its potential success.

11.2 Part I: A Problem with Ecological Significance and a Method with Ecological Validity

11.2.1 Overcoming Barriers

An estimated half-million persons are added to the population of the handicapped each year through illness or injury (Arthur, 1967). Of the estimated 36 million Americans with disabilities (Disabled USA, 1984), many achieve education and, later, employment or self-employment because there are specially seated workplaces available for them in accessible facilities (for example, homes, schools, offices, and factories). Unfortunately, many other disabled Americans are excluded from these opportunities because standardized design guidelines fall short of the ideal and do not accommodate their particular disability. Current minimal standards sometimes fall short of the ideal for the disabled for two reasons: First, because the practices used to implement the standards may be underconstrained by the current specifications, or, second, because design tradeoffs are used to strike a practical and economical balance between the architectural dimensions needed to accommodate both the disabled and the nondisabled.

A significant reason for these problems is that, at present, it is impossible to measure the dynamic fit of goal-directed activities (e.g., moving wheelchairs) to the environments in which they must be performed. Consider the special case of designing environments for wheelchairs. Only recently have methods begun emerging to determine design guidelines for the fit of such activities performed from relatively stationary chairs and wheelchairs (see Dainoff, 1987, 1991a, 1991b; Abdel-Moty & Khalil, 1989). However, no significant work has yet been done on ways to measure dynamic maneuvers of wheelchairs over accessible routes through functional spaces toward goals. The proposed project offers a way to remedy this shortcoming.

In its specifications, the American National Standards Institute (ANSI, 1986) emphasizes the need to recognize that persons with disabilities that confine them to wheelchairs are no longer "average" persons. They are shorter and wider, rolling instead of walking, being unable to climb stairs. They require ramps or oversized elevators, require more space to turn around, and more clearance under tables and other equipment. Nor can they see over or cross over barriers that others might more easily. Moreover, they require, on the average, more time to egress through cluttered environments (such as furniture arrangements and milling crowds).

This list comprises but a few characteristics that distinguish the wheelchair-bound individual from ordinary individuals. Wheelchair users have fewer opportunities for actions than the nondisabled. Consequently, a question of great interest is how functional spaces for wheelchair users can be designed to afford freedom for them to act in ways comparable to nonwheelchair users. Consider other important characteristics of wheelchair users that make the designing of functional spaces explicitly for them even more critical.

11.2.2 Determining Dynamic Tolerances of Fit

The Uniform Federal Accessibility Standards (UFAS) sets the specification for a clear opening width for doorways. When the approach is head on, the recommendation is for widths of 30 in (76 cm), whereas turning a wheelchair to enter an opening requires greater clear widths. For most approaches, the addition of an inch leeway on either side suffices—making for a minimum clear width of 32 in (81.5 cm). However, to accommodate the likelihood that control of straightline travel of the wheelchair will be imperfect adds at least 2 in tolerance. For instance, to accommodate traversing passageways that are more than 24 in (61 cm) long increases the minimum clear width to at least 36 in (91.5cm) as compared to doorways with unrestricted approaches. For similar reasons, the minimal clear widths of checkout aisles in stores must also be 36 in (91.5 cm). However, the specifications must be even more generous in libraries. Because their greater length makes larger meanders from the straightline path even more likely, aisles between library bookcases require greater tolerances. In this case, the UFAS recommends a clear width of 42 in (106.5 cm) where possible. Where a wheelchair user might meet other wheelchair traffic or pedestrian traffic, then the clear width of passages must be adjusted to even greater tolerances (66 cm and 48 cm, respectively). Ultimately, our project aims at dynamic in-the-field testing to see if such standards remain realistic under a variety of wheelchair velocities, approach angles, and intentions.

These alterations to the specified clear widths of doorways or passageways, as a function of direction of approach, distance to traverse, or type of traffic, are only estimates. Dynamic measures might verify whether they are adequate adjustments to existing codes. Clearly, wheelchair speeds may increase when going down ramps, under emergency egresses, keeping up with traffic flow, and so forth. The possibility of varied speeds in approaching doorways or moving through passageways, therefore, requires dynamic measures to set their clear width tolerances. Wheelchair velocities may vary as a function of the layout of the environment, the circumstances, and the intentions of the wheelchair user.

Physical mechanics dictates that wheelchairs moving at even moderate speeds will have momentum characteristics that make them more demanding on space requirements than slow-moving ones. Momentum influences both maneuverability and control—sometimes in a positive way and sometimes in a negative way—depending on circumstances. The formula for linear and angular momentum involves the multiplication of mass by velocity. For every unit of increase in speed or in the mass of the wheelchair user, there is a dramatic, multiplicative increase in the minimal requirements for stopping and turning. For this reason only dynamic measures can determine the limits on the safe and comfortable fit of the active wheelchair to architectural spaces.

In preparation for the rest of this chapter, it is well worth rereading the Gibson and Crooks' (1938) quotation given at the beginning of this chapter. Clearly, its full appreciation entails an innovative approach to accommodate the facts of dynamic measurement as discussed.

11.2.3 Limitations of Current Measurement Techniques

The Americans with Disability Act (ADA) provides an important incentive to actualizing the commitment of our society to allow every member of this community to participate, fully, in all aspects of life. Designing for architectural spaces that are accessible and usable by wheelchairs is a fundamental step toward equal rights. Legislation can do no better than to implement the best design guidelines that exist. If these are inadequate, then the resulting standards and codes will be equally wanting. In most cases, current standards have evolved from practical experience, legal precedent, and the intuition of experts. They have not been set nor verified by scientific methods. Without dynamic measurements there is no alternative to current practices. The simplicity of the proposed method can be better appreciated.

As observed earlier, the field needs dynamic measures of active wheelchairs to determine the safest, most comfortable, and efficient paths for traveling between points in the environment. Current methods may measure the total time to travel a fixed path, but not the selection of the path, nor the continuous time accumulated at each point along the path. A few experimental studies investigate wheelchair use in simulated environments. They use treadmills or dummies attached to wheelchairs, rather than human wheelchair users in actual workplace settings. None of the existing methods can make direct and continuous measurements of wheelchair use over paths that connect different areas in the workplace. We now appraise the existing techniques for dynamic investigations of this problem.

One might use aerial movies to measure wheelchair paths through natural environments. Unfortunately, aerial techniques prove impractical for several reasons: One must mount several cameras at heights greater than the ceilings usually available. Filming or taping the wheelchair's path requires an extremely wide angle lens that introduces gross distortions. This makes exact measurements problematic. Finally, the filmed or videotaped record from each camera must be digitized and their data somehow combined—a costly, time-consuming, and tedious job. Current automatic digitizing programs prove very expensive and are not, indeed cannot be, entirely automatic. Alternatively, automatic sonic digitized recordings are possible. This technique mounts sound emitters on the wheelchair and distributes an array of microphones widely over a prearranged path. As the wheelchair moves its location, speed, and direction are recorded. Such devices work best in limited spaces. They are also subject to serious problems of acoustic reflectance in nonstandardized environments. Neither audio nor video techniques work well in environments cluttered with furniture and other architectural barriers. These methods restrict the wheelchair routes to those within range of the camera or emitters. Consequently, they are unable to measure the user's natural preferences for routes that fall outside the predetermined set. These methods are thus less practical, more costly, and more restricted than the method to be proposed.

In short, we know of no current ecologically valid method for "in-the-field" measurements. Such a method is indispensable to the design of functional spaces in which wheelchair users perform daily a wide variety of diverse activities. A chief concern is that although ordinary environments may be sufficiently clear for walkers, they are usually not barrier free for wheelchair users. Hence, design criteria must differ for environments that allow these different modes of locomotion. Dynamic measurement of clear movement through such environments requires the development of new tools and techniques. We have built a prototype of a new measurement instrument and tested its feasibility which we describe next.

11.2.4 A New Technique for Measuring Dynamic Fit of Active Wheelchairs

For this project we have developed a unique tool and associated methods for determining the relevant variables for the fit of wheelchair activities following paths through functional spaces. Following Gibson and Crooks (1938), one of our aims is to discover how users perceptually select fields of comfortable (safe and efficient) travel within fields of possible travel. A current series of experiments are designed to uncover the informational basis and the stable styles of control by which wheelchair users navigate successfully through doorways and passageways. In the heart of this research is a typical wheelchair that has been modified to gather data online, while running diverse routes through such environments. A prototype of the computerized wheelchair has been built and tested. These exploratory experiments are described later. Here we use the computerized wheelchair to examine a range of doorway and passageway width tolerances under a variety of experimental conditions with different velocities, widths, distances, and intentions. Our ultimate aim is to discover the relevant information variables, control parameters, and values or goals, whose interrelationships define the dynamic field that moving wheelchair users carry with themselves as they move about (Gibson & Crooks, 1938). For convenience, we might call these three kinds of parameters—observables, controllables, and valuables—the *affordables*[1] of the task situation.

This prototype computerized wheelchair has an on-board, laptop computer that reads data from optically encoded accelerometers. The accelerometers are attached to a pair of measurement wheels that have been added to the undercarriage of a standard wheelchair between the large wheels by which the vehicle is steered and driven. This instrumentation allows the wheelchair's changing locations, orientations, and velocities to be automatically recorded at each point along the route traversed (within tolerances of approximately 2 mm for each 100 cm of lineal distance traveled). The existence of this first prototype demonstrates the feasibility of the concept, the soundness of

---

[1]Gibson (1979) introduces the term *affordance* for those objective properties of environments which, taken in reference to an actor, provide opportunities for that actor's actions (i.e., goal-directed behaviors). It is one thing, however, to have an opportunity for action and another to have the means to seize that opportunity; hence, the necessary distinction between affordances *as opportunities to be seized* and effectivities as the means by which such opportunities may be seized. Similarly, it is still another thing for an actor to value one opportunity more than others so that it and not they is intended and acted on—that is, to select, on a given occasion, one affordance as being more valuable than another; hence, the need for the term *valuable* to pick out the intended affordance. For these reasons, we introduce the term *affordables* to refer to the component parameters (observables, controllables, valuables) by which affordances, intentions that select them, and the effectivities that realize them, may be conveniently referred to under a common rubric.

the hardware and software designs, the practicality of the engineering and production standards used, and a reasonable cost-effectiveness.

For these reasons, we believe that the computerized wheelchair has more face validity for the purposes of relatively unrestricted dynamic measurements in natural settings than any other currently available method. Furthermore, in field testing, the first prototype has functioned successfully, although improvements might be made. For instance, in the future we hope to develop a second prototype that will have strain gauges added to the wheel hubs which are connected to the circular handrails. This will allow manual forces applied to the wheels to register the torques applied to each wheel. Because it is these manually applied wheel torques that both drive and steer the wheelchair, these modifications will allow direct measures to be made of both the kinematics and kinetics of the vehicle's navigation through naturally cluttered environments. However, in what follows, we base our analyses solely on the kinematic description of wheelchair activities because these data are all that is currently available to us.

The approach to be taken clearly recognizes that adequate environmental architectures for wheelchair users can not be merely normal designs with add-on allowances for a "special population" of users. Rather, moving wheelchairs require environments with fundamentally different designs than those that accommodate nondisabled walkers. Clear passage through shared functional spaces for the latter is not necessarily clear passage for the former. The economics of design and construction, however, must remain realistic. If minimal standards are too excessive, then costs of construction will be wasteful, perhaps, even exorbitant—setting up but another barrier to society's commitment to provide realistic accommodations for wheelchair users.

To reiterate: Our aim is to develop accurate techniques for dynamic measures of functional spaces for wheelchair activities that avoid intuitive overestimations. Such techniques could contribute significantly to lowering the economic barrier to mainstreaming wheelchair-bound disabled individuals into society at large. Of course, merely having an instrument that is appropriately designed to measure the required dynamic fit is not enough. One must also have a theory that clearly indicates which measurements should be made.

In the next section, we develop the background for the scaling (ecometric) problem and the transduction (ecomechanics) problem. With this background in place, we are better able to formulate an answer to the question of how to measure the similarity of information and control required for goal-directed actions to be successful.

11.3   Part II: Theoretical Background: Dynamic Fit as an Ecometric Problem

In this section we introduce some of the basic concepts that underlie ecometrics and ecomechanics. Such concepts will be involved in any attempt to solve the scaling and transduction problems associated with actors (e.g., wheelchair users) navigating successful goal paths.

Consider, first, some properties an environmental situation must have to afford an actor navigating through a cluttered environment: The medium that surrounds the actor must permit freedom of movement, and the actor—who may be a walker, flyer, swimmer, or vehicle user—must maintain mechanical contact with a surface of support. For swimming and flying creatures, the medium and surface of support are the same—water and air, respectively. For arboreal and land creatures, the medium is air or water, and the surface of support is tree, rock, ground, floor, stairs, or sidewalk. If the actor is neither arboreal nor aquatic, say a human, horse, or dog, then the surface must be sufficiently rigid to support the actor's weight, have sufficient friction to allow traction, and be sufficiently level to prevent falling over. Environmental properties that afford opportunities for actions—that is, goal-directed activities—are called *affordances* (Gibson, 1979).

Examples of environmental affordance properties are the *graspability* afforded by certain objects, the *supportability* afforded by certain surfaces, and the *edibility* afforded by certain substances. In general, properties of objects or surfaces count as affordances if they provide appropriate structural and informational support for the action capabilities of properly attuned actors, that is, actors who have the means, opportunity, and motivation to carry out the relevant actions. Hence the definition of an affordance necessarily implicates corresponding action skills, called *effectivities* (Shaw & Turvey, 1981; Turvey & Shaw, 1979). Such effectivities determine whether a specific class of actors, for whom information specifying the relevant affordance property is available, can use that information to realize that affordance property, that is, to guide its behavior successfully toward an intended goal.

Table 11.1 shows examples of how affordances, effectivities, and actions have an underlying dimension of abstract similarity. Affordances and effectivities are functionally defined and functionally similar under the action of a sufficiently skilled organism relative to appropriate environmental structures. The affordance refers to the scale-dependent aspects of the physical situation that support an opportunity for a definite action; effectivities refer to the commensurate means available to the actor for realizing that definite action in that given situation. The actor's intention to act toward realization of that affordance goal in the given situation provides the motive to act. As in a court of law, when all three conditions—means, motive, and opportunity—are met, then we may conclude that the agent has a realistic intention to act. Under this interpretation, an action is necessarily goal-directed and intentional, involving the goal-specific affordance and the intention-specific effectivity.

TABLE 11.1: *The Interrelationship of Affordances, Effectivities, and Goal-Directed Actions.*

AFFORDANCE OF E (opportunity for action)	EFFECTIVITY OF O (means for acting)	ACTION OF O ON E (realizing intended affordance)
graspable	able to grasp	O grasping E
climable	able to climb	O climbing E
catchable	able to catch	O catching E
passability	able to pass through	O passing through E
sit-on-able	able to sit	O sitting on E

A wheelchair is a vehicular tool, that is, a tool that aids locomotion. Hence, whatever similarities carry over from affordance information to effectivity control must do so through the tool that interfaces the actor-as-perceiver with the environment-as-perceived and acted on. But what happens to the affordance description of the environment and the effectivity description of the actor when tools as objects and tools as functions enter the story? We address this important question next.

11.3.1 Tools as Ecological Interfaces

Tools enhance, extend, or restore the action or perception capabilities of humans or animals. They may assist in manipulation, locomotion, or exploration. Tools may extend existing capabilities or restore lost functions either partially or completely. Tools may be simple machines (e.g., levers, ramps, pulleys, springs) or complex tools (e.g., trucks, bulldozers, automatic assembly lines, refineries) that amplify the capacity for work, or they may be devices that amplify information detection (e.g., microscopes, telescopes, sonar, radio, television) or amplify its usage (e.g., computers, libraries). Currently, it is still an open question whether computational tools may also amplify intelligence. They surely carry out, in many ways, the tasks that would otherwise take an intelligent person to do. In short, tools may focus or extend effectivities of humans, thereby providing greater access to affordances in the environment and, consequently, new opportunities for action.

Tools may also serve a prosthetic function by restoring the effectiveness of lost or infirmed sensory abilities (e.g., eyeglasses or hearing aids), or they may restore lost or infirmed action capabilities. Some action prostheses restore manipulatory abilities, such as the artificial hand or arm; others restore lost or infirmed locomotory abilities, such as crutches, walkers, artificial legs, and wheelchairs. Manually driven or motor-driven wheelchairs fall into the category of vehicular tools. Little if any work, to our knowledge, has been done on the formal description of such tools and how they alter the dynamic fit of the disabled user to that environment in which they are typically used. Moreover, achieving empirical validation of the dynamic fit of vehicular tools might be furthered by considering how people navigate other vehicular tools, such as automobiles. We consider such comparisons later. A natural question is how does an object with a tool function relate to the affordances of the environment and the effectivities of the actor? Does the tool act as a mediator between the environment's affordances and the actor's effectivities. Or, if the tool does not sit in the seam between the environment and the actor, does it belong more to one component of the ecosystem than the other?

Figure 11.1 shows the relationship that a tool might have to the user and the user's environment in which it is applied. Notice that the tool might be treated as a mediating device which belongs neither to the organism nor to the environment, but interfaces the two. This interpretation seems to overly complicate the relationship between organisms and their environments and to potentially destroy the theoretical balance that is needed to keep affordances and effectivities mutual and reciprocal (i.e., mathematically dual) as illustrated in Table 11.1. This theoretical balance is desirable because it allows the relationship between information and control to remain epistemically direct for all the reasons that we have discussed at length elsewhere (Shaw & Turvey, 1981; Turvey & Shaw, 1979; Turvey, Shaw, Reed, & Mace, 1981). Mathematically, this directness in the coupling of the components of the ecosystem allows us to use some powerful theorems from (self-adjoint) information/control theory that permit an economy of description for modeling psychological ecosystems, affordances and effectivities, information and control, and the perceiving-acting cycle that would not otherwise be possible (Shaw et al., 1992; Shaw, Kugler, & Kinsella-Shaw, 1990). There seems to be another alternative which keeps intact our fundamental assumptions.

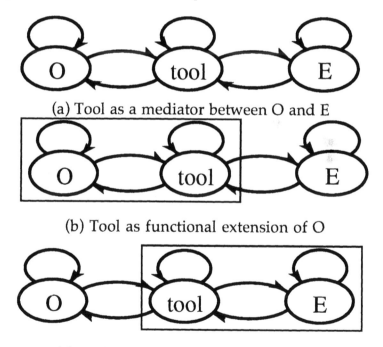

(a) Tool as a mediator between O and E

(b) Tool as functional extension of O

(c) Tool as functional extension of E

Figure 11.1. *Possible ways to partition the role of a tool in an ecosystem: (a) as a mediator, where E<=>T<=>O; (b) or as an extension of environmental affordances, where (E<->T)<=>O; or (c) or as an extension of an actor's effectivity, where E<=>(T<->O).*

We prefer to distinguish between tools-as-objects and tools-as-functions. As objects, tools must be treated as contributing to the affordance structure of the environment. That is, before the tool is used, like any object it has its own affordances, thereby inviting certain actions. But tools-in-use are no longer just objects but play an intrinsic role in extending, replacing, or restoring the user's repertoire of effectivities—that is, the actor's capabilities for realizing affordances of the environment. Tools, therefore, have a *dual* function—as objects within the environment's affordance structure and as components within the actor's effectivity system. Through the dual function, a tool both scales and transduces the information detected and the control executed in the course of the task involving the tool use.

For instance, on the ground a steel rod may afford grasping, lifting, and wielding. But once grasped and properly braced under a boulder, the rod becomes a lever and is part of the effectivity for leveraging the rock causing it to roll down a hill. Hence, to call this object a rod is to distinguish its role in the affordance structure of the environment, whereas to call it a lever is to distinguish its function as an integral part of an effectivity system. There is clearly a logical distinction between the affordance structure of an object that invites a goal-directed function and the effectivity, or goal-directed function, it invites. These concepts should not be confused. The former refers to the environment as a source of goals that an organism might achieve, whereas the latter refers to the organism as a source of intentions that achievement of environmental goals might satisfy. Hence, there is no ambiguity introduced by dually partitioning the tool, first, as a part of the functionally defined environment with affordance properties and, then, as part of the functionally defined organism with effectivity properties, as shown in Figure 11.1b and 11.1c.

Consider the methodological benefits of recognizing the dual functional roles that tools play: As objects they have affordances like any other, but as tools they engage effectivities, thereby permitting actors to perform tool-specific, goal-directed functions. Thus, a tool function of an object contributes, in a unique way, to the affordance-effectivity compatibility underlying an actor's fit to its environment. Furthermore, and here is the main point, because tool use requires a high degree of coordination between perception and action, *tools focus behaviors*. When theoretical description and empirical study might be intractable at the microscale level of the enormous number of degrees of freedom exhibited by neuromuscular acts, both theory and experiment become more tractable when only the macroscale degrees of freedom of the tool using system is considered.

This suggests a way that one might develop a manageable inventory of solutions to the degrees of freedom recognized by Bernstein (1967). The scientific study of tool functions involved in manipulatory and locomotory acts might identify more tractable analogues of their corresponding, more complicated, unaided goal-directed actions. A study of these "free-action" analogues might provide a theoretical basis for better understanding the underlying effectivities that underwrite them.

If tools promote the fit of actors to their environments in ways more visible than tool-independent behaviors, then studying activities involving wheelchairs may help us understand the coordination requirements for successful locomotory navigation in general. For describing the behavioral interface of the wheelchair to the floor surface is much simpler than describing the free actions of a walker. Of course, so far as specifics are concerned, the study of one is not a substitute for the other. Even so, as related locomotory cases, they are equally interesting examples of how similar systems may exhibit intentional dynamics under different navigational demands.

For these reasons, it would be important to have a way of measuring the degree to which tools promote the common fit of actors to the task environment, as compared to the same actors without the tool. It seems reasonable to assume that use of the same tools by different actors performing the same tasks increases the task constraints so that they become dynamically more similar as intentional systems. We turn next to a discussion of various ways in which systems (actors plus tools) might be similar in performing goal-directed behaviors.

11.3.2 Similarity across Physical Systems

Assume there are two systems, each with the same tool interface with the environment but different masses. If the tools and tasks are similar, then their behaviors should be similar—differing by only a scale factor reflecting their different masses. For instance, imagine two wheelchair users of different weights who must select efficient paths for navigating through a room cluttered with furniture so as to exit through a doorway. If they travel at the same speed (are kinematically similar), the heavier actor (greater mass) will require larger turning radii because he or she will have the greater momentum (mass x velocity). In spite of the fact that the two cannot travel the same curvilinear paths at the same velocity, there are still numerous ways that they may perform their tasks similarly. They may cover similar distances to get to the doorway, take similar amount of time to do so, go in similar directions, use similar work (i.e., proportional to their masses), and so forth.

Table 11.2 shows the dimensions involved in the various kinds of similarity that one system might have to another with respect to the paths they follow in navigating through an environment. (Similarity means alike up to a scale factor). Geometric, kinematic, and kinetic path similarity hold for systems that have proportional relationships among state variables involving the dimensions of L, LT, LTF, or DTF, respectively. Likewise, work, impulse, and torque similarity hold if the two systems exhibit proportional path functions (integrals) involving these same dimensions in which each dimension is multiplied by force (i.e., $\int FdL$, $\int FdT$, $\int FdD$, respectively). Specifically, geometric path similarity means the two systems cover similar distances between homologous environmental locations; kinematic path similarity means the two systems arrive at homologous places on a similar temporal schedule; and kinetic path similarity means the two systems apply similar forces to move over similar distances (similar work forces), times (similar impulse forces), and directions (similar torque forces).

Table 11.2: *Various Kinds of Physical Similarity Two Systems Might Exhibit.*

Similarity over	length (L)	time (T)	angle (D)	force (F)
distance	√			
time		√		
direction			√	
work	√			√
impulse		√		√
torque			√	√

Perceptual similarity involves a common proportionality defined over information variables detected by the systems, whereas actional similarity involves a common proportionality over control variables used by the two systems to produce their respective behaviors. Finally, intentional similarity means the two systems share similar goals (i.e., target parameters, manner parameters, or both). When the goalpaths pursued by the two systems are similar, then we say that they are similar with respect to their intentional dynamics. We now describe how this works.

Table 11.3 indicates the intentional dynamical similarities that two systems might exhibit in their goal-directed performance, in which the physical similarities defined in Table 11.2 refer to measures indicating how the path is generated from the initial condition (start-up) to some final condition. This path may or may not reach the target nor do so in the manner of approach intended. By contrast, the intentional dynamics similarities, defined in Table 11.3, refer to measures indicating how the path ought to be generated to conform to an intended final condition (reaching the intended target in the manner intended). Hence, Table 11.2 refers to measures of the causal generation of the mechanical path indifferent to the goalpath, whereas Table 11.3 refers to measures of the intentional control required to generate the ecomechanical path, that is, the intended goalpath. The former is a "blind push from behind" by

Table 11.3: *Various Kinds of Intentional Dynamical Similarity Two Systems Might Exhibit.*

type of similarity between systems	information (target parameters)			control (manner parameters)		
	distance to contact (δ)	time to contact (τ)	direction to contact (γ)	work to contact ($F \times \delta$)	impulse to contact ($F \times \tau$)	torque to contact ($F \times \gamma$)
target	√	√	√			
manner				√	√	√
goal-path	√	√	√	√	√	√

forces determined retrospectively by physical law as a function of the past states of the system; the latter is a 'directed pull from in front' determined prospectively by the actor's control law being provided with (perceptual) information about how the observed future states of the system reckon with those intended. For example, *time* as a variable for physical dynamics tells us how much time has elapsed since the system left its initial start-up state given where it is; whereas *time-to-contact*, as a variable for intentional dynamics, tells how much time should elapse before the system reaches its intended final state, given where it is. A similar prospective interpretation should be given to the other "contact" variables.

The importance of extending physical similarity theory to include intentional dynamical similarity is to allow measures of the similarity between intentionally driven systems analogous to the similarity measures possible for causally driven systems. Dimensionless analysis, an outgrowth of dimensional analysis and similarity theory, provides a way to construct measures of physical similarity between the observed characteristics of two systems as well as between a single system's observed and intended characteristics.[2] Both of these similarities will be of interest to comparing systems with both physical as well as intentional dynamics. The general nature of these measures is discussed next.

11.3.3 Dimensionless Ratios as Measures of Similarity

Dimensionless analysis developed from dimensional analysis: the classic parameter approach introduced by Buckingham (1914) and Rayleigh (1915), systematically developed by Bridgman (1922), reformulated by Drobot (1954) to recognize dimensionless functions, and used by Schuring (1977) to emphasize law-based similarities. Dimensionless analysis serves two important functions for model construction and theory. First, it puts a limiting framework around the model to be constructed. This framework is based on the necessity of any functional relation to remain invariant if the units are changed. A second important function capitalizes on the fact that any physical system can be analyzed into functions of a limited number of

---

[2]The notion of ecowork and affordance-effectivity duality as dimensionless quantities, were introduced in a series of lectures on ecometrics given in the late 1970's at the University of Connecticut by R. Shaw, first reported in Warren & Shaw (1981), and given experimental application by Warren (1982).

dimensionally fundamental variables that define its equation of state. The number of independent dimensionless groups, as proven under the π-theorem, is equal to the difference between the number of variables that make them up and the number of dimensions involved (Ipsen, 1960). We explore these two functions of dimensionless analysis in the context of the tasks involving the perceptual control of actions required to perform wheelchair navigation tasks and other locomotory tasks. Precedents for using such an approach in psychology have been set by Warren and Shaw (1981), Kugler, Kelso, and Turvey (1980, 1982), and especially applied to ecological psychology by Shaw and Warren as reported in part by Warren (1982).

The physical dimensions of interest, as discussed earlier, are those of length (L), time (T), direction (D) and force (F). The L, T, D, F system of variables are fundamental in that they can be used to define any other variables, called *derived variables*. Using these fundamental dimensions, the respective functions describing a pair of systems can be compared so as to reveal any similarities that might exist. Such similarities may exist at various levels of analysis with respect to the state equations of the systems

In order for a comparison of the equations for two systems to be legitimate, however, certain formal criteria must be satisfied. One important criterion, discovered by Fourier in 1822, is that such equations be *dimensionally homogeneous*, that is, be expressed in the same dimensions. (Consequently, a solution to the scaling and transduction problems introduced earlier depends on this property holding between the equations describing the information and those describing the control as exhibited by the perceiving-acting cycle involved in a given task.) On initial inspection, however, two equations may not appear to satisfy *dimensional homogeneity* because they may be expressed in different derived variables whose symbols are quite distinct. However, by reducing derived variables to the fundamental dimensions of F, L, D, T (or the corresponding fundamental dimensions, F, δ, τ, γ, for contact variables), it is possible to determine if two systems satisfy this homogeneity property. For example, in Newton's famous law (stated in a form to emphasize its dimensional analysis), $F = MA$, although mass and acceleration appear as derived variables, they can be expressed as a function of the fundamental dimensions so that $A = LT^{-2}$ and $M = F(LT^2)^{-1}$.

The most important progress in the study of similarity among systems resulted in the development of the so-called π–theorem (Buckingham, 1914), one of the great achievements of similarity theory.

This beautiful theorem asserts that any analysis of a physical phenomenon in dimensional terms can always be reduced to a simpler functional relationship among dimensionless variables (Drobot, 1954; Rosen, 1978; Stahl, 1963). An appreciation of this theorem is necessary if one is to understand how systems too complicated to be described in terms of mathematically closed formulae (e.g., differential equations) might nevertheless be compared in terms of more abstract levels of similarity. The worth of this approach was summarized by Stahl (1963) as follows:

> "It is reiterated that similarity criteria may always be obtained in a simple manner from governing differential equations, when such equations are available. When clear differential formulations are not at hand much progress can be made by the study of simplified model formulations of the problem, followed by an apparent dimensional 'integration' which gives the pertinent similarity criteria without actually performing the numerical integration process. In other cases there may be no differential equation which appears appropriate and one can choose similarity criteria on the basis of prior experience and direct manipulation of dimensional variables. " (p. 369)

This is not the place to go into details regarding this approach, but fortunately several lucid accounts are available that the interested reader might consult (Birkhoff, 1960; Duncan, 1953; Johnstone & Thring, 1957; Langhaar, 1967; Sedov, 1959; and in the area of similarity of biological systems, see especially, Günther, 1975; Rosen, 1978; Stahl, 1963). To give the spirit of dimensional analysis and similarity theory, and to show how dimensionless numbers naturally arise in the former and have application in the latter, consider a classical example—the case of the generalized Reynolds number.

An Example: The Reynolds Number. In 1638 Galileo made initial contributions to similarity theory by noting that static objects, such as pillars, tree trunks, and animal's legs, should scale area of support as a function of volume of the mass supported. Newton went still further by recognizing that laws of nature should be formulated so as to be independent of certain dimensions, such as the overall size of objects. Finally, similarity theory moved to a new plateau of abstraction and generality when it was recognized that one might avoid specific dimensions altogether by composing variables from the ratios of two variables which were not just identical in dimensionality, but whose dimensions canceled. Such variables, when evaluated, came to be known as "dimensionless numbers" —the most famous of which is the Reynolds number.

Specifically, a variable is *dimensionless* if it is formed by a ratio of two quantities measured in units of the same dimension so that these units cancel out of the numerator and the denominator, leaving a pure ratio. A so-called π-number is a numerical evaluation of a dimensionless variable. The numbers receive their name from the fact that the well-known geometric ratio π= *3.1415* is itself a dimensionless number, one achieved by dividing the circumference of a circle by its radius, where both are measured in the same units. Hence, the units cancel leaving a dimensionless π-number. Quantities in which this is so are said to satisfy the property of *dimensional homogeneity*.

The Reynolds number, perhaps the most famous example, derives from the Navier–Stokes equation which describes the behavior of fluids. In this context, the Reynolds number expresses abstractly the ratio of inertial forces to viscous forces acting on a small volume of fluid. To be more specific, the Reynolds number is composed in the following way (Here, for convenience, mass, M, is used as a fundamental rather than a derived dimension):

$$vL\rho/\eta$$

in which v is velocity (LT^{-1}), L is characteristic length, η is viscosity ($ML^{-1}T^{-1}$), and ρ is density (ML^{-3}). If we substitute these fundamental dimensional quantities in this formula, it confirms that Reynolds number is dimensionless; namely,

$$(LT^{-1})L(ML^{-3})/(ML^{-1}T^{-1}) = LLLMTL^{-3}MT = MTL^{3}M^{-1}T^{-1}L^{-3} = [1].$$

(Brackets indicate a dimensionless quantity rather than an integer.)

The use of dimensionless numbers in science is exemplified by applying the Reynolds number in biology. For instance, it has been shown that hemodynamical systems (blood flow) are similar across the circulatory systems of a variety of animals because they all share fundamental similarities with hydrodynamical systems in general (Stahl, 1963). In general, a rather large set of invariant dimensionless numbers have been found that seem to hold for complex multivariable biological systems. These numbers specify relationships of these otherwise diverse systems that are independent of specific parameters such as size. One theorist goes so far as to claim inductively that "behind the complexity and astonishing variety of forms and functions quantitative criteria of similarity for all living systems have been disclosed" (Günther, 1975, p. 660).

11.3.4 Dimensionless Measures of Similarity Between Information and Control Processes

In the same spirit, the Gestalt psychologists claimed that abstract structure can be recognized, even though transposed from one situation to another (Koffka, 1935). Similarly, Gibson (1979) has spoken of invariant information as being both "timeless" and "formless," that is, as being independent of the specific parameters of a given environmental situation. Recent evidence suggests that information underlying perception of environmental structure must be very abstract. If what comes in on the perception side is to be of use to what goes out on the action side, then there must be a similarity of goal-specific information detected to the control which is guided by that information. For example, a cat must detect the target information, the direction and distance to be jumped, in order to land on a perch, say, the top of a fence. In the information picked up both from the environment and from its own body in that environment, there must be a reasonably precise specification of the torque required to aim itself and the impulse forces required to do the necessary work of transporting its body mass to the target. These general requirements for any successful action by any actor can be summarized by using the concept of affordables introduced earlier.

The observables present in the perceptual information specify the affordance goal to be selected (e.g., the fence's perchability). The affordance's observables are then transduced into ecometrically scaled, ecomechanical values of control parameters governing the biomechanical degrees of freedom, or controllables. If these action–control parameter values define a successful goal-directed function, or effectivity, for accomplishing the intended goal (e.g., jumping to the fence), then they comprise the set of values, or valuables, sought. More generally: If the mapping of the observables over the dually scaled controllables allows an actor to transduce its energy into the intended valuables, then an opportunity for action is successfully seized. This is the condition that allows the affordance and the corresponding effectivity to be duals, and as such it is guaranteed that the information and control are sufficiently similar so that the former can be both transduced and scaled into the latter.

Indeed, a target's observables must be sufficiently abstract to be invariantly transposed from the particular form of information they must take to be detected to the particular form they must take to be the controllables that allow a successful action toward that target. This is the basic assumption of ecometrics (Shaw & Kinsella-Shaw, 1988). Mathematically, this assumption entails that a "flow" of information and the "flow" of control over the perceiving-acting cycle together define a *dynamic invariant*, or what physicists call a *conserved quantity*. A formal argument can be made that this generalized abstract quantity, called the *total action potential* (Appendix A), must be conserved if one is to explain how action skills acquired in one situation can generalize to other similar situations (see Shaw et al., 1990, for details).

Furthermore, the conserving of this total action potential whenever a goal-directed behavior is successful has been mathematically shown to entail that an inner (scalar) product invariant (Appendix B) must hold between the dynamic flow of goal-specific information and the dynamic flow of the control of goal-relevant energy expenditures (for details, see Shaw et al., 1992). In other words, the information detected over the path to a goal must determine the path over which the actor controls its manner of approach to the goal—otherwise, either the target moved toward is not the intended target or the manner of approach is not the one intended. In either case, if there is an error in information detection or control, the path traveled will not be the intended goal path.

Proving the existence of a conserved quantity, or dynamic invariant, as a condition for success of goal-directed behavior might be called *the fundamental theorem of ecometrics*. Showing empirically that this theorem does in fact hold in a variety of experimental contexts for different actors is very important for motivating research into systems that exhibit intentional dynamics, that is, goal-directed behaviors. For the theorem asserts that an abstract similarity must hold over the perceiving-acting cycle whenever an actor successfully uses perceptual information to achieve an intended action goal—a statement tantamount to the claim that a dimensionless quantity exists which is a kind of *intentional dynamical invariant* and may be constructed over the relevant task affordables. These are clearly π-numbers at the ecological scale rather than an arbitrary physical scale, as illustrated in the following example.

Consider two cats who satisfy their respective intentions equally well by jumping to the top of a fence. Let one cat be more massive than the other. If we construct the equations for each performance (analogous to Reynold's equations for flow), the control and information variables should be similar (just as the inertial and viscous forces were similar). The pair of resulting ratios of these similar quantities will produce a π-number that is the same for both cats, being indifferent to the difference in their masses. The total action for a given cat comprises certain functions of the information detection and energy control that maintain an invariant relationship just in case the cat's behavior is successful. When the two cats' behaviors satisfy similar goals, then other similarities should hold as well. For instance, the parabolic paths through the air to the fence top should be geometrically similar in shape and kinematically similar in the time to traverse the paths; likewise, the work-to-contact with the fence top should be kinetically similar, and so forth. The fact that they also have the same targets and similar manners of approach to those targets and hence similar goal paths should be reflected in the dimensionless ratio. Hence, the similarity of their goal-directed tasks with respect to both ordinary causal dynamics and intentional dynamics should be revealed by their respective π-numbers being identical. The degree to which these numbers are not identical provides a measure of the dissimilarity of their actions. This is the promise of dimensionless analysis that we wish to explore in tasks involving different actors performing the same or similar locomotory tasks.

Our research project, therefore, ultimately aims at determining the empirical validity of the fundamental theorem of ecometrics, as sketched earlier. As a means for exploring this thesis of intentional dynamics, there is no problem more scientifically interesting nor socially significant problem than wheelchair navigation. In the next section, we outline the method of attack on this problem and present some recent encouraging results. It should be clearly noted, however, at the outset that although these results obtained encourage the use of π-numbers, they also indicate that the current use has been too limited to solve the prospective control problem in its most general form (e.g., locomotory navigation toward a goal through a cluttered environment). Consequently, after reviewing the promising but limited results of current research by ourselves and others, we shall discuss the mathematical foundations of π-numbers. Our goal will be to suggest the ways of removing these limitations so that π-numbers might be generalized to measure similarity of any actions by systems operating at the ecological scale under the aegis of intentional dynamics.

11.4   Part III: Recent Investigations of the Fit of Actors to their Environments

11.4.1 π-numbers of Different Orders

Anthropometry is the study of the design of environments, furniture, tools, and equipment in units proportional to the intrinsic measurement of the human body (Panero & Zelnik, 1979). The architect Corbusier employed anthropometric principles, on a large scale, in the design of cities and buildings and, on a smaller scale, in the design of furniture and equipment. The famous Bauhaus initiated exploratory investigations into such design principles in the 1930s. Since then, the anthropometric approach has become an integral part of all design disciplines, especially architecture, human factors, and human engineering. Anthropometric π-numbers are those dimensionless ratios based on body-scaled measurements that play a role in the design and evaluation of environmental structures, furniture, tools, and equipment. Clearly, anthropometric π-numbers constitute an important class of geometric π-numbers. Very recently the empirical investigation into the validity of such geometric π-numbers has grown in popularity.

Specifically, in our laboratory and elsewhere, we use π-numbers to express similarities between energy control and information detection. As argued earlier, under the basic ecometric theorem of intentional dynamics, such similarities should exist whenever actors must make compensatory motor adjustments, given goal-specific information, to carry out successful goal-directed activities. Such compensatory motor adjustments, under the control of perceptual information, specify the actor's fit to the relevant environmental structures. This is the pragmatic meaning of the perceiving-acting cycle (Gibson, 1979).

A decade ago, as his dissertation, Warren carried out a now famous study exploring the kinetic basis for anthropometric (geometric) π-numbers (Warren, 1982, 1984). He showed that for people of different heights the perceived optimal stair design was specific to the person's individual height defined as a function of leg length. (Here the measure of optimal stair design is the riser-to-tread ratio taken relative to the climber's leg length.) Nevertheless, the ratio of leg dimensions to stair dimensions for all people (riser height in cm/leg length in cm $= r/l$)—short and tall ones—were identical ($r/l = \pi_p = .88$). These people then had to climb a variable motorized stairmill. The stairmill ran at speeds requiring a wide range of step frequencies (30 to 70 cycles/min), while the subjects climbed as comfortably as possible. Warren obtained an important result. Subjects produced an invariant optimal work measure (π_a = .26, as measured in calories/kg-cycle by oxygen utilization methods) only for those designs that matched the dimensionless number obtained from their original perceptual judgments.[3] This research showed convincingly the existence of anthropometric π-numbers that hold predictably over both perceptual and action measures (π_a and π_p). How general are these π-numbers?

Konczak, Meeuwsen, and Cress (1992) investigated stair designs with maximum riser heights. They asked whether leg strength and hip–joint flexibility provide additional relevant constraints on both the perceptual judgment and action capability of subjects regarding such stairs. Thus, their work generalized Warren's conditions. Warren measured only anthropometric factors, omitting the effect of different dynamic conditions (such as different speeds) and used only young adults of the same approximate ages. Konczak et al. (1992), by contrast, showed that younger and older adults differ in the perception of maximum climbable riser height. In other words, adults can accurately perceive the relative limitations and, presumably, the change in their action capabilities that aging brings. Specifically, their analysis showed that one's action capability (in stair climbing) is subject to multiple biomechanical factors (that is, kinetic π-numbers). These factors, taken together with anthropometric constraints, are better descriptions of the action capability than the anthropometric constraints (geometric π-

---

[3]The action π_a-number, on first consideration, may not seem to be truly dimensionless as it must be in order to be a π-number since it involves a kinetic quantity (minimum energy expenditure per vertical meter measured in units of calories/kg-m) taken in reference to a geometric quantity (leg length). However, π_a is found by, first, plotting energy expenditure, E_d, as a function of riser height, R, and, then, taking R_0, the riser height value at which E_d is minimal, as numerator and leg length, L, as denominator to obtain $\pi_a = Ro/L = .26$ for all people regardless of height so long as the stair design is anthropometrically optimal for them. This was Warren's (1982) method. A more general method exists for computing π-numbers from dimensionalized measures that are dimensionally inhomogeneous. If one takes the measures in two different situations, as over learning trials for an individual, a test-retest, or over different individuals, dimensionless ratios can be constructed from measures that fail to satisfy the property of dimensional homogeneity. Let the two measures taken in the first situation be k_1 and k_2, dimensionalized as MLT and LT, respectively; and where k'_1 and k'_2 are the same measures taken on a different occasion. We can then form a dimensionless quantity by taking their cross-ratio as follows: $k_1 MLT/k_2 LT$.

numbers) taken alone.

By using π-numbers, these experiments make an exciting discovery. They show that the control of energy expended for action (kinetic π-numbers) is formally similar to the information available through visual perception (geometric π-numbers). But where is the dimensional homogeneity required? It must be supplied in some way, for it seems that people can "see" the work to be done, the impulse forces to be scheduled, and the torques by which to direct their behaviors toward selected goals in the manner intended. Perhaps, the underlying ecometric commensurability required for this to be so is a similarity relationship existing between something felt and something seen. Could it be the similarity between the "felt effort" experienced during the act of climbing the stairmill and the information detected in seeing the stair design? Is this not the source of the commensurability of haptic information with visual information required for their dimensional homogeneity? We return to this point later.

Other approaches have also successfully constructed π-numbers for different activities than stair climbability. Each of these is constructed from a ratio of an anthropometric variable with an environmental variable, defining what we call an *ecological π-number*. In principle, such ecological π-numbers may be anthropometrically based (geometrical π-numbers), biomechanically based (kinematic π-numbers), bioenergetically based (kinetic π-numbers), or intentionally dynamically based (ecological π-numbers). Researchers have identified a range of π-numbers in different scientific domains. In psychology, following Warren (1982, 1984), Mark (1987) found geometrical π-numbers, whereas Lee (1974) and Todd (1981) found kinematic ones. As pointed out, in at least one case these numbers have been validated both for perceptual judgment and for physiological measures of action under normal conditions. These findings suggest that the human perceptual system is a fine measuring device for different dimensions of locomotions, such as climbability (Warren), sittability (Mark), catchability (Todd), brakingability (Lee), and passability (Warren & Wang, 1987).

In addition, we have extended this methodology to wheelchair passability studies. We review these results next.

11.4.2 Wheelchair Passability Studies

With the exception of the work on driving (e.g., Gordon, 1966) or braking automobiles (Lee, 1976), or flying and landing of airplanes (e.g., Warren, 1991), research on the dynamic fit of actors to their environments has been restricted to the investigation of natural biological motions, that is, movements unassisted by prosthetic or vehicular tools. Humans, and even some animals, use tools successfully as substitutes to increase their innate or learned abilities. The success of these tool-using activities naturally suggests the possibility that π-numbers might be used to construct scales that measure the dynamical fit of human activities to their functionally (affordance) defined spaces.

The General Hypothesis: Function Unit Scaled Actions. In the initial research, we restricted our efforts to the careful measurement of some of the chief dynamic variables that affect the observed value of π–numbers for wheelchair passability with respect to doorways and passageways. Our working assumption is that such π-numbers will be most stable when the behavior engaged in is most comfortable. The orientation we take to this problem is a functional one and was anticipated by Gibson (1979). (Please reread carefully the extracted quotes at the beginning of is chapter.)

What one derives from these passages is that the measure of fit of an actor to its environment is not rigid but plastic and functionally variable; the actor's egospace, or field of safe and comfortable action, whether it be manipulation or locomotion, is not restricted to a fixed region of space-time. Because a tool extends the actor's body so that the boundary between the animal and the environment is not fixed at the surface of the skin but can shift. Similarly, this action field is not stationary relative to the environment but moves with the actor, shifting and changing, continually, bending and twisting, and also elongating or contracting to accommodate obstacles that encroach on it and limits its boundaries. Thus, no dimensionless constants may be based merely on static structural units of measure but must be based on functionally defined units.

In the same vein, we wish to argue that a wheelchair, as a vehicular tool, may be considered an extension of the user's body; moreover, it carries with it a dynamically plastic field of safe and comfortable travel—a field that functionally tailors its boundaries to fit whatever constraints that might arise from the actor's dynamics or from the environment. This means tool changes the functionally defined action capabilities of the user, or more briefly, the user's *effectivities* by enabling the user to do things in a way that otherwise could not be done (Flascher & Shaw, 1989). The success of tool using activities naturally raises the possibility of generalizing body-scaled relationships to the environment so that anthropometric measures might be generalized to ecological measures. When the former are geometrically defined, the latter must be defined as units of measure of the functional fit of the user to the environment—a fit that usually is dynamically scaled. Hence, we call such measures *function unit scaled*. Such measures are to be contrasted with extrinsically defined, static anthropometric measures of fit. But what difference should we expect tool use (e. g., wheelchair use) to make on the calculation of π-numbers?

As indicated earlier, a tool can be defined as an environmental structure the control of which aids an actor in achieving a goal-directed activity. As shown in Figure 11.1, we are interested in π-numbers that are identical when the two systems, S_1 and S_2, are similar, have similar tool interfaces, T and T', with similar environmental structures, E and E' (recall Figure 11.1). That is, where S_1: O+T->E yields π and S_2: O' + T' -->E' yields π', then we expect π, ideally, will be identical to π'. For example, where Warren (1982) found identical π-numbers for people of different body-scales climbing stairs of different but anthropometrically proper designs, so, for the reasons given earlier, we would expect that the π-number yielded by one wheelchair user passing through a doorway of proper clearance will be identical to the π-number yielded by another user in a different wheelchair passing through a different but properly designed doorway.

Several studies of passing through apertures (that is, clear widths) revealed the possibility of a universal ratio of body size and minimal passable gap size. Humans as well as frogs perceive an aperture as affording passage if it is at least 1.3 times their body width (Ingle & Cook 1977; Warren & Whang, 1987). As pointed out, Warren (1984) found that the perceptual category boundary between climbable and unclimbable stairs corresponds to a critical riser height as a function of riser height/leg length ratio. Interestingly, Warren was able to compute a critical, dimensionless number describing this relationship.

By analogy, our laboratory uses π-numbers to discover control similarities among different wheelchair users who pass through the same openings at different velocities or through different openings at the same velocities. Warren's research showed convincingly a π-number that holds predictably over perceived dimensions of the environment (a stair design) can be used to predict the minima of a kinetic measure (oxygen consumption). We try to generalize these measures even further so that ultimately they might include intentional dynamic π-numbers.

In pursuing these practical goals, we assume that:

1. The dynamic fit of a wheelchair user is a function of the

efficiency, comfort, and safety associated with carrying out a specific task in a given environmental setting.

2. That geometric, kinematic, kinetic, or intentional dynamic π–numbers provide scaling coefficients that can index the dynamic fit of wheelchair activities to the functional environment.

3. That critical values of dynamic π–numbers can be defined and used in qualifying guidelines for the design of wheelchair environments.

11.4.3 Passability by Walkers, Automobiles, and Wheelchairs

Ecologic π-numbers and their generalized forms provide measures for characterizing the functional architecture of spaces designed to accommodate active wheelchairs. The experiments reported next suggest the main themes of future investigations. The main purpose of these preliminary experiments was twofold:

1. To assess whether passability is body scaled (i.e., indexed by a geometric π-number) or whether it is dynamically scaled (i.e., indexed by a kinematic π-number).

2. To compare passability π-numbers for walking, car driving, and wheelchair using and other activities.

We investigate these questions and methods in both the laboratory and the field. Future endeavors will focus more broadly on characterizing accessibility routes through functional workspaces of different designs (offices versus homes).

Experiment 1. Automobile Passability Judgments. (Flascher, Shaw, Carello, & Owen, 1989).

Method. Ten subjects (all experienced drivers) were asked to judge the minimum clear width passable by a car. Using the limits method borrowed from psychophysics, subjects judged aperture clear widths regarding the maximum width of their own car and a smaller or larger experimental car. Subjects made judgments under six conditions for each size car from a distance of 12 m: with subjects (1) sitting in their own car; (2) sitting in the experimental car; and (3) standing outside. Hence, there were 2 x 6 = 12 conditions in all.

Results. We calculated the dimensionless π-number for passability by dividing the judged minimal passable gap by the target car width. The passability π-number in this task was 1.22. Judgment position had no influence on judgment $F_{(2, 9)} = .747$; however, there was a difference between the judgment for the experimental and own car $F_{(1, 18)} = 7.008$, $p < .05$). The π-number for the subjects own car (1.24) was larger than the π-number for the experimental car (1.20).

Conclusions. These results support the assumption of a functionally defined scale for the perception of passability, of which body-size scale (that is, geometric dimension) is a limiting case when no dynamics are involved. Even though the passability ratio for cars (1.22) was different from that for walking (1.3), they seem close enough so that differences may be accounted for by different task demands due to different modes of locomotion.

Experiment 2. Dynamic Choices of Wheelchair Passability (Flascher et al., 1989; Flascher & Shaw, 1989).

Method. Each of four subjects, while sitting in a wheelchair, was confronted over 14 trials with a row of 11 apertures using a forced-choice method. The apertures were ordered by size (either ascending or descending order from the subject's point of view). Subjects were asked to roll themselves along a line parallel to the aperture row, stop when they saw the smallest passable gap, and then race through the chosen aperture as fast as they could. Apertures were varied in 2 cm increments to create 7 ranges: 59–79, 61–81, 63–83, 65–85, 67–87, 69–89, and 71–91 cm. Each subject approached each row from the left and from the right in a counterbalanced design. The size of the passed aperture and the direction of approach and speed of passage were recorded.

Result. The average passability π-number in this task was found to be 1.18.

Conclusions. From these results it can be concluded that in active selection as in passive judgment the same conclusions as in Experiment 1 are warranted.

Experiment 3: Perceived Passability Judgments for Walkers and Wheelchair

Users. (Flascher & Carello, 1990).

Method. Fourteen naive subjects were asked to judge the minimum aperture width that was passable under two different styles of locomotion: walking through or wheeling through in a wheelchair. Using the limits method, subjects judged aperture passability under four different conditions (two in each style) from a distance of one eyeheight: standing or sitting in a wheelchair. None of the subjects had ever used a wheelchair before.

Results. The passability π-number for wheelchairs in this experiment was found to be 1.22 and for walking 1.12, $F_{(1, 13)} = 8.564$, $p < .05$. There was no significant difference between the judgment positions (standing or sitting; $F_{(1, 13)} = .521$), and there was no interaction effect, $F_{(1, 13)} = 2.966$).

Conclusions. In this case, a direct comparison between the perceived widths for wheelchair passability and walking reveals a significant difference in π-numbers.

Discussion: How General are π-numbers?

In these experiments we were able to identify a passability π-number for car driving (1.22), for wheelchair using (1.22 and 1.18 for different tasks), and for walking (1.12). The discrepancy in these results generally suggest a need to understand perceptual scaling in terms of function units (dynamic π-numbers) rather than merely in terms of body dimensions (static π-numbers), as has previously been the case (e.g., Warren, 1982).

Although a significant difference was found to hold between dynamic π-numbers for walking and for wheelchair wheeling, one cannot argue for a basic difference in their perceptual characteristics. It now seems likely that the value of a π-number is changed (possibly reduced) with experience. For instance, in the first experiment we found that subjects scaled differently to their own cars than they did to an experimental car (which was a different size from their own, being either bigger or smaller). In the wheelchair experiments, none of our subjects had any prior experience with wheelchairs, so no conclusions could be drawn regarding learning and the possible reduction in the size of the π-number. In future research, of course, the hypothesis might be tested by comparing novice versus expert users. We intend to do so.

On the one hand, these experiments confirmed the usefulness and general applicability of the π-number methodology. On the other hand, they suggest that limits might be placed on the generality of these numbers. In other words, the assumption that a passability π-number might be a fixed universal constant now seems less likely. However, this limitation, if proper, still does not reduce their usefulness as the basis for a generic function scale that might be used for measuring the perceived or actual fit of actors to their environments (i.e., their use as ecological π-numbers). Recent experiments support this appraisal.

An important issue is whether the π-number of passability is "body scaled" or function-unit scaled (where the "body" scale may be defined for geometric dimensions of a wheelchair, automobile, or walker). In this regard, tasks involving different geometric ratios between diverse bodies and apertures, relatively close π-numbers for passability were nevertheless found. This presumably resulted from the tasks placing similar functional requirements on relatively independent body dimensions. Consequently, it seems likely that functionally scaled π-numbers may exist rather than geometrically scaled ones, as originally thought (Warren, 1982, 1984). However, to ratify this conclusion, further investigations across a wide range of task categories are called for to determine the exact relation of perceptual π-numbers to action π-numbers.

The experiments discussed also suggest that, although π–numbers are useful for gaining a handle on a problem, they should be treated as functional and dynamic variables that may change rather than as fixed universal constants—what we later shall call "dimensionless functions". Nevertheless, over a wide variety of modes of locomotion and passability conditions, π-numbers provide useful measures for both specific dynamic conditions (e.g., comfort mode or maximum speed) as well as for limiting cases (e.g., minimum passable gap size, maximum riser height, etc.). The differences in π-numbers over different tasks further suggest that, because of their abstractness, they may provide a task-dependency measure of great generality. Such a scale could be used to index the differential properties affecting the degree of fit of different types of behavior to different kinds of situations. In this way, one might scale the dynamic fit of wheelchair users to their environments, whereby making reference to effort and fatigue to which they may be subjected in obtaining or maintaining the respective fit.

These findings suggest that the investigation of the various π–numbers identified should be extended in new directions. For instance, dimensionless analysis should be carried out in a wide variety of work-spaces. To do so, however, requires moving beyond π-numbers to more abstract descriptions of dimensionless quantities. We need these more general mathematical procedures for coupling the affordables intentionally—that is, for coupling the function units in perceptual information (observables) to the dual function units in the control of actions (controllables) as constrained by the goals (valuables) intended.

For dynamic tasks, dynamic π-numbers are most relevant. Thus, a method is needed to determine how expert actors (e.g., wheelchair users) select those routes that are minimally clear for travel from among those that are not, and from among the minimally clear routes how they select those that are also most safe, efficient, and comfortable. Only in this manner will the most adequate design principles emerge. Our future project has as its chief aim to propose such mathematical methods whereby continuing to develop and evaluate the instrument we have for making the required dynamic measurements.

11.5   Part IV: Generalized Dimensionless Analysis

11.5.1 Overcoming Traditional Failures to Address the Ecometric Scaling Problem

Earlier we discussed the seminal experiments on stair climbing conducted by Warren in his dissertation carried out in our laboratory. This research showed convincingly the existence of anthropometric π–numbers that hold predictably over both perceptual (information) and action (control) measures—the use of geometric π-numbers to index kinetic efficiency and comfort. However, to reach into the realm of intentional dynamics, even more general dimensionless concepts than these must be used.

By using π-numbers to express stair-climbability, Warren's experiments made an exciting discovery. They showed that the control of energy expended for action is formally similar to the information available through visual perception so that actors can *see* geometric design information as if scaled in terms of energy demands. In the second phase of the same experiments Warren validated these perceptions. Here subjects actually climbed stairmills of designs more or less like the stairs judged in the first phase of the experiment. Although actors of different heights found different stairmill designs easier to climb, measurements showed they expended the least amount of energy on those stairmills that they had selected a priori in the first phase. Hence the actors were solving the ecometric scaling problem. Because, as a function of anthropometric fit, geometric information (the stair design) constrained the kinetic activity of climbing, we can also conclude that the actors were solving the ecomechanics transduction problem as well.

Earlier we suggested that the ecometric commensurability underlying Warren's climbability task was a similarity between something felt and something seen. We speculated that the similarity needed holds between the "felt effort," experienced during the act of climbing the stairmill, and the information detected in seeing the stair design before climbing. Hence, haptic information must somehow be commensurate with geometric visual information. This suggests a need for understanding at a more general level how ecological π-numbers work.

An ecological π-number is specific to a rule for the perceptual control of action as defined for a particular affordance goal (e.g., stair climbability, aperture passability). Like any other π-number, it must satisfy the physical property of dimensional homogeneity mentioned earlier. Ecological psychology typically treats information and control as being dimensionally inhomogeneous, thus creating the problem of how kinematics can specify kinetic values. That is, information is treated kinematically (e.g., L,T), being defined over dimensions of geometry and time; whereby control is treated kinetically (e.g., F, L, T), being defined over dimensions of force (or mass), length, and time. For example, we speak of visual information as kinematic, as in the case of time-to-contact with a target (Lee, 1974). Yet, such force-free information somehow specifies forceful control, as in the application of braking forces to avoid colliding with the target seen. Runeson has identified the problem of how kinematic information specifies dynamics (kinetics) as the search for the KSD principle (Runeson & Frykholm, 1983). Clearly, there is no dimensional homogeneity under this interpretation. In contrast, here we argue that the KSD principle is, in fact, a DSD principle with kinetic based information detection specifying kinetic-based energy control. Thus, dimensional homogeneity is satisfied, and ecological π-numbers may be constructed. Before presenting this DSD, view as one way to rid ourselves of the incommensurability problem, consider another way. The traditional approach attempts to avoid this incommensurability problem by taking a different tack. But, as we shall see, this approach has major shortcomings.

In standard information theory, as developed by Shannon, the concept of information is dimensionless, being measured in "bits." (Shannon & Weaver, 1949) (*Bits* are binary units obtained by taking the logarithm to the base 2 of the total number of exclusive disjunctive choices possible under a given choice-set of items.) Likewise, from the development of abstract machine theory we have a dimensionless concept of control. Consider Turing's notion that state transitions a machine makes (reading symbols on tape, writing symbols on tape, erasing tape, moving tape right, moving tape left) are treated as determined by rule-directed symbol manipulations independent of the forces required to control a real machine. Just as information variables are not bound to the semantic dimensions of a message's content, so are control variables not bound to the physical dimensions of a machine's plant components. By both being dimensionless, the problem of dimensional inhomogeneity is avoided from the start by assumption.

Does not the existence of dimensionless descriptions imply the widest possible theoretical generalization across a wide class of analogous systems because their specific semantic and physical details can be ignored? So why not build π-numbers directly from these classical vacuous notions of information and control? If only dimensionless variables go into the ratio, then there can be no problem of their inhomogeneity. But is a deeper problem of incommensurability really avoided by this tactic?

Unfortunately, ignoring the dimensional variables and attempting to work directly with dimensionless ones does not solve the commensurability problem, so that information can scale control constraints. Because information and control variables are not bound to qualitative dimensions, their scale is left free floating, undefined, and hence potentially incommensurate. For instance, the information value of any nested (fractal) pattern is infinite (undefined) unless an ad hoc, extrinsic measurement constraint is applied to fix the level of analysis. Likewise, the control of state transitions in a nested state space is undefined until the level of control is extrinsically constrained, say, by an ad hoc programmer's ploy or physical limitations of the machine or organism.

How, then, can we be sure that the ratio of information to control is the appropriate one for the perceiving-acting cycle involved? An answer to this question comes in two parts: First, the concept of π–numbers must be generalized to a mathematical level of abstraction sufficient to allow information and control to be treated commensurately if indeed they can be. Second, it must be shown that under the intentional dynamics approach, one is, indeed, justified in treating them commensurately. Such commensurability would guarantee dimensional homogeneity of these dynamical processes over the perceiving-acting cycle; thus, dynamical π-numbers might then be actually constructed to reflect the degree of success of attempted goal-directed behaviors.

11.5.2 Higher Order Dimensionless Invariants

To illustrate the generalization process needed to pursue to get higher order invariants, it is useful to scrutinize further the hydrodynamic example already introduced with the Reynolds number. As mentioned earlier, one specific Reynolds number is specific to a family of flows, namely, to flows in which the ratio of inertial forces to viscous forces is the given Reynolds number. In the literature these flows are called *geometrically similar steady state flows*, referring primarily to the constant velocity of the flow along a given section. It is important to note that flows characterized by the same Reynolds number are often called *dynamically similar flows*. In other words, a Reynolds number is deemed necessary and sufficient to capture dynamically similar, steady flows.

An analogy can be drawn between a geometrical π-number for a wheelchair "flowing" through a passageway and a Reynolds number for a liquid flowing through a passageway. However, the analogy is not simple; although in the former case we refer to a geometrical π-numbers, in the latter case we refer to a different, more general, dynamical π–number. Consequently, a closer look is needed to clarify what is meant by dynamical π-numbers as a generalization of geometric π–numbers. Earlier studies on climbability and passability revealed a geometric π-number which was specific to the "minimal" dynamics involved. In other words, subjects were instructed to apply a controlling force (analogous to inertia force) to maintain a normal, comfortable, and constant speed. Each subject belonged to a population with specific attributes affecting the effort required to maintain that comfort mode activity (similar to overcoming viscous forces). From attempts to control (inertial forces) one's movement while remaining in a comfort mode (overcoming viscous forces), a Reynolds number interpretation seems to follow naturally.

Using terminology from hydrodynamics, we might consider the "normal speed" to be a steady-state mode of motion. However, these steady-state modes are ideal and, in that sense, represent only a minimal dynamical solution. We say ideal, because real flows are not constant but change as their boundary conditions change. For instance, the flow velocity of a river is not invariant but changes as the river's cross-section changes, or the speed of a car changes according to the changing road conditions (e.g., roughness), even if road width remains the same. Thus, for the general case of flow in natural settings, the term dynamical means continuously changing velocity, or continuously changing viscosity, or whatever. For our "dynamical" examples, the analogy sought is between a Reynolds number for real rather than ideal dynamic flow, that is, for flow that is continuously changing between cross-sections and also in time at the same cross section. Thus, similarly, the passability π-number must also be capable of reflecting change between cross-sections and in time for each cross-section as a function of changing environmental and subject conditions.

In Figure 11.2 we have an example of how the Reynolds number's effect on the mode of flow depends on the environmental boundary condition that is present. For flows of equal velocity, equal viscosity, and so on, but different boundary conditions (roughness of the pipe), different frictional values arise causing a bifurcation into flows exhibiting different modes of turbulence. Notice how the Reynolds number range (R) of the joint critical-transition region increases as friction (f) decreases. This is a function of roughness on the boundary of the flow (which effects the growth of the boundary layer and therefore the size of the region containing the inviscid core). The analogy to intentional dynamics is straightforward: Intentions, defined on the initial condition, not only select target parameters, defined on the final conditions, but select manner parameters as well which, in turn, determine the manner of flow toward the target from the point of initial intent. The value of this manner parameter is a boundary condition, like roughness, which selects the particular path over which the actor moves. In fluid mechanics, the path that minimizes fluid action is that which allows flow to remain most laminar, just as in ecomechanics, the goal-path that minimizes the actor's need for self- control (while maximizing target direction information) is that which allows behavior to remain most comfortable.

Consequently, the most natural generalization is from dimensionless invariants that are constant π-numbers (constant dimensionless functions along a steady flow) to nonconstant dimensionless functions (along varying flow). At this level of generality, instead of using invariant dimensionless constants (π-numbers) to express the dynamical similarity between different flows (fluid particles or moving vehicles), we should use π-functions—that is, invariant dimensionless functions. A mathematical strategy for achieving the generalization process is heuristically outlined next.

11.5.3 Searching for the Laws Responsible for Dynamical

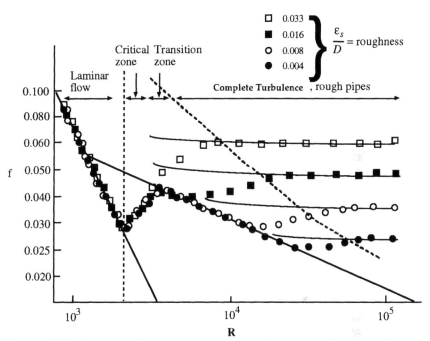

Figure 11.2 *The effect of the environmental condition of roughness on flow as a function of Reynolds number.*

Similarities

The heuristic outline sketches a popular technique by which the laws underlying dynamical similarities might be discovered and characterized. (For more detail and definitions please consult the appendices.)

Step 1: *Identify the dynamical dimensional space (so called base space) in which the laws governing the phenomena might be formally expressed.* The dimensional space in most cases captures a certain amount of redundancy in the measured variables used. The derived variables of velocity, acceleration, or momentum can be expressed in terms of the fundamental M (or F), L, T dimensions. In most cases, the dynamics of this dimensional base space is expressed in differential equations.

Step 2: *Characterize the redundancy of the dimensional base space that expresses the dynamical similarities of interest.* This redundancy can be characterized in terms of the symmetry structure of the base space. Symmetry structure is best characterized by the continuous Lie group of infinitesimal differential operators derived from the original differential equations modeling the relevant system dynamics (for details see Shaw et al., 1990; Stephani, 1989). The Lie group symmetry technique is especially useful for simplifying the analysis of complex dynamic systems because it reveals the participating laws (invariant subgroups) responsible for creating the symmetry structure of the base space in the first place.

Step 3: *Using the Lie-group technique, factor the dynamical space into its invariant subgroups.* These will identify the participating laws. In other words, by unfolding the factor structure of the Lie-group description of the base space, we reveal the laws (invariant subgroups) that are operationally identified with the base space redundancies (symmetries). By doing so, dimensionally homogeneous, invariant functions are obtained (see Appendix A).

Step 4: *By using Drobot's more general reformulation of Buckingham's π-theorem (Buckingham, 1914), obtain π-numbers (if the dynamics is simple) or invariant functions (if the dynamics is complex).* Invariant functions are those functions that do not change when the system of measurement is changed. Thus, similar properties remain similar over two systems, even if one of the systems is rescaled. Generally, a two-step procedure is required to obtain invariant functions: One first constructs dimensionless constants and then generalizes them to the dimensionless invariant functions. There is a more direct way to obtain invariant functions. If one uses Drobot's reformulation of the π-theorem (Drobot, 1954; Kasprzak, Lysik, & Rybaczuk, 1990), then there is no need to construct the dimensionless constants first.

It is worth noting that Drobot reformulated the π-theorem so as to obtain invariant functions that are directly defined over the π–numbers of Buckingham's π-theorem (see Appendix A for details): From Buckingham's π-theorem we are guaranteed that it is always possible to reduce the number of independent parameters in a problem by compacting the dimensional parameters into a set of dimensionless parameters. From Drobot's reformulation and generalization of this theorem, we are also guaranteed that the Buckingham procedure will work for more abstract dimensionless functions. (Usually the more recent reformulation is given without crediting Drobot with the generalization; e.g., Gerhart & Gross, 1985, p. 364, whose treatment we follow.)

Assume any physical process governed by a dimensionally homogeneous function whose argument consists of n-dimensional parameters, such as

$$x_1 = f(x_2, x_3, \ldots, x_n),$$

where the x's are dimensional variables. Then there exists an equivalent function whose argument consists of a smaller number (n-k) of dimensionless parameters, such as

$$\pi_1 = F(\pi_2, \pi_3, \ldots, \pi_{n-k}),$$

where the π's are dimensionless groups constructed directly from the x's. The amount of reduction in the parameter set, signified by k, is usually equal to, but never more than, the number of fundamental dimensions involved in the x's. Here the dimensionless parameters are constructed as products of powers of the dimensional parameters (see Appendix A for details).

Essentially this important theorem asserts that a complex system with redundant structure can be made nonredundant and therefore simplified. This amounts to finding the smallest number of parameters for describing a system by discovering the minimal generator set for all the paths through the system's state space.

Step 5: *Model the way and degree to which two systems have similar intentional dynamics (i.e., goal-directedness).* Drobot's dimensional space can be interpreted as a *fiber bundle space.* This suggests that differential geometry can be used to provide a fiber bundle interpretation of the invariant functional associated with the flow of the relevant potential (i.e., the generalized Hamiltonian representation of the total action potential). The similarity between the two systems can then be identified with that connection in the fiber bundle which gauges the flow of information on one manifold to the flow of control on the dual manifold. A third lower dimensional manifold, called the *quotient manifold,* defines a connection in the fiber bundle which carries the scale factors for relating the two flows. The parametrically simpler flow of the generalized action potential on the quotient manifold is obtained through group factorization of the flows on the other two manifolds (see Appendices A and B).

Earlier we asked how one can be sure that the ratio of information to control is the appropriate one for the perceiving-acting cycle involved? It was suggested that an answer to this question comes in two parts: first, that the π-number concept must be generalized to a level of abstraction sufficient to allow information and control to be treated commensurately; steps 1–5 suggest an approach to answering this first part of the question. The second part requires showing that an intentional dynamics approach justifies treating information and control flows as being commensurate. Intuitive and formal arguments supporting this claim have been made by us elsewhere (Shaw & Kinsella-Shaw, 1988; Shaw et al., 1990; Shaw et al., 1992). These earlier arguments were the primary motivation and justification for the generalized dimensionless invariant story told here.

11.6 Conclusion: The Explanatory Value of Dimensionless Analysis

This chapter has been organized around the notion of the π-number and its generalization to dimensionless functions. The search for these dimensionless parameters promises the following benefits to psychological theory: (a) π-numbers and their generalizations can be used as the basis for intrinsic measures that relate on a single scale quantities that might otherwise be taken as unrelated. For example, they might show intrinsic measurement relationships holding between environmental variables and organismic variables, affordances and effectivities, perceptual and action variables, and so forth. We have concentrated on how π-numbers might be used to reveal intrinsic ecometric relationships between information and control variables. (b) π-numbers and their generalizations provide means to reduce less tractable problems involving systems with high-dimensional state spaces to problems involving systems with low-dimensional state spaces. (Recall that the reduction in complexity of state space will always be proportional to the number of relevant invariants, or π-numbers, discovered.) Finally, (c) through the use of principal π–numbers (or principal π-functions), that is, π-numbers based on laws, the functional relationship of psychological variables can be shown to imply lawful rather than arbitrary bases for explanations. This should have the effect of moving psychology into close alignment with other law-based sciences, such as physics and biology. (Readers interested in pursuing this point might consult Kugler & Turvey, 1987; Rosen, 1978; Schuring, 1977). Our main concern in this chapter has been to suggest ways in which dimensionless analysis might serve the interests of ecological psychology by providing a formal interpretation of intentional dynamics. We return to this issue next.

11.6.1   Reducing Intractable Degrees of Freedom

The intentional dynamical approach to ecological psychology is a search for laws that relate perception to the control of goal-directed actions. Since Noether's pioneering work in applying group invariant solutions to characterize the classical conservations as symmetries (see Goldstein, 1968), one assumes that the discovery of such lawful invariants will help reduce the degrees of freedom of intractably complex systems to something more tractable. The search for invariants is the search for fundamental symmetries in the relationship between energy control and information operations as they interact to form the perceiving-acting cycle. How might this search proceed?

Although algebra is needed to solve problems, geometry is used to clarify them. An explicit geometrical setting for this approach is provided by the notion of a *quotient manifold, M/G*, where a manifold *M* is "divided" by the invariants of some regular group of transformations. For our purposes, these invariants will derive from a "compounding" of control and information detection operations into groups of operations. The coupling of these groups is expressed in their common (dual) action at conjugate points along a goal path on the ecological manifold. (In the Appendices we show that this is a gauge manifold defining a connection in the fiber bundle, and that it is of lower dimension than the control and information base manifolds that it relates.) Why is the ecological manifold, obtained from coupling the information and control manifolds, simpler (more tractable) than either manifold taken alone? The answer lies in showing that the ecological manifold can be constructed as a *quotient manifold* from these dual groups characterizing the two base manifolds. Each point on the dual quotient manifold will correspond to an orbit of each group.

Two main points should be made regarding the relevance of these remarks to the current approach: First, the Lie-group descriptions of the control and information detection operations capitalize on the existence of such invariant group solutions. When the goal-path is successfully traversed by the perceiving-acting cycle, we reveal the major group invariant that all the orbits (possible paths) share. Any other quantities that commute with a dynamical invariant also reduce the complexity of the original manifold, M. The discovery of such invariants greatly reduces the complexity of the original information and control manifolds, when treated separately, to a simpler quotient manifold, M/G, when treated jointly. From this joint treatment a single ecological field manifold is derived on which the orbits of the separate group operators are coupled into a single dual prescribed goal path. The motivation for constructing the quotient manifold is but a more geometric way to present the algebraic approach pioneered by Buckingham in his famous π-number theorem (see Olver, 1986, for details of this relationship).

The use of π-numbers helps to resolve the crucial problem facing psychological theorists—how to reduce the complexity of a given system to the lowest possible level. For instance, the intentionally dynamical systems to which we have referred may be difficult to model because they have equations of state that posit a very complicated, high dimensional state space. This problem has been referred to in control theory as the "curse of dimensionality" (Bellman & Dreyfus, 1962), in action theory as the "degrees of freedom problem" (Bernstein, 1967), and in ecological psychology as "intractable nonspecificity" (Shaw, Turvey, & Mace, 1982). The original manifold, on which the system's goal-directed behavior is defined by superposing information and control paths, may be extremely complicated. The use of dimensionless numbers reduces the dimensionality of this manifold to a new description using quotient manifolds. This move significantly reduces the degrees of freedom of the original dual state-space manifold. Hence, the search for law-based ecological π-numbers and their generalizations is clearly mandated.

11.6.2 Affordances as Dimensionless Invariants

Gibson held that perceptual information was scale-free, consisting of formless and timeless invariants (Gibson, 1979). Affordances are compounded from invariant relationships among environmental variables taken in reference to organisms as acting perceivers. Information is specific to these invariants of invariants. That affordances are dimensionless invariants is exemplified by Lee's generalized tau numbers specifying time-to-contact (i.e., surface contactability) and the tau derivatives specifying the marginal values for braking to avoid hitting objects (the negative affordance of collidability) or accelerating to intercept objects (interceptabilty) (Lee, personal communication, July 1992). Affordances as scale-free properties hold across species. The invariant ecometric properties that make one surface, say a bridge, afford support for one creature, say an elephant, makes another surface, such as a leaf, afford support for another creature, say an ant. The affordance of supportability is an abstract dispositional property which, by definition, must be scaled to the animal in question. Thus, the ratio of elephant-relevant properties to the bridge will reveal a value that is dimensionless and equivalent to the ratio of ant-relevant properties to the leaf. Warren (1982, 1984) showed that the climbability of stair designs by actors of different size nevertheless yielded the same π-number. Lee's tau-derived π-number for a collidability is the same regardless of the different distances and velocities of approach; hence, it is the same, regardless of the type of animal approaching or the type of surface being approached.

In all cases of affordances, supportability, edibility, collidability, graspability, passability, or whatever, the affordance is nonspecific or scale-free. It picks out an equivalence class of actor/environment ratios which, if dimensional homogeneity is satisfied, are dimensionless numbers or dimensionless functions. Not to see this is to miss one of Gibson's most elegant insights. For if affordances were merely specific measures of an individual animal's fit to its environment, there would be as many affordances of a given type as there are individual animals and individual environmental situations. Gibson aspired to a general theory of perception that could be applied to all kinds of creatures in a variety of econiches, and not one restricted to particular environmental situations, individuals, or species. This generality of the affordance concept allows ecological psychology to lay claim to being the study of the lawful aspects of perception and action.

Effectivities by which affordances are realized are no less scale–free than the affordances to which they are dual. Mathematically, the isomorphism that takes an affordance into the corresponding dual effectivity and vice-versa is called a *duomorphism*. The fundamental ecometrics postulate asserts that the duomorphism which, by definition must hold between affordances and their effectivities, guarantees that the measurement of goal-specific information and of the control of goal-directed action must be the same. (Mathematically, they must share Lie–group orbits that are similar up to isomorphism.) Hence, by the group factorization theorem, the same ecological dimensionless invariants always exist when similar goal-directed actions are successful—regardless of the goal, the creature, or the environmental situation.

The ratio of an affordance to its dual effectivity will yield a dimensionless invariant and have a value specific to the degree of success in attaining a goal—a value independent of the semantics of the situation. Information detection and energy control are operations that coordinate the effectivities with the affordances they serve. The intentionally coordinated perceiving-acting cycle can, therefore, be thought of as a "gauge" group with orbits that run through the environment and the organism thereby generating a goal path. If intentional dynamics is as lawful as we suspect, then modeling complex systems with such dynamics at the ecological scale will yield the simplest explanations possible, and the curse of dimensionality will have been lifted.

11.7 References

Abdel-Moty, E., & Khalil, T. (1989). Computer-aided design and analysis of the sitting workplace for the disabled. In A. Mital (Ed.), *Advances in Industrial Ergonomics and Safety I, Proceedings of the Annual International Industrial Ergonomics and Safety Conference.* (pp. 863-870). New York: Taylor & Francis.

American National Standard Institute. [ANSI]. (1986). A117.1-1986.

Arthur, J. (1967). *Employment for handicapped.* New York: Abington Press

Bellman, R., & Dreyfus, S. (1962). *Applied dynamic programming.* Princeton, NJ: Princeton University Press.

Bernstein, N. (1967). *The control and regulation of movements.* London: Pergamon Press.

Birkhoff, G. (1960). *Hydrodynamics: A study in logic, fact, and similitude.* Princeton, NJ: Princeton University Press.

Bridgman, P. (1922). *Dimensional analysis.* New Haven, CT: Yale University Press.

Buckingham, E. (1914). On physically similar systems. *Physical Review, 4,* 345–376.

Dainoff, M. (1987). Some issues related to seated posture and workstation design. In G. Salvendy, S. Sauter, and J. Hurrell Jr. (Eds.), *Social ergonomic and stress aspects of work with computers.* Amsterdam: Elsevier.

Dainoff, M. (1991a). Ergonomic comparison of VDT workstations. In S. Sauter, M. Dainoff, and M. Smith (Eds.), *Promoting health and productivity in the computerized office.* London: Taylor & Francis.

Dainoff, M. (1991b). Reducing health complaints in the computerized workplace: The role of ergonomics in education. *Journal of Interior Education and Research, 16,* 31–38.

Drobot, S. (1954). *Studia Mathematica, 14,* 84.

Duncan, W. (1953). *Physical similarity and dimensional analysis.* London: Ewald.

Flascher, O., Shaw, R., Carello, C., & Owen, D. (1989, July). *Perceiving aperture passability under ego-extension conditions.* Paper presented at The Fifth International Conference on Event Perception and Action, Miami University, Oxford OH.

Flascher, O., & Shaw, R. (1989, October). *Issues regarding tools, ego-extension, and perception of affordances.* Paper presented at The Fall Conference of the International Society for Ecological Psychology, Dartmouth College, Hanover, NH.

Flascher, O., & Carello, C. (1990, June). *Visual and haptic perception of passability for different styles of locomotion.* Paper presented at The Spring Conference of the International Society for Ecological Psychology, The University of Illinois at Champaign-Urbana.

Gerhart, P., & Gross, R. (1985). *Fundamentals of fluid mechanics.* Reading, MA: Addison-Wesley.

Gibson, J. (1950). *Perception of the visual world.* Boston: Houghton-Mifflin.

Gibson, J. (1979). *The ecological approach to visual perception.* Boston, Houghton–Mifflin.

Gibson, J., & Crooks, L. (1938/1982). A theoretical field analysis of automobile driving. In E. Reed and R. Jones (Eds.), *Reasons for realism: Selected essays of James J. Gibson.* Hillsdale, NJ: Lawrence Erlbaum.

Gordon, D. (1966). Perceptual basis of vehicular guidance. *Public Roads, 34*, 53–68.

Günther, B. (1975). Dimensional analysis and theory of biological similarity. *Physiological Reviews, 55*, 658–699.

Ingle, D., & Cook, J. (1977). The effect of viewing distance upon size preference of frogs for prey. *Vision Research, 17*, 1009–1019.

Ipsen, D. (1960). *Units, dimensions, and dimensionless numbers.* New York, NY: McGraw-Hill.

Johnstone, J., & Thring, M. (1957). *Pilot plants, models, and scale-up methods in chemical engineering.* New York, NY: McGraw-Hill.

Kasprzak, W., Lysik, B., & Rybaczuk, M. (1990). *Dimensional analysis in the identification of mathematical models.* Singapore: World Scientific.

Koffka, K. (1935). *Principle of gestalt psychology.* New York, NY: Harcourt Brace.

Konczak, J., Meeuwsen, H., & Cress, E. (1992). Changing affordances in stair climbing: The perception of maximum climbability in young and older adults. *Journal of Experimental Psychology of Human Perception and Performance, 18*, 691-697.

Kugler, P., Kelso, K., & Turvey M. (1980). On the concept of coordinate structures as dissipative structures: I. Theoretical lines of convergence. In G. Stelmach and J Requin (Eds.), *Tutorials in motor behavior.* Amsterdam: North Holland.

Kugler, P., Kelso, K., & Turvey M. (1982). On the control and coordination of naturally developing systems. In J. Kelso and J Clark (Eds.), *The development of movement control and coorrdination.* New York, NY: Wiley Press.

Kugler, P., & Shaw, R. (1990). Symmetry and symmetry-breaking in thermodynamic and epistemic engines: A coupling of first and second laws. In H. Haken (Ed.), *Synergetics of cognition.* Heidelberg, Germany: Springer-Verlag.

Kugler, P., Shaw, R., Vicente, K., & Kinsella-Shaw, J. M. (1990). Inquiry into intentional systems I: Issues in ecological physics. *Psychological Research, 52,* 98-121.

Kugler, P., & Turvey, M. (1987). *Information, natural law, and the self-assembly of rhythmic movement.* Hillsdale, NJ: Lawrence Erlbaum Associates.

Langhaar, H. (1967). *Dimensional analysis and theory of models.* New York, NY: Wiley Press.

Lee, D. (1974). Visual information during locomotion. In R. MacLeod and H. Pick (Eds.), *Perception: Essays in honor of J. J. Gibson.* (pp. 250–267). Ithaca, NY: Cornell University Press.

Lee, D. (1976). A theory of visual control of braking based of information about time-to-collision. *Perception, 5,* 437–459.

Mark, L. (1987). Eyeheight-scaled information about affordances: A study of sitting and stair climbing. *Journal of Experimental Psychology: Human Perception and Performance, 13,* 361–370.

Olver, P. (1986). *Applications of Lie groups to differential equations.* New York, NY: Springer-Verlag.

Panero, J., & Zelnik, M. (1979). *Human dimension and interior design.* New York, NY: Whitney Library of Design.

Rayleigh, J. (1915). The principle of similitude. *Nature, 95,* 2368, 66-68; and *95,* 2389, 614.

Rosen, R. (1978). *Fundamentals of measurement and representation in natural systems.* New York, NY: North-Holland.

Runeson, S., & Frykholm, G. (1983). On visual perception of dynamic events as an informational basis for person and action perception: Expectation, gender recognition and deceptive intention. *Journal of Experimental Psychology: General, 112,* 585–615.

Schuring, D. (1977). *Scale models in engineering.* New York, NY: Pergamon Press.

Sedov, L. (1959). *Similarity and dimensional methods in mechanics.* New York, NY: Academic Press.

Shannon C., & Weaver, W. (1949). *The mathematical theory of communication.* Urbana, IL: University of Illinois Press.

Shaw, R., Kadar, E., Sim, M., & Reppenger, D. (1992). The intentional spring: Strategy for modeling systems that learn to perform intentional acts. *Journal of Motor Behavior, 24,* 3-28.

Shaw, R. (1987). Behavior with a purpose. *Contemporary Psychology, 32,* 243-245.

Shaw, R., & Kinsella-Shaw, J. (1988). Ecological mechanics: A physical geometry of intentional constraints. In P. Kugler (Ed.), *Self-organization in biological workspaces.* Berlin: Springer-Verlag.

Shaw, R., Kugler, P. & Kinsella-Shaw, J. (1990). Reciprocities of intentional systems. In R. Warren and A. Wertheim (Eds.), *Perception and control of self-motion.* (pp. 579-619). Hillsdale, NJ: Lawrence Erlbaum Associates.

Shaw, R., & Turvey, M. (1981). Coalitions as models of ecosystems: A realist perspective on perceptual organization. In I. M. Kubovy and J. Pomerantz (Eds.), *Perceptual organization.* Hillsdale, NJ: Lawrence Erlbaum Associates.

Shaw, R., & Turvey, M., & Mace, W. (1982). Ecological psychology: A consequence of a commitment to realism. In W. Weimer and D. Palermo (Eds.), *Cognition and the symbolic processes, II.* Hillsdale, NJ: Lawrence Erlbaum Associates.

Stahl, W. (1963). The analysis of biological similarity. In J. Lawrence and J. Gofman (Eds.), *Advances in biological and medical physics: Volume 9.* New York: Academic Press.

Stephani, H. (1989). *Differential equations: Their solution using symmetries.* Cambridge: Cambridge University Press.

Todd, J. (1981). Visual information about moving objects. *Journal of Experimental Psychology: Human Perception and Performance, 7,* 795-810.

Turvey, M., & Shaw, R. (1979). The primacy of perceiving: An ecological reformulation of perception for understanding memory. In L. G. Nillson (Ed.), *Perspectives on memory research: Essays in honor of Uppsala University's 500th anniversary.* Hillsdale, NJ: Lawrence Erlbaum Associates.

Turvey, M., Shaw, R., Reed, E., & Mace, W. (1981). Ecological laws of perceiving and acting: In reply to Fodor and Pylyshyn. *Cognition, 9,* 237-304.

Warren, R. (1991). Preliminary questions for the study of egomotion. In R. Warren and A. Wertheim (Eds.), *Perception and control of self-motion.* (pp. 3-32). Hillsdale, NJ: Lawrence Erlbaum Associates.

Warren, W. (1982). *A biodynamic basis for perception and action in bipedal climbing.* Unpublished Doctoral Dissertation, University of Connecticut.

Warren, W. (1984). Perceiving affordances: Visual guidance of stair climbing. *Journal of Experimental Psychology: Human Perception and Performance, 10,* 683–703.

Warren, W., & Shaw, R. (1981). Psychophysics and ecometrics. *Behavioral and Brain Sciences, 4,* 209–210.

Warren, W., & Whang, S. (1987). Visual guidance of walking through apertures: Body-scaled information for affordances. *Journal of Experimental Psychology: Human Perception and Performance, 13,* 371–383.

Appendix A

Concepts and Theorems Required for Dimensionless Analysis and to Build a Fiber Bundle Geometry

Here we present the basic definitions and theorems used in this paper. We restrict ourselves to a pure mathematical language. The discussion follows the major steps of dimensional analysis as expressed by Kasprzak et al. (1990). Furthermore we also used Shaw et al. (1990) and Shaw et al. (1992).

Definition: Group $G = (G, *)$ is an algebraic structure on a set G with an operation $*$ called multiplication if
   (a) for any $a, b \in G$ there is a unique $c \in G$

$$ab = c$$
   (b) for any $a, b,$ and $c \in G$

$$a(bc) = (ab)c \quad \text{(associativity)}$$
   (c) there exists $e \in G$, for which

$$ea = a \quad \text{for any } a \in G$$
   (d) for any $a \in G$ exists $a' \in G$

$$a'a = e$$
Furthermore if for any $a, b \in G$

$$ab = ba, \text{ then the group is commutative (Abelian).}$$

Definition: We call G_1 group an *invariant group* under transformation $s \in G$ if for any $a \in G_1$

$$G \supseteq G_1 \text{ and } sa = as.$$
In other words, s is a commutator of G.

Definition: We call S subgroup an *invariant subgroup* of group G if for any $a \in G$ and any $s \in S$

$$sa = as.$$
An invariant group is sometimes called *a normal subgroup*.

Definition: If N is a normal (invariant) subgroup of G, then $aN \ a \in G$ classes are *compatible classes*. That is, if

$$a \neq b \ a, b \in G, \text{ then } aN \cap bN = 0.$$

Definition: If N is a normal (invariant) subgroup of G, and $a \in G$, then we can define the multiplication of aN *compatible classes:*
$$aN \blacklozenge bN = abN \quad \text{for } a,b \in G.$$
These classes with this multiplication constitute a group which is called the *factor group* of G with respect to N. The standard notation of this group is G/N.

<u>Example</u>: If G is the quaternion group $\{\pm 1, \pm i, \pm j, \pm k\}$, and $N = \{1, -1\}$, then G/N is the Klein group $\{1, i, j, k\}$. The Klein group is commutative and $ij=k$, $ik=j$ and $jk=i$.

Definition: Group G is called a *Lie group* if it has the structure of both a manifold and a group. To be more specific, G is a Lie group if

- the manifold is a topological space
- the group multiplication is a continuous operation on the manifold.
- differentiations (infinitesimal transformations) are also defined on the manifold.

Definition: A G Lie group is called a *Lie algebra* if the infinitesimal operators are represented by vector fields and satisfies a *Lie bracket operator* defined as

$$[B,A] = (BA - AB),$$

which provides us another derivative (Lie-derivative) on the manifold. The bracket product satisfies the Jacobi identity, that is

$$[A,[B,C]] + [B,[C,A]] + [C,[A,B]] = 0.$$

If the bracket product vanishes for any A, $B \in G$, then the algebra is said to be *commutative*, because $AB=BA$.

Definition: The M/Q equivalent classes of M with regard to a G transformation is a *quotient manifold*, if M is a smooth manifold and G is a local group of transformations (e.g., infinitesimal differential transformations). Any class of M can be identified for any $x \in M$ by the orbits of G passing through x.

Example: Let M be a manifold of an arbitrary physical system with Z_1, Z_2, Z_3, ..., Z_s fundamental physical quantities, and G be the corresponding group of scaling transformations a_1, a_2, a_3, ..., a_s. Then the so-called π *theorem* (see later) for this physical system provides an example of the significance of the M/G quotient manifold.

Theorem: Let A be an operator of a second-order differential equation and X be the generator of a group of local (point) transformations. Then the differential equation of operator A is invariant under the symmetry operation dX/dt if and only if

$$[dX/dt, A] = kA.$$

Example: Kepler's problem (Stephani, 1989, pp. 96–97): motion of planets around the sun is governed by

$$\ddot{x}^i = - M x^i / r^3$$

which satisfies the conditions of the theorem for the following 5 point symmetry transformations:

$$X_n = a_{n,m}^k \left(x^m \frac{\partial}{\partial x^k} + \dot{x}^m \frac{\partial}{\partial \dot{x}^k} \right) \quad n,m,k = 1, 2, 3$$

$$X_4 = t \frac{\partial}{\partial t}$$

$$X_5 = t \frac{\partial}{\partial t} + \frac{2}{3} x^m \frac{\partial}{\partial x^k} - \frac{1}{3} \dot{x}^m \frac{\partial}{\partial \dot{x}^k} \quad m,k = 1, 2, 3$$

The first three generators X_n constitute the three-dimensional rotation group according to the spherical symmetry of the gravitational field of the sun. The invariance of temporal translation expressed in X_4 reflects the stationarity of the gravitational field. X_5 is an implicit expression of Kepler's third law (spatiotemporal relationship of any two orbits of planets):

$$t_1^2/r_1^3 = t_2^2/r_2^3$$

In sum, these invariant transformations express the underlying spatio-temporal symmetry structure of the gravitational field of the sun.

Definition: Assume that the sets X, Y have a group G defined on them, and there exists an $A:X \to Y$ mapping, then we call A an *invariant function* if for any $x \in X$ and for any $g \in G$ $A(\phi(g)x) = \psi(g)A(x)$.

Without proof we present some of the important properties of an invariant function.
  - Function A maps orbits into orbits.
  - A can be decomposed into the sum of invariant functions on the orbits.
  - The invariance property of function A is equivalent to the commutativity as shown in Figure 11.A1.
  -The symmetry value of an invariant function (Figure 11.A1) cannot be smaller than the symmetry of the argument.

On the basis of these properties there is a method to simplify the determination of an invariant function (for more details see Kasprzak et

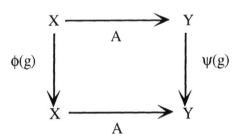

Figure 11.A1: *Diagram for an invariant function.*

al., 1990).

Definition (by Drobot): An algebraic structure $P = (P, \bullet, [\ ]^\circ)$ (P is the base set on which two operations, "$\bullet$ = group multiplication" and "$[\ ]^\circ$ = real power function" are defined) is called a *dimensional space* if
      (a) $(P, \bullet)$ is a commutative (Abelian) group.

(b) The power function satisfies the following identities:
For any $a, b \in R$ and $X, Y \in P$

$$X^{a+b} = X^a \bullet X^b$$
$$(X \bullet Y)^a = X^a \bullet Y^a$$
$$((X)^a)^b = X^{ab}$$
$$X^1 = X$$

(c) $P_0 = R_+ \in P$

(d) On $P_0 = R_+$ the group multiplication $\bullet$ = multiplication of real numbers, and

(e) The powers of elements of the subset $P_0 = R_+$ are identical to the ordinary powers of the numbers.

Definition: In a dimensional space every system of dimensionally independent elements is called the *base* of this space.

Definition: $Z = F(Z_1, Z_2, Z_3, ..., Z_s)$ is called a *dimensional function* of $Z_1, Z_2, Z_3, ..., Z_s$ if $Z, Z_1, Z_2, Z_3, ..., Z_s \in P$.

Definition: Z is called a *dimensionally homogeneous and invariant function* of $Z_1, Z_2, Z_3, ..., Z_s$ if

(a) for any $\alpha_1, \alpha_2, \alpha_3, ..., \alpha_s \in P_0$ and $Z_1, Z_2, Z_3, ..., Z_s \in P$ there exists an $\alpha \in P_0 = R_+$ for which

$$F(\alpha_1 Z_1, \alpha_2 Z_2, \alpha_3 Z_3, ..., \alpha_s Z_s) = \alpha F(Z_1, Z_2, Z_3, ..., Z_s).$$

(homogeneity condition).

(b) for each dimensional transformation Θ

$$F(\Theta Z_1, \Theta Z_2, \Theta Z_3, ..., \Theta Z_s) = \Theta F(Z_1, Z_2, Z_3, ..., Z_s).$$

(This guarantees the invariance of a dimensional function, that is, the independence from the system of units used.) Note: These are the only functions relevant in physics, and in that regard we do not want to make intentional dynamics an exceptional case.

Theorem π: If Z is dimensionally homogeneous and invariant function depending on $z_1, z_2, z_3, ..., z_m$ among which the first $z_1, z_2, z_3, ..., z_n$ are dimensionally independent and the rest $z_{n+1}, z_{n+2}, z_{n+3}, ..., z_m$ are dimensionally dependent arguments, then

$$Z = f(\pi_1, \pi_2, \pi_3, \ldots, \pi_k) \prod_{i=1}^{m} Z^a,$$

where $k = n - m$.

Definition: A *fiber bundle* is a pair of manifolds *(E, M)* and a projection *Pr: E -> M*. *M* is called a *base space* on which a dynamic structure (field) is defined. For any $m \in M$ $Pr^{-1}(m)$ is called the *fiber* over *m*. In most cases fibers have further geometric structure. For instance, if they are vector spaces then we call *Pr: E -> M* a *vector bundle*.

If the fibers are the cotangent spaces on the manifold *M*, we call it the *cotangent bundle, T^*M*.

If the fibers are the tangent spaces on the manifold *M*, we call it the *tangent bundle TM*. Physicists call it velocity space (see Figure 11.A2: Examples of different fiber bundles). A fiber bundle can also be viewed simply as a bunch of fibers projected from the same base space.

Definition: A *Hamiltonian* is a function H, defined in phase space (q,p).

$$H(p,q) = \sum_{i=1}^{n} \dot{q}(q,p)p_i - L(\dot{q}(q,p))$$

In other words, a Hamiltonian is represented by the vector field

$$v = \frac{\partial H}{\partial p}\frac{\partial}{\partial q} - \frac{\partial H}{\partial q}\frac{\partial}{\partial p}$$

on a cotangent bundle (q,p). The cotangent bundle is its natural structure.

Definition: A *generalized Hamiltonian, G,* is defined in a generalized (complexified and compactified) phase space

$$\{(Q_1, \ldots, Q_n; P_1, \ldots, P_n), i(Q^*_1, \ldots, Q^*_n; P^*_1, \ldots, P^*_n)\}$$

where *Q* and *P* are the generalized (target-specific information) coordinates and generalized momenta (manner-specific movements), respectively, and both are observable in the exterior frame. In the interior frame, accessed by the complex operator, we have dually Q^* and P^*—the generalized (manner-specific information) coordinates and

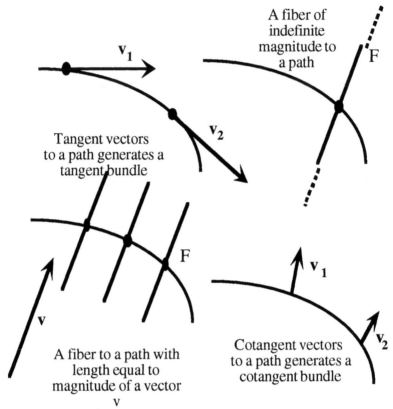

Figure 11.A2: *Examples of different fiber bundles.*

generalized momenta (metabolically produced control impulses), respectively. The generalized Hamiltonian, G, therefore, is defined (complexly) adjointly over the exterior, and interior phase spaces by

$$G(\dot{Q}, P; Q^*, P^*) = \{ \sum_{i=1}^{n} \dot{Q}(Q,P)P_i - L(\dot{Q}(Q,P)) \} + \{ \sum_{i=1}^{n} \dot{Q}^*(Q^*, P^*) P^*_i - L(\dot{Q}^*(Q^*, P^*)) \}$$

Just as an ordinary $H(q,p,t)$, over a time invariant path in phase space conserves the time-forward flow of total energy (kinetic + potential energy), so the generalized Hamiltonian, $G(Q, p; Q^*, p,t; +t, -t)$, conserves the relationship between time-forward and time-backward flows of energy and information that comprise total (goal-specific) action. Total action, G, is the sum of total control information, $(Q, Q\sim, +t, -t)$, and total useful energy $(P, P^*, +t, -t)$ over both frames (i.e., the interior and exterior phase spaces). (Shaw et al. 1991, pp. 607-608).

Appendix B

A Sketch of the Fiber Bundle Geometry Needed for the Intentional Dynamics

A *fiber bundle geometry* is the appropriate mathematics in which to express a potential solution to the ecometric (information) scaling problem and the ecomechanic (energy) transduction problem. Using the generalized Hamiltonian and its natural fiber bundle structure, we can define two different fiber bundles on both the control field and the information field construed as base spaces:

$$Pr_{info}: C \to I \quad and \; Pr_{contr}: I \to C$$

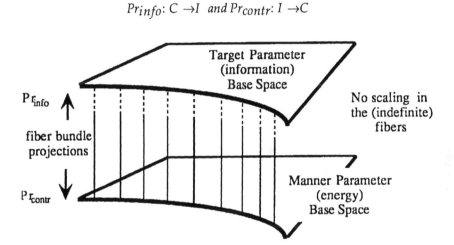

Figure 11.B1: *Indefinite Ecological Fiber Bundles without Gauging.*

Definition: A *fiber* is defined as a projectivity from one vector field over a base space, containing information flow paths to another vector field over another base space which contains control paths.

Fibers are parallel to one another and independent unless connected by some function. For instance, a goal-path integral which scales information detection flow paths in one base space to energy control paths in another base space. The ecometric scaling problem is to discover the definite magnitude (gauging) to place on the fiber defined over the information field. The ecomechanics transduction problem is to discover the definite magnitude (gauging) to place on the fiber defined over the control field. These two gauging procedures have no necessary correspondence or commensurability to each other unless the intended action is successful, then the gauging determines the same curve—the ~oal path (see, e.g., Fig 11.B2.).

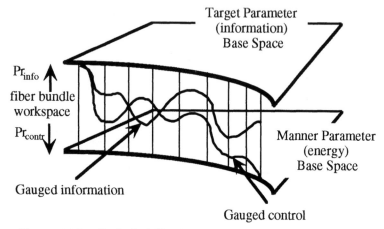

Figure 11.B2: *Ecological fiber bundles with non-ecological gauging.*

Definition: Assuming the proper gauging in two fibers the function that maps values of one fiber onto the value of adjacent fibers is called a connection in a fiber bundle (Fig 11.B4.).

In the case of navigation by a vehicle the connection is a vector space, that is, the tangent space defined over the *contact elements* to the fibers (i.e., defined as the direction of the curve as it crosses the fiber). The *goal path* in an ecological workspace is a connection in a fiber bundle that *gauges* (or scales) one region of a field to another region of the same or different field. We use a connection to relate the flow of information to the flow of energy within the same field or across different fields, such as, across the interior (organismic) field and the exterior (environmental) field.

Definition: When the connection is itself a field it is called a gauge field.

Definition: *Ecological physics* is the study of the gauge field between energy flow fields and information flow fields that couples an environment and organism into an ecosystem.

Definition: *Intentional dynamics* studies the properties of this dual-information-control gauge field for goal paths through a workspace.

Definition: *Goal paths*, therefore, are curves of dually compact points of contact elements defining a gauge field coupling the interior and exterior fields. Each point along the path is a dual pair of numbers—one number specifying the location of the fiber in each base space, and the other number specifying the value of the projection along the fiber to the path, moreover:

—When the projection pushes the value forward from the information base space to the curve, then it is the gauge value for solving the ecometric scaling problem.

—But when the projection pulls the value back from the curve to the control base space, then it is the gauge value for solving the ecomechanics transduction problem. Together they determine the gauging of the information and control flows to the goal path by means of the complex involutional group of reciprocity maps characterizing the perceiving acting cycle (as described by Shaw et al. 1990).

In other words, if we treat each fiber as unity, then the connection (curve) partitions each fiber into complementary magnitudes (length) whose ratio is proportional to the scaling of the first base space to the second space, and whose reciprocal ratio is proportional to the scaling of the second base space to the first. This is shown below.

Energy flow (+t)

Start   Target

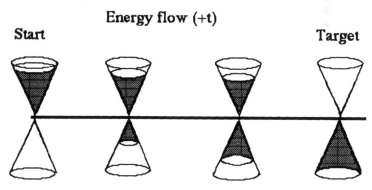

Figure 11.B3: *The information-energy potential along a goal path in the ecological workspace exhibits an inner-product invariant relationship.*

The dual cones represent projectivities (fibers) from a goal-path in the environmental workspace. Each fiber is two dimensional (a cone) consisting of a tangent vector to the goal-path and the magnitudes of a dual-potential specific to each goal-path point (The two base spaces are suppressed). The upright cone, incident at a goal-path point, represents a quantity of energy potential, whereas its dual inverted cone represents a quantity of information potential. The fact that the sum of the control and information complements remains constant over the goal-path integral illustrates the meaning of the *inner-product invariant*. Notice the mini-max relationship between the dual cones.

The points in either field's base space are dually compact because they each hide reciprocal degrees of freedom (or constraint) that reside in the other field's base space. These hidden degrees of freedom are represented by the fact that the fibers incident at each point along the goal-path are projections from each base space.

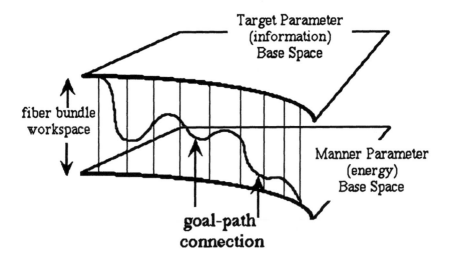

Figure 11.B4: *The goal-path connection with an ecological gauging in a fiber bundle workspace.*

The three most important properties of intentional dynamics:

1   Since the goal-path connection (curve) partitions each fiber into complementary magnitudes of energy control and information detection, then their inner-product remains invariant over the goal-path integral. This dynamical invariant is to be expected if generalized action is a conserved potential.

2   This inner-product invariant, therefore, provides us with a
        metric criterion against which to measure the degree of success
        of an intentional system in reaching its goals.

3   Also, because these energy and information fiber bundles exist
        at each point in the workspace, then the coupling between the
        two fields is both "soft" (transactional, or informational) and
        "hard" (interactional, or energetic) (in the sense of Kugler &
        Turvey, 1987, and Shaw & Turvey, 1981). Thus each point along
        a goal path in the connection has conjugate values in the control
        and information fields vis-a-vis the fibers projecting from each
        of the dual base spaces

Chapter 12

"Center of Mass Perception": Affordances as Dispositions Determined by Dynamics

Geoffrey P. Bingham and Michael M. Muchisky
Indiana University

12. 0   Introduction

Affordances are dispositional or relational properties. They consist of properties of objects in relation to properties of a perceiver/actor in the context of a task or a goal to be achieved by way of the affordance (Gibson, 1979). The basis of the relation unifying object properties into an affordance property is the task goal. Thus, understanding affordances requires some understanding of the manner in which physical properties of object and actor are invoked by tasks.

The relation between affordances and goals is problematic. On the one hand, affordances are task specific. On the other, a given goal might be achieved via various means involving different affordances respectively. The way to resolve this apparent contradiction is via the process of goal specification. Tasks can be analyzed into a sequence of alternative subgoals corresponding to the various ways that the superordinate goal might be achieved. Actual performance of the task requires the successive determination of subgoals. How might such goal specification proceed? Goal specification entails the perception of affordances corresponding to subgoals. The implication is that there are superordinate and subordinate affordances. The problem is to understand the relation between the former and the latter. For instance, a goal might be to transport an object from a table to a shelf. Suppose that one recognizes that the object affords being so transported by oneself. That is, the size and weight and surface properties are such that one could control the trajectory of the object from the table to the shelf via a variety of means, using either one hand or two or perhaps using two elbows if both hands are occupied.

The particular means need not be specified and, in certain circumstances, are unlikely to be specified until the action is actually being performed. For instance, suppose a collection of differently shaped objects, all of similar sizes and weights, is to be transported to the shelf. Possible transportation using a one handed grasp for each might be apprehended. Nevertheless, assuming that the shapes are different enough to require different types and locations of grasping, the particular form and locus of the grasp in each instance is unlikely to be determined at the time when the overall sequence of actions is initiated. What structure might underlie the progressive specification of the affordance for grasping each object? Imagine further that the objects must be replaced on the table from the shelf. The way that each object is grasped this second time may be related, but not identical to the initial grasp. What structure is the basis for this constrained variability? Finally, imagine that the objects must be placed the second time at orientations other than their initial orientations. The grasp types and locations are very likely to be different from the previous ones, yet they will remain related. What structure is the basis for both the new organization and its relation to the old?

Performers anticipate the ways in which they can combine physical properties of themselves with physical properties of objects to achieve specific outcomes, meaning trajectories and/or states[1] of self and objects. The detailed relations that can be established and employed are simultaneously constrained and indefinitely variable. The basis of this generativity is the physical dynamics of actors and objects, and it is on these dynamics that the focus of our study must be placed. As perceptible properties rooted in the dynamics of objects and actors, affordances entail KSD or Kinematic Specification of Dynamics. Informative patterns within energy distributions impinging on the sensory apparatus are kinematic in their dimensions, that is, the patterns vary in lengths and times. The problem is to describe how such spatiotemporal properties provide information about dynamically determined or mass related properties of objects and events and to discover which spatiotemporal properties provide information about affordances in given instances.

---

[1]Although trajectories and states are usually construed as kinematic, they need not be necessarily. We do not here intend to assign goals to a kinematic level of description.

Affordances are dynamically determined dispositions that cannot correspond to simple properties or variables taken from classical dynamics. Due to inalienable requirements for identification, scaling, and orientation of affordances, more than one classical dynamical variable is required for the determination of any given affordance property. Accordingly, an affordance exhibits continuity of structure (including dynamical variables) shared with neighboring affordances. The continuity of such structures is the coherent basis for progressive goal specification. The shared structure and/or variables provides a basis both for individuating the entire structure at a superordinate level as well as for progressing to more distinct subordinate alternatives for action.

In what follows we investigate "graspability" as one of the most oft-cited examples of an affordance (Gibson, 1979; Mark, 1987, Mark, Balliett, Craver, Douglas, & Fox, 1990; Turvey, Shaw, Reed, & Mace, 1981; Warren, 1984). First, we attempt to provide a general description of "graspability," one that is indifferent to the wide variety of tasks in which grasping is involved. We succeed only in illustrating the task specificity of affordances that are, nevertheless, related by common dynamical characteristics. Next we briefly address the problem of goal specification. We suggest that, underlying and enabling the process of goal specification, there should be a continuous perceptible structure that is rooted in the dynamics of objects and actions and that contains related alternatives for action. We advocate the search for a perceptible layout of dynamically determined alternatives. Accordingly, we describe investigations of "center of mass perception" for the visual guidance of precision grasping which reveal a very small, yet distinctly structured region within the layout of affordances for grasping.

12.1   "Graspability" and the Task Specificity of Affordances

We might describe *grasping* as "taking on and supporting an object's weight by enclosing object surfaces with hand surfaces." This accords with the object properties typically listed as determining *graspability* namely, the shape, size, and weight of the object taken with respect to corresponding properties of the hand. However, in grasping a pole on the subway or in grasping someone's hand to shake it, we do not support the object's weight. Although the former might involve supporting a portion of one's own weight, the latter involves no weight bearing whatsoever. Certainly, mere contact of hand and object surfaces together with some weight bearing is not sufficient because we do not "grasp" a wall when we "lean" on it with our hand. Nevertheless, such "leaning" is very closely related to the function performed when one "grasps" the pole on the subway. Enclosing with hand surfaces might seem to capture what we mean, but one encloses water with hand surfaces when swimming without really "grasping" the water. Also, one might enclose a cricket with one's hand surfaces to trap it without really 'grasping' it. Although it might be in contact with hand surfaces that support its weight, room would be left for it to move within the hand. We might say that one is merely "trapping" or "imprisoning" the cricket without really "grasping" it.

Perhaps, we might refine our definition by requiring that the surface contact be of a sort that prevents the object surfaces from moving relative to hand surfaces. Surface texture or compliance properties are sometimes listed as relevant to "graspability" because they determine frictional forces. We note again, however, that this is relevant as well to "leaning" on a wall with one's hand. Previously, we invoked enclosure by hand surfaces to distinguish "grasping" from "leaning," but certainly complete enclosure of an object is not necessary for "grasping." Napier (1956) distinguished a number of distinct types of grasps, depending on the amount of enclosure from "power grasps" to "precision grasps." In a *power grasp,* an object is wrapped and nearly enclosed by palmer hand surfaces. However, in a *precision grasp,* an object is merely pinched between the distal segments of the thumb and fingers, often only the index finger. This latter grasp involves relatively small areas of contact between object and hand surfaces. Nevertheless, to fix an object with respect to the hand, precision grasps require contact on opposite sides of an object so as to produce oppositely directed and mutually canceling forces yielding a stable configuration at equilibrium[2]. Thus, we might revise *enclosure* to mean pinched between such oppositely directed forces from hand contact. Still, difficulties remain.

Does one grasp an object when one supports its weight on the flat surface of an upturned palm, or does one merely support it? Is this "supportability" as opposed to "graspability?" It's similar to the case of

---

[2]With misgivings that we subsequently make explicit, we will use "center of mass perception" as shorthand for "perception of grasp locus on an object affording a neutrally stable precision grasp." We use the shorthand because the perception of this affordance happens to coincide more or less with "center of mass perception."

"leaning," but with a change in direction (that is, palm facing the gravitational direction) and with the difference that the object's weight is now being supported by hand surfaces. The difficulty is that a continuum can be established between a "power grasp" of an object and a "support grasp" of this sort.

Imagine grasping a soda can by wrapping the cylindrical shape with fingers, palm, and thumb as one normally would. This illustrates the essential *power grasp*. Now, imagine grasping the can by placing its end on the palm and contacting the cylindrical sides with the fingers and thumb. This is another form of power grasp. Orient the palm of the hand upright so that the weight of the can rests on the palm. Gradually, extend the fingers and thumb at the metacarpal–phalangeal (or knuckle) joints, one at a time. The result is a gradual, continuous, and benign transition from a strict power grasp to a support grasp. The transition is benign because an equilibrium state was maintained throughout the transition and the object remained on the hand, its trajectory still determined by the hand. Despite the change in the relation between the hand and object, the function was preserved, namely, the trajectory of the object with respect to the hand was fixed. Thus, the support grasp must be a proper type of grasp that does not involve enclosure or pinching between hand surfaces. Of course, a similar transition can be described between the grasp of the pole on the subway and the mere "lean." The implication is that the 'lean' is a type of grasp as well, despite the fact that this departs from what we usually mean by a grasp. Certainly, we would not refer to leaning on a table top with our hands palm downward as "grasping the table." Nevertheless, such a posture is not disjoint in character with postures normally referred to as "grasping." Consider, for instance, "palming" a ball which is a type of grasp. One could not lift a table in this way , but the table could perhaps be slid sideways.

We have suggested that the essence of grasping might be the fixing of relative positions of hand and object surfaces in contact with one another. But this is not consistent with the active manipulation of objects within a grasp. Ultimately, stable grasping involves dynamic equilibrium (Raibert, 1986) as much as (or more than) static equilibrium. A ball or a pencil might be manipulated within a stable grasp by momentarily fixing and rolling it between two digits, free of other hand surfaces. Alternatively, one might simply allow the ball or pencil to roll along the palmer hand surface from wrist to finger segments. An excellent example that illustrates the difficulties and complexities of dynamic grasping is that seen in the movie *Blow Up* in which the protagonist fidgets with a coin by flipping it end over end across the back sides of the proximal-most segments of his fingers. One might protest that this was not grasping because the coin was situated on the back of the hand. But during each flip, the end of the coin was lodged between two fingers. Does one not grasp a cigar when one holds it between two fingers? Actions like this seriously challenge and defy any straightforward scheme for defining an affordance property like graspability in a categorical fashion distinct from given tasks.

There is a tremendous amount of functional variability underneath the categorical terms, *grasping* or *graspability*, and to the contrary, there is functional similarity despite variation in terms appropriately or preferentially applied, as in *leaning, supporting, trapping, grasping* , and so on. What determines the variability, on the one hand, or the similarity, on the other, is the physical properties that are marshaled in achieving specific goals and the manner in which they are marshaled, including the geometries according to which they are configured and the values associated with the particular properties.

12.2   The Problem of Goal Specification

Even if we focus on what would seem to be an indisputable and relatively uncomplicated grasping task, for instance, passing a soda can from one person to another, the relation between the goal and affordances for grasping is one to many. For instance, various types of precision grasps, power grasps, or support grasps all might satisfy and be admitted by the relevant constraints in the task.

The proliferation of means to an end is a familiar problem in the study of human action. For instance, within the trajectory-formation tradition, goals have been described as kinematic entities, that is, particular end positions or trajectories to be achieved via the appropriately organized dynamics of the motor system (e.g., Hollerbach, 1982). The problem has been that a good deal of kinematic variability can occur in goal directed limb movements without significant functional repercussions. A desired end point can be reached without hesitation in reaching movements despite significant mechanical perturbation of the hand on its way to that end point. In recognition of this and other related stabilities commonly exhibited in human limb movements, Saltzman and Kelso (1987) formulated an approach in which tasks were described explicitly in terms of dynamics. This general approach, called *task dynamics*, has been adopted widely (Beek,

1989; Beek & Bingham, 1991; Bingham, 1988; Bingham, Schmidt, & Rosenblum, 1989; Bingham, Schmidt, Turvey, & Rosenblum, 1991; Feldman, 1986; Hogan, Bizzi, Mussa-Ivaldi, & Flash, 1987; Kugler, 1986; Kugler & Turvey, 1987; Riccio, Martin, & Stoffregen, 1992; Solomon & Turvey, 1988; Stoffregen & Flynn, In Press; Stoffregen & Riccio, 1988; Thelen, 1989; Warren, 1988).

Nevertheless, even at the dynamical level of description, there may be many ways to achieve a given goal. For instance, in passing the soda, one might specify the palmer side of the recipient's hand as the equilibrium point or "point attractor" of a critically damped mass-spring task dynamic defined with respect to the center of mass of the soda can as the mobile. This abstract organization for the multilink arm, hand, and can system would leave unspecified, however, the use of either a power grasp, a variety of precision grasp, or even a support grasp to achieve control over the center of mass of the can. Fortunately, as the performer approaches and begins to move through the actual performance of the task, additional constraints may come to bear on the possible subgoals. For instance, transfer to the recipient's grasp might be facilitated by use of either a support or precision grasp which would leave greater can surface exposed for contact by the recipient. On the other hand, desire for stability in the face of significant lateral accelerations might rule out the use of a support grasp.

The remaining subgoal would be to establish a stable precision grasp. The problem is that this subgoal remains ill specified. A precision grasp can be established at a continuous variety of locations on the can. The locations are not equivalent in their effect on the stability of the resulting precision grasp. Formulating the subgoal requires that a coherent and connected subset of affordance properties be specified so that an appropriate member of the set might be selected.

12.3   The Center of Mass in Precision Grasping

We consider the types of equilibrium postures in precision grasps in which the stability of the equilibrium configuration is determined by the location, relative to the center of mass, of an axis running between the two pinching hand surfaces. For an equilibrium posture, the forces applied to opposite and (approximately) parallel object surfaces by the index finger and thumb pads must be oppositely directed, co-linear, and equal.[3] We call the line along which they are directed the *opposition axis*

[3]This also is not precisely true. Fearing (1983) has shown that strict opposition of contact forces is not required. The angle between the force vectors from opposing segments of the hand can vary within a tolerance determined by the frictional characteristics of the object and hand surfaces. Thus, a grasp can be less than enclosing in this sense as well, that is, the hand surfaces would surround an arc of less than 180°.

(Iberall, Bingham, & Arbib, 1986). Such equilibrium configurations can vary in their stability from stable, to neutrally stable, to unstable configurations as the opposition axis is varied in its position with respect to the center of mass.

When this axis passes above the object center of mass, the equilibrium configuration is stable. If perturbed, the object will return to its original orientation when the perturbing force is removed. Such a configuration is optimal when trying to keep a hot cup of coffee upright in one's grasp.

When the opposition axis passes directly below the object center of mass, the configuration is unstable. When displaced by perturbation, the object will continue to rotate away from the original orientation. This configuration is useful for passively reorienting an object in one–handed manipulation, for instance, picking up a rod lying on a surface so that the end nearest the grasper is finally oriented upward within a grasp placed just below the top end.

When the opposition axis passes directly through the center of mass, a neutral stability is achieved. The object exhibits no preferred orientation around the opposition axis. The object will remain at any given orientation and can be reoriented to arbitrary orientation with a minimum of effort. This configuration might be preferred in a "peg-in-a-hole" task in which fine manipulation of orientation is required. This posture would also be desirable in situations in which the final desired orientation is not apparent at the moment of initial grasping.

As a generator, the center of mass produces a continuous set of related although distinct dispositions. The transition from strongly stable, to weakly stable, to neutrally stable, to weakly unstable, and finally to strongly unstable can be affected by continuously varying the position of the opposition axis with respect to the center of mass. Although the center of mass is a property found in all objects, the dispositions are not. For instance, one cannot grasp a banana in a precision grasp so as to pass the opposition axis through the center of mass, because the center of mass lies outside the banana's skin. Nevertheless, if the banana's weight is to be supported in a precision grasp at equilibrium, then the opposition axis must lie along a vertical through the center of mass.

Because affordances are perceptible properties, these dynamical dispositions only remain as possible affordances until we establish that they are perceptible. The focal question becomes whether and how observers can apprehend dynamically determined properties of objects and events. Over a decade ago, Runeson introduced and discussed the notion of "dynamic event perception" and the perception of "dynamic properties" (Runeson, 1977). Runeson suggested that many perceptible properties might be understood or modeled as built from, and therefore continuous with, properties of standard mechanics. They need not be identical to any familiar or named properties in mechanics, although they might happenstantially be close in specific circumstances.

One prerequisite for dynamic properties to be construed as affordance properties is that the dynamic property be a dispositional property of objects. Some standard mechanical properties can indeed be construed as dispositionals. For instance, moments of inertia are dispositions to turn with specific rotational accelerations given certain torques applied about axes corresponding to those inertial moments. Another example, pursued in the research to be reported in this chapter, is the center of mass. The center of mass can be construed as a disposition for an object to accelerate along a straight line in proportion to the amount of force applied along a line passing through the center of mass. Whether these dispositions are perceptually salient is as much to be doubted as is the possibility that torques or forces are explicit control variables in human action (Bingham, 1988; Feldman, 1986; Hogan, et al., 1987; Stein, 1982). Nevertheless, the center of mass is usefully approached as the generator of a family of dispositions that are of relevance to human activity.

12.4 Perception of the Center of Mass

Dynamic properties are physical properties and, as such, are described in units of mass as well as length and time. On the other hand, patterns detectable by the perceptual systems are described only in length and time units. Thus, in the context of events, information about dynamics must be found in motions described, using kinematics, in length and time units. Information about dynamics involves a series of relations, first between dynamics and kinematics and then, between kinematics and optics (Bingham, 1987a, 1987b, 1988, In Press; Bingham, Rosenblum, & Schmidt, In Press). Kinematics or trajectories as information are properties of events as are the dynamics that generate them. The event trajectories must next map into spatiotemporal patterns in distributed energy fields that can be detected by the perceptual apparatus.

The trajectories of an object held in a grasp might provide information about the dynamical disposition for stable, neutrally stable, or unstable equilibrium. The trajectories would map either into the transforming distributed structure of the tissues of the haptic array (Bingham, et al. 1989; Solomon, Turvey, & Burton, 1989a; b) or into flow patterns in the optic array (Bingham, 1987a, 1987b, In Press; Bingham et al., 1989, In Press; Gibson, 1979; Todd, 1981). People performing assembly tasks while blindfolded have been seen to test an object by lifting it briefly on first encounter with the hand and then to go for a grasp established about the center of mass (Iberall, et al. 1986).

However, only visual information about the object would be available before an object has been contacted by the hand and, most typically, before coming into contact with the hand, an object lies immobile on a supporting surface. Would visual information about the center of mass be available in such situations? If so, then information would have to be found in geometric properties of the object. Thus, when the perception of dynamics in events has been called a KSD problem (Kinematic Specification of Dynamics), the perception of dynamical properties in static situations might be called a GSD problem, that is, Geometric Specification of Dynamics (Muchisky & Bingham, 1991). On the other hand, geometric properties are a subset of those studied in kinematics, so we might dispense with the extra terminology. Furthermore, to be apprehended visually, object properties must map to informative properties of the optic array which would be spatiotemporal properties generated by observer motion, even if the object itself was unmoving. We studied center of mass perception as a KSD problem in which we addressed the question of information in object geometry but did not attempt to reveal or describe the corresponding optical information.

We studied the simplest of object manipulation tasks, that is, to grasp an object using a precision grasp yielding neutral stability. Essentially, we asked observers to locate the center of mass (henceforth, CM). We asked if only visual information were available before contact, would it be sufficient for observers to locate the CM of an unmoving object? If so, then what would the information be? Consideration of the potentially relevant informational factors is instructive because this task provides an indication of the minimum complexity that we should expect to encounter in describing affordances as dynamically determined dispositions.

Three general factors are relevant, namely, form, scale, and orientation. These are fundamental to the description of any affordance property as follows. Form is entailed by the need to recognize an affordance, that is, for a solution to the identification problem (Bingham, 1987b, In Press, Bingham et al. In Press). Such form might be either geometric or kinematic. Scale is important for two related reasons: First, it contributes to a determination of identity (Bingham, 1987bBingham, Rosenblum, & Schmidt, In Press), because objects and events are scale specific; they occur within definite scale ranges. More precisely, as studied in allometry (e.g., Hildebrand, Bramble, Liem, & Wake, 1985), in scale engineering (e.g., Baker, Westine, & Dodge, 1973), and in similarity theory (e.g., Szücs, 1980), scale and form are co-determinate (Bingham, 1993a, 1993b; Bingham & Muchisky, 1992; Shaw, Mark, Jenkins, & Mingolla, 1982). Second and relatedly, the functional value and/or repercussions of an object or event is determined by its scale relative to the scale of the observer or actor (Warren, 1984; Warren & Whang, 1987). Finally, orientation is important because gravity is a dynamical component that contributes significantly to the form of every terrestrial event at the scale of human activity, including human activity itself. Bingham, Rosenblum, and Schimdt (In Press) have shown that orientation must be specified to determine identity when using kinematics as visual information about events. To the extent that functional repercussions are intrinsic to affordances as perceptible properties, orientation also must be relevant to the identification of any affordance. The implication of this last factor is that the recognition of affordances is an inherently multimodal affair, because kinesthetic or haptic information about the gravitational direction would be an inalienable component of visual recognition.

Accordingly, in studying CM perception, we manipulated three geometric object properties: shape, size, and orientation. The shape of an object is relevant because the CM is a symmetry property (Becker, 1954; Sears, Zemansky, & Young, 1987). The CM falls on any axes of reflective symmetry or at the center of a radial or rotational symmetry in an object with a homogeneous mass distribution, because the CM is the point around which the mass is balanced. Three noncoplanar axes of reflective symmetry or a center of radial symmetry about two axes uniquely specify the location of the CM.

The detection of symmetry properties would locate the CM. Might greater amounts of symmetry increase the accuracy in locating the CM? If so, we should expect random error in locating the CM to decrease with increasing symmetry. (Note that we have assumed a homogeneous mass distribution. This assumption is valid for a large class of objects which are made of a single material such as wood, plastic, metal, ceramic, stone, glass, or organic fibers.)

The size of an object of a given shape determines, for any given error distance, the repercussions of missing the CM. Ignoring torques around the opposition axis created by frictional forces, a miss of a given distance will produce less rotational acceleration of a larger object than of a smaller object. The greater the rotational acceleration, then for a given amount of rotation, the smaller the time in which to respond, or alternatively, within a given response time, the greater the amount of undesired rotation. However, shape, in addition to size, affects rotational acceleration. Both shape and size determine an object's moment of inertia or resistance to rotation. Might shape and size interact in determining the accuracy of judgments of the center of mass? We should expect random errors in locating the CM to increase both as objects increase in size and as shapes become more elongated and less compact.

Finally, orientation is potentially relevant for two reasons. First, the consequences of inaccurately locating the CM are a function of orientation with respect to gravity. The repercussions of missing the CM are potentially less severe for misses along the vertical direction than for those along the horizontal. If the pattern of errors reflects this, we should expect more errors along the vertical than along the horizontal direction from the CM. Second, orientation is purported to affect the recognition of symmetry in an object. According to Rock (1973), predominant axes and especially reflective symmetry axes are more easily recognized when parallel to gravity. The further away from a vertical orientation a symmetry axis was, the less often it was recognized as such. When reflective symmetry axes are parallel to gravity, might accuracy in locating the CM increase?

We have run a number of experiments investigating these and related questions (see, e.g., Bingham & Muchisky, 1993a, 1993b). We describe two of them at length. In the first, we investigated the use of object symmetry properties as visual information about the location of the CM in an object. In the second, we investigated the use of approximations to symmetry as visual information about CM location.

12.5   Symmetry, Size, and Orientation in Center of Mass Perception

A set of 7 planar objects was designed to vary the number of axes of reflective symmetry from 0 to 4. We used planar shapes cut from 1 cm thick plywood to simplify the problem while retaining much of the variation in object shape relevant to precision grasps. Rotational symmetry (or periods of rotation in the plane of the object required for self congruence) also varied. Furthermore, some of the shapes possessed radial symmetry, that is, reflective symmetry through a point in contrast to a line as in axial reflective symmetry. The shapes and symmetries of the objects are shown in Figure 12.1. Each of the planar shapes was created in three sizes of 100 sq. cm., 200 sq. cm., and 300 sq. cm. The largest set allowed a pair of tongs to reach far past the centers

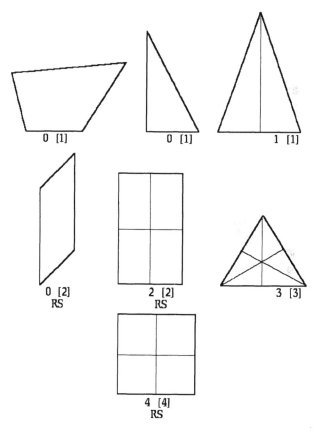

Figure 12.1. *The planar shapes used in the experiment included a right triangle [0 reflective axes, period=360°], a quadrilateral [0 reflective axes, period=360°], a parallelogram [0 reflective axes, period =180°, radial symmetry], an isosceles triangle [1 reflective axis, period=360°], a rectangle [2 reflective axes, period=180°, radial symmetry], an equilateral triangle [3 reflective axes, period=120°], and a square [4 reflective axes, period=90°, radial symmetry].*

of the figures from almost any direction in the plane of the objects.

Fifteen undergraduates at Indiana University participated in the experiment for pay. The objects were presented to observers with the plane of the figure parallel to gravity. Objects were held upright in a transparent spring loaded clamp affixed to a wooden base as shown in Figure 12.2. The side of the figure facing observers was an unfinished smooth-wood surface. The side facing the experimenter had polar coordinate paper attached to it with the origin of the coordinates fixed at the CM. The Archimedean method was used to determine the location of the CM in each object and to place the polar graph paper accordingly. This was accomplished by suspending each object from two different points along its perimeter. A plumb line was hung from each point in turn and marked on the object. The intersection of the two lines marked the location of the CM.

Observers used a set of tongs to express their judgments of the CM. The tongs were held and manipulated in one hand like a large pair of scissors. The point of the tongs that contacted the surface of the object viewed by participants was padded to prevent indentation of the surface. A sharp point contacted the side observed by the experimenter to allow precise determination of contact coordinates.

Observers were asked to judge where they felt the "stable point" was. We explained that the stable point referred to the point at which an object would remain stable without rotating about the point of contact when held upright with the thumb and index finger. Furthermore, if the object were to be rotated to another orientation, it would remain in the orientation in which it had been placed. This was demonstrated using an object other than those used in the judgment trials. The phrase *stable point* was used for two reasons: First, to avoid using the word *center* in the task description; second, stability was the disposition that we wished our observers to achieve. During each trial observers were asked to close their eyes while the object in the clamp was changed. In this way, they were prevented from obtaining information about the CM by witnessing the experimenter's handling of the objects.

Participants indicated their judgments of the stable point by lightly grasping the object with the tongs at the appropriate location. Participants never actually picked up the objects. The experimenter measured the error in estimation by noting the angle and the radial distance of the point of contact (in millimeters) in the polar coordinates on the back of the object. Observer's viewed all 6 of the objects at 4 different orientations at 0°, 90°, 180°, and 270°, from initial orientations similar to that shown in Figure 12.2. Each participant saw each of the 7

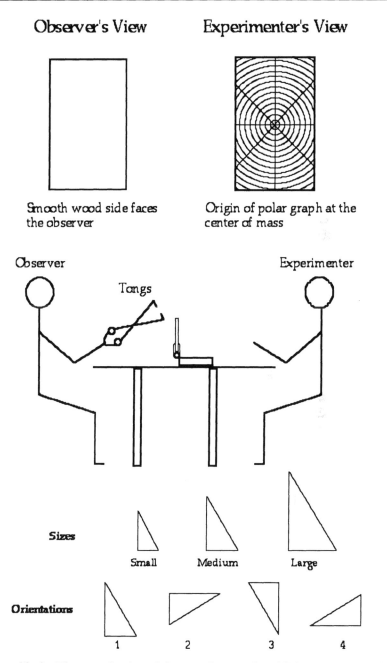

Figure 12. 2. *The organization of the experiments. In addition to variation in shape, objects were varied in size and presented in 4 different orientations.*

shapes in 3 sizes at each of the 4 orientations 3 times each for a total of 252 presentations in 3 blocks of trials. Presentation order within each block was randomized. An experimental session lasted approximately 75 min. Two sessions were required for each participant. All factors were within-subjects.

12.5.1   Results

Measurements in polar coordinates were converted to Cartesian coordinates with the origin at the CM. Zero degrees of polar angle was designated as the positive X direction. The spatial distribution of the data for each object exhibited an elliptical shape including distinct major and minor axes as illustrated in Figure 12.3a. Major axes in the distributions generally aligned with X axes. The X axis had been aligned with distributions exhibited in previous experiments. As shown in Figure 12.3b, for analysis of the systematic error along both the X and Y axes, we used the means of the X and Y data points across trials. Standard deviations calculated for each participant across trials (as well as across participants and trials) were used to analyze the random error along the X and Y axes.

Systematic error was analyzed by performing repeated measures analyses of variance (ANOVA) on the X or Y data with shape, orientation, and size as factors. For the X data, the size and shape factors were not significant, but the orientation factor was, $F(3, 27) = 11.5$, $p < .001$, as was the size by orientation interaction, $F(12, 108) = 3.8$, $p < .001$. In a simple effects test, size levels were significantly different, $p < .04$, at orientations 1 and 4 whereas orientation was significant, $p < .001$, at all levels of size. Other interactions were significant, but the means in each case varied only by $+/-1$mm and the patterns seemed random.

On the Y axis, the shape factor was significant, $F(6, 54) = 5.5$, $p < .001$, but all means were within $+/-1$mm except for the quadrilateral which was at 1.8 mm. Both orientation, $F(3, 27) = 9.3$, $p < .001$, and size, $F(2, 18) = 28.8$, $p < .001$, were significant. The interaction was not.

X and Y means were affected primarily by orientation. The overall trend was for the centroids of the distributions to be located below and to the left of the CM by about 2–3 mm. Given the fact that fingerpads are on the order of 15 mm in width, this variability should have minimal functional consequence, that is, all of the mean judgments can be counted as accurate.

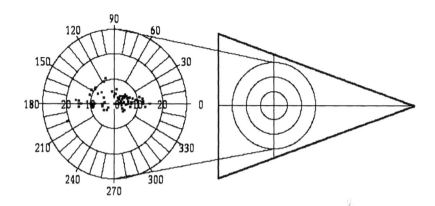

Figure 12.3a. *The elliptical distribution of the data is illustrated. The polar coordinate paper used in the experiment had a larger number of coordinate lines for precise measurement.*

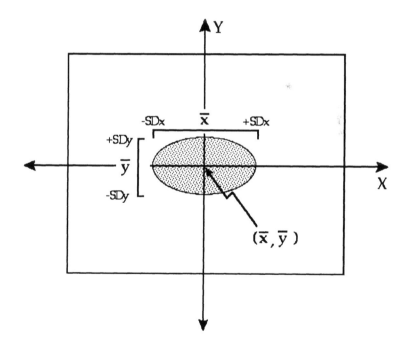

Figure 12.3b. *Modeling the elliptical distributions simply in Cartesian coordinates for statistical analysis. X and Y means represented systematic errors while X and Y standard deviations represented random errors.*

Random errors were analyzed by computing X and Y standard deviations across trials within shapes, sizes, and orientations. A repeated measures ANOVA was performed on X SDs with size, orientation, and shape as factors. There were only main effects of size, F (2, 18) = 37.9, p <.001, and shape, $F(6, 54) = 33.2$, p <.001. As shown in Figure 12.4, the X SDs increased as size increased in most objects. The exceptions were the square and the rectangle in which size changes produced no difference. The size by shape interaction was significant, F (12, 108) = 2.61, p <.004. In simple effects tests, size was significant, p < .01, for all shapes except the square and rectangle, p >.1, while shape was significant, p <.05, at all size levels. The main effect for shape reflects the decrease in the X SDs as symmetry increased. The equilateral triangle was an exception to this trend, exhibiting random errors comparable to the parallelogram.

All three main effects were significant for the Y SDs, including size, $F(2, 18) = 14.74$, p < .003; orientation, $F(3, 27) = 5.90$, p <.004, and shape, $F(6, 54)=7.94$, p<.001. As the symmetry of the objects increased, the random error on the Y axis decreased. The Y SDs increased for the less symmetric objects as the size of the objects increased. The size by shape interaction was marginal, p <.08. The reason that this interaction was only marginally significant in this case was that random errors in the three most symmetric shapes were affected by orientation. The Y SDs for these objects were higher when the Y axis was parallel to gravity. The same trend occurred in the X SDs although there it failed to reach statistical significance. The effect can be seen in Figure 12.4 in which mean SDs for X or Y vertical versus horizontal were plotted separately for each size. Means for the most symmetric objects (rectangle, equilateral triangle, and square) at vertical orientations of both X and Y axes were greater than those at horizontal orientations. The orientation by shape interaction was significant, $F(18, 162) = 1.66$, p <.05. There was a trend in this direction for other shapes as well, but it was clearest in the most symmetric shapes. This effect was especially curious in the case of the square, because nothing specific to the shape changed with changes in orientation of the square. Thus, to account for this effect we must go beyond consideration of shape.

The random errors for the equilateral triangle consistently deviated from a trend of decreasing random error with an increase in the number of axes of reflective symmetry. The equilateral triangle, which has three axes of reflective symmetry, exhibited more random error than the rectangle, which has only two axes of reflective symmetry. Furthermore, the parallelogram, which has no axes of reflective

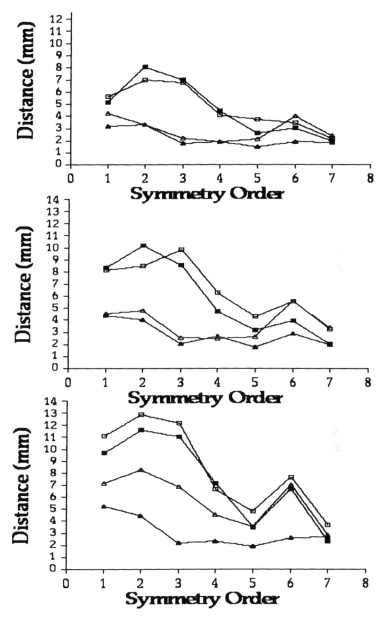

Figure 12.4. *Mean random X and Y errors for 3 object sizes, 7 shapes, and 2 axis orientations either horizontal or vertical. X axis vertical (open squares); X axis horizontal (filled squares); Y axis vertical (open triangles); Y axis horizontal (filled triangles). The object shapes were in order of total symmetry: 1) quadrilateral; 2) right triangle; 3) isosceles triangle; 4) parallelogram; 5) rectangle; 6) equilateral triangle; and 7) square. Top: 100 sq cm. Middle: 200 sq cm. Bottom: 300 sq cm.*

symmetry, exhibited less random error than the isosceles triangle, which has a single axis of reflective symmetry. Clearly, the amount of reflective symmetry can not be the only determinant of random errors.

The inclusion of other types of symmetry, however, may account for the results. The objects used in Experiment 1 varied in the presence or absence of radial symmetry and in amounts of rotational symmetry as well as in amounts of axial reflective symmetry. The triangles, in particular, possessed no radial symmetry, whereas the parallelogram as well as the rectangle and square did (see Figure 12.1). A multiple regression regressing the number of reflective axes, the presence or absence of radial symmetry, and size on X SDs accounted for 80% of the variance with beta weights of $-.49$ for axial reflective symmetry, $-.47$ for radial symmetry, and .43 for size. A multiple regression on Y SDs including the number of reflective axes ($\beta = -.20$), whether the orientation of the Y axis was horizontal or vertical (B=.36), whether the object possessed any symmetry or not ($\beta = -.44$), and size ($\beta = .55$) accounted for 72% of the variance. Similar but slightly weaker results were obtained using rotational symmetry in place of axial reflective symmetry. A number representing the total amount of symmetry (number of axes of reflective symmetry plus number of self–congruences in 360° of rotation plus 1 for radial symmetry) accounted for 67% of the variance in X SDs, but only 26% of the variance in Y SDs in simple linear regressions.

12.6   Reflective Versus Rotational Symmetry

In an attempt to parse out the effects of axial reflective symmetry versus rotational symmetry, we performed additional experiments with two new sets of objects. The first set contained both types of symmetry, and a second comparable set contained only rotational symmetry. The addition of semicircular protrusions along the edges of circles restricted the infinite symmetry of a circle down to the amounts of reflective symmetry exhibited by objects in the initial experiment. In the second set of shapes, the reflective symmetries were eliminated and the rotational symmetries preserved by the addition of small half-oval protrusions just to the side of each larger semicircular bump.

Overall, the results were similar for the two sets, although superior for the set with both types of symmetry. The implication was that observers could use either type of symmetry or both.

In all of the objects except the first and least symmetric of each set, the CM of the object happened to fall very close to the center of the underlying circle upon which all of the objects were based. The CM in the first asymmetric objects in each set fell to the negative side on both the X and Y axes due to the grouping of the protrusions added to eliminate symmetry. In the absence of the symmetry exhibited by the other objects, observers overestimated the amount to which the CM was moved away from the center of the underlying circle. This result leads to our next question.

What do observers do in the absence of symmetry? Do they use an approximation to symmetry? The results with the perturbed circles would indicate that this was not the case. Approximation to symmetry would have been reflected in a tendency to err in the direction of the center of the underlying circle. Participants erred in exactly the opposite direction. The result suggests, nevertheless, that the underlying approximate symmetry structured the estimates. Did participants use the approximate symmetry to base estimates of CMs? We investigated this question in the next experiment.

12.7 Center of Mass Perception and Approximation to Symmetry

The location of the CM in a planar object that has two axes of reflective symmetry is determined readily because the CM must lie on both axes. An object with only one reflective symmetry axis leaves the location of the CM along that axis undetermined. If a rectangular planar object (with two axes of reflective symmetry) were perturbed by attaching another small rectangular piece to one side, the piece might be added so as to preserve reflective symmetry about one axis while leaving the near symmetry about the remaining axis apparent. If observers used an approximation to symmetry in judging the location of the center of mass, then we would predict that the judgments would tend toward the location specified by the symmetries of the original perturbed rectangular shape. If the size of the added rectangle were to be gradually increased, a new larger rectangle with two symmetry axes would be approached accordingly. As the shape approached the larger symmetric figure, the perturbation would become effectively a perturbation away from the larger rectangle by the removal of pieces. In such circumstances, judgments based on approximation to symmetry should err in the direction of the location determined by the symmetry of the larger rectangle as shown in Figure 12.5a.

As is also shown in Figure 12.5a, we created a continuum along which the location of the CM changed linearly across objects. We added successively larger square figures to one side of an initial square-shaped object. The CM moved 1.25 cm along the X axis in each successively larger object. The series consisted of 9 objects beginning with a square (100 cm^2) and finishing with a rectangle (300 cm^2). We predicted that the occurrence of a second symmetry in the square (object 1) and in the rectangles (objects 5 and 9) would structure the pattern of errors as shown in Figure 12.5a. We predicted that the symmetries would attract

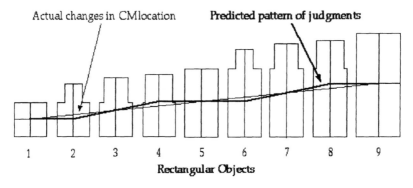

Figure 12.5a. *The rectangular objects used to move the CM along a line along with predictions of how judgments might vary about the line.*

judgments, depending inversely on the deviation from approximate symmetries.

Ten undergraduates at Indiana University participated in the experiment for credit in an introductory psychology course. Participants were required to perform the same grasping task as before. All participants viewed each of the 9 objects at 3 orientations (0°, 135°, and 270°), 3 times each in a session that lasted 1 hour. The main reflective symmetry axis was vertical in the 0° orientation just as shown in Figure 12.5a. All factors were within-subjects.

12.7.1 Results

Systematic errors were analyzed in terms of mean errors along X and Y axes. Repeated measures ANOVAs were performed on the X and Y distances with shape and orientation as factors. Along the X axis, orientation and shape were both significant, $F(2, 24) = 11.34$, $p < .004$, and

$F(8, 96) = 7.55$, $p <.001$, respectively. Along the Y axis only shape was significant, $F(8, 96) = 3.29$, $p <.002$. We will focus on the X means for the 0° orientation in which the symmetry axis common to all of the objects was vertical. As shown in Figure12.5b, the results exhibited exactly the inverse of the pattern predicted by direct approximation to symmetry. Instead, the X means followed a pattern consistent with overestimation of the perturbations as was found with the circular figures of the previous experiment. (The mean errors on the Y axis, though significant across shapes, showed too little range -.75 mm to .75 mm to be of any real significance. Observers stayed very close to the single common symmetry axis.)

Random errors were analyzed via ANOVAs performed on the SDs along the X and Y axes with orientation and number of symmetry axes as within-subject factors. Rather than analyzing each object separately, we grouped the objects into those that had a second reflective symmetry axis and those that did not. For the X SDs, the amount of symmetry was significant, $F(1, 12) = 53.63$, $p <.001$. Judgments were much less variable with both reflective symmetries than with only one. The amount of random error also increased linearly with increases in the size of the objects as can be seen clearly for objects 1, 5, and 9.

The combined pattern in Figure 12.5b and 12.5c demonstrates that symmetries, whether exact or approximate, strongly influence judgments of CM location in objects. The crossing and recrossing of the 0 distance axis in Figure 12.5b exhibits a pattern clearly defined in terms of the points at which the second axis of reflective symmetry appears. In addition to being on the 0 distance axis at the symmetry points, the data cross at points midway between the symmetry points, that is, the crossing locations are symmetrically distributed with respect to the underlying points of symmetry in the set of objects. Rather than staying close to the CM location of the nearest symmetric shape, observers seem to overestimate the strength of the perturbation, whether it be material added to a symmetric form or material subtracted from a symmetric form. Halfway in between the two balance.

12.8   The Role of Dynamics in the Patterning of Estimations

We have described this task in terms of the perception of a dynamic property—the center of mass. One might argue, however, that this is inappropriate and that to interpret these results in terms of perceiving

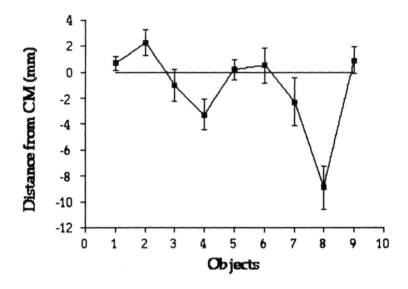

Figure 12.5b. *Mean judgments (with standard error bars) for the upright orientation. Object numbers correspond to those shown in Figure 5a.*

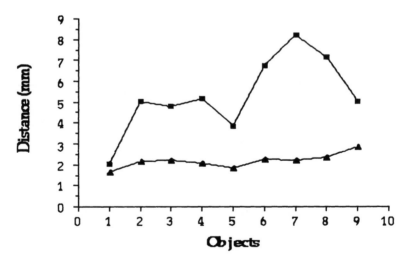

Figure 12.5c. *Mean standard deviations along the X (squares) and Y (triangles) axes.*

dynamic properties or affordances is to read into the data what is not properly there. We reduced our object variations to variations in planar shapes. This suggests that the participants might only have been judging centroids of areas as such, that is, a strictly geometric property with no dynamic content whatsoever.

The difficulty for such an account is in handling variations in judgments produced by variations in object orientation. The judgments reflected the gravitational direction. This certainly implied dynamic content. Systematic errors tended to fall below and to the left of the CM. Rather than undershooting, we might have expected to see overshooting so as to produce a stable equilibrium. However, because the grasps were established from the right side of the object, contact below and to the left of the CM would allow the object to rotate to the right coming to rest against the hand. The result would be an especially stable configuration.

We also found an orientation effect in random error patterns in the first experiment. In the most symmetric objects, random errors were greater along an axis aligned with gravity. The interesting aspect of this effect was that it occurred when no shape related changes accompanied changes in orientation, that is, with the square. In fact, the whole effect seems to have emerged as the objects became more symmetric and shape variations became less of a factor. How might we account for this pattern?

The nature of the pattern is shown in Figure 12.6a and 12.6b, in which contour plots of judgment frequencies are shown as distributed about the CM on object surfaces. Plots for an equilateral triangle appear in Figure 12.6a and for a square in Figure 12.6b. For each object, the results for each of the 4 orientations are shown with the gravitational direction indicated. As previously noted, the distributions tended to be elliptical with the long axis of the ellipses aligned with the gravitational direction. Why should this have been so?

What are the functional repercussions of missing the CM by various distances in various directions? In particular, what is the difference between missing above or below the CM as opposed to missing off to the side? The analysis is shown in Figure 12.7. The object acts as a physical pendulum, so the relevant equation of motion is that for the pendulum. We used the parallel axis theorem to compute the moment of inertia around the contact point. The inertia was computed as the inertia around the CM plus the object mass times the square of the distance from the CM. The inertia about the CM was computed, in turn, as a function of two shape specific constants and the squares of object

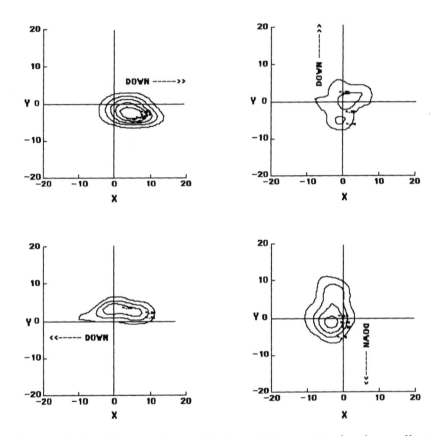

Figure 12.6a. *Contour plots of judgment frequencies for the smallest equilateral triangle plotted separately for each orientation. The direction of gravity is indicated as "down." The X–Y origin is at the CM.*

dimensions. When these were substituted into the original expression, the masses canceled meaning that the rotational acceleration is a function only of the geometry of the object and gravity; q and r can be thought of as polar coordinates on the object surface with the origin at the CM. Thus, for each object of a given shape, the rotational acceleration can be computed for each potential contact point on the object surface using the derived function as a single valued function (yielding rotational acceleration) in two variables (q and r). All of the objects produced acceleration surfaces of the same basic shape shown in Figure 12.8. Only the relative height of the hills and the steepness of their slopes forming the valley varied. The valley always aligned with gravity. We suggest that the elliptical distributions in the data were

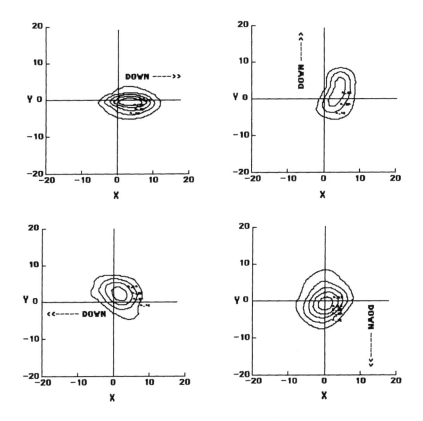

Figure 12.6b. *Same as 12.6a for the medium-sized square.*

constrained to lie along the valley in this plot. Greater errors occurred along the vertical direction where the repercussions were less strong.

Note that this plot is the product of a very local analysis in time, which does not reflect the asymmetry in the stability of points located above versus below the CM. These are the accelerations generated instantaneously at the given location. The graph is relevant when a complete lack of rotation is desired (as opposed to rotations that take one into desirable configurations) and when corrections or responses to rotations can only be provided in some finite time.

The implication of our analysis was that yet another dynamical property—the moment of inertia—contributed to the dispositional property perceived by our observers. This possibly was lent further support when we addressed a remaining aspect of the orientations of

Calculation of Rotational Acceleration

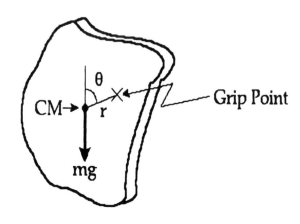

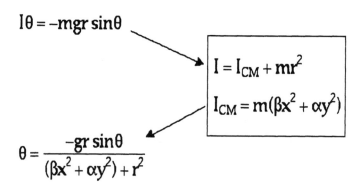

Figure 12.7. *Dynamics generating rotational acceleration around contact points displaced from the CM. I is the moment of inertia; m is object mass; g is gravity; r is the distance from the CM to the contact point or axis of rotation. Remaining variables are described in the text.*

judgment patterns. We have noted that in objects with maximum symmetry, the elliptical data distributions tended to align with gravity. Most often, however, and especially for asymmetric objects, elliptical data distributions tended to align regularly within the object geometry, whatever its orientation. However, the major or minor axes of the distributions did not align with the longest axes that one might find in

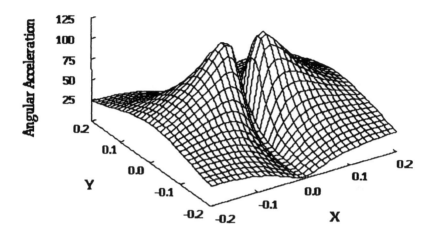

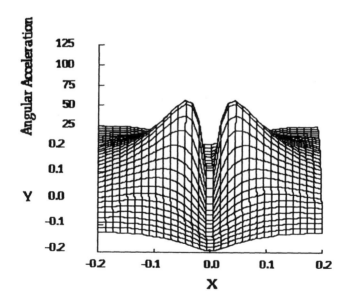

Figure 12.8. *Two perspectives on an angular acceleration surface plotted over X and Y object coordinates centered with the origin at the CM. The Y axis is parallel to gravity.*

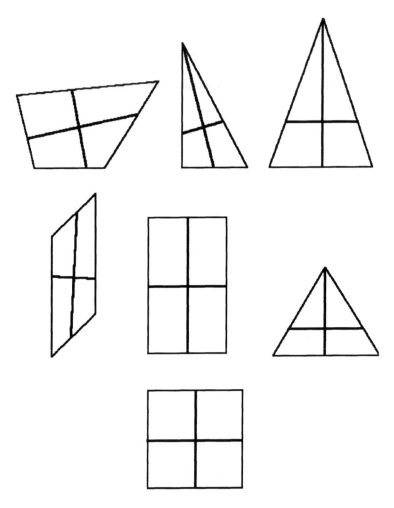

Figure 12.9. *The axes of the principle moments of inertia for the objects used in the first experiment shown together with the major and minor axes of the elliptical data distributions.*

the objects. (In the triangles, such a strategy would place one at an edge of the object.) Rather, in objects with axes of reflective symmetry, the axes of the distributions tended to align with the symmetry axes. (Symmetry axes in the squares and rectangles, for instance, did not correspond to the longest axes which run diagonally from corner to corner.) But what of the asymmetric objects? A possibility would be to use the longest axes through the CM, but the axes of elliptical judgment distributions did not correspond to these[4]. How were axes determined in judgments? Did their means of determination relate to the use of symmetry axes where possible?

In search of an account, we turned to the moment of inertia as a dynamic property related directly to the CM. Without elaborating on the dynamical role of the moment of inertia (see Solomon, Turvey, & Burton, 1989a, 1989b), we simply note that the principle moments of inertia form a Cartesian coordinate system, intrinsic to an object and its inertial dynamics, a coordinate system with an origin at the CM. We computed the locations of the principle moments of inertia for the objects used in the CM perception experiments. As shown in Figure 12.9, the axes of the elliptical data distributions conformed very well to the axes of the principle moments. This was true in symmetric and asymmetric objects alike because the axes of the principle moments of inertia correspond to axes of reflective symmetry when they are present. The regular coincidence of judgment axes and inertial axes could not be mere coincidence. We inferred that the principle moments were determining perceptible dispositions of use in grasping, but these dispositions remain to be fully understood.

12.9   Understanding KSD: "CM Perception" Versus Dynamically Determined Dispositions

Our results provide conclusive evidence that people can perceive the location of contact points for precision grasps that yield a neutrally stable equilibrium about the opposition axis. Our observers performed with impressive accuracy, certainly sufficient to any act of grasping or object manipulation. Within a tolerance, this locus coincided with the location of the center of mass in an object and so, for convenience, we

---

[4]A suggestion that has appeared in the robotics literature is to use symmetry axes where possible and the longest axis through the CM otherwise. The problem is that as a rectangle is slightly perturbed into a parallelogram, the perceived axis would have to jump discontinuously from an axis running between the midpoints of the sides to an axis running diagonally corner to corner. The scheme is fundamentally unstable. Given the inevitable noise and stochastic fluctuations in any measurement system coupled with the fact that perfect Platonic rectangles are nonexistent, such instability could well yield indeterminacy in perception, that is, robots paralyzed by indecision.

have referred to this as the "center of mass perception." However, this reference is strictly incorrect for two very important reasons. Understanding these reasons will help to resolve confusions concerning KSD that have arisen in recent discussions.

First, as demonstrated by the complex patterning of judgments in our studies, *observers perceive* not the center of mass, but a *dispositional property of objects that is determined by a collection of dynamical factors* including, in addition to the center of mass, gravity, moments of inertia, frictional and compliance properties of the hand and object surfaces, and response times of the grasping system in response to perturbations of the object. That perceptible dispositions or affordances must always be determined by such a collection of dynamical factors is ensured by the character of the three attributes essential to any affordance, namely, form, scale, and orientation. Regarding form, we investigated geometric symmetries as form properties essential to the identification of a stable grasp locus because the CM is a symmetry property. But we found that symmetries of kinematic form also might play a role as shown in Figure 12.8. Furthermore, we found that the latter kinematic form was orientation-specific while the use of the former geometric forms was also conditioned by orientation. (Note once more that the orientations were determined by gravity which entailed the use of multimodal perception in performing the task.) Regarding scale, we were led to discover the relevance of moments of inertia by the necessary consideration of object size in relation to the size, speed, and perceptual skill of the performer.

Second, *the center of mass* in an object *contributes* as a dynamic factor to the determination of *a continuously related family of dispositions*. The locus of the center of mass is equally relevant to dispositions for stable, neutrally stable, and unstable equilibria in precision grasps. Furthermore, although these types would appear to be categorically distinct, they are in fact all part of a single continuous structure. Unstable equilibria surround and continuously grade into any point of stable equilibrium in which the graded magnitude of instability might be indexed by the distance to the nearest attracting stable equilibrium. The significance is that apprehension of a neutrally stable grasp locus entails the apprehension of a coherent set of alternative affordances for precision grasping. Which affordance is actually used can be determined as additional constraints are brought to bear in the process of progressive goal specification. For instance, returning to the task of transporting a set of objects from table to shelf, suppose that the actor notices upon approaching the third object that it is inverted. Instead of a stable grasp, an unstable precision grasp below the object center of mass might then be used to place the object upright on the shelf. The specification of this subgoal in the task could only be based on the apprehension of the entire structure comprising the set of alternatives. Additional research on the role of the moments of inertia in grasping and manipulation of objects might reveal the more extensive nature of this structure.

We note that the use of these dispositions was conditioned by the availability of perceptual information as shown by the results of our having perturbed object symmetry. The fact that estimates were affected by the perturbation of symmetry showed both that symmetry is indeed used as information about the locus of neutrally stable contact points and that the use of the disposition depends on the availability of that information, at least before contact. When symmetry was perturbed, the use of approximate symmetry (or neighboring symmetries) as visual information allowed the performer to contact an object within a ballpark which could then be improved via haptic information that would become immediately available upon contact. An affordance is as much conditioned by the availability of perceptual information as by other dynamical factors, but that information will evolve over time and the course of an act and will involve a number of different modalities.

So, we close with the emphatic warning that the "D" in KSD should be understood as a reference to dispositional properties determined by the interplay of dynamical factors. The exact nature of those properties is an empirical problem. We must not confuse our hard won analytical apparatus, namely, dynamics, with the phenomena incidentally under study, namely, perceptible events and affordances. Classical mechanics was not developed in a week as it would have been had the dynamics underlying everyday events been immediately accessible through perception. Nevertheless, those dynamics generate the regularities that comprise and are detectable as affordances.

12.10   References

Baker, W.E., Westine, P.S., & Dodge, F.T. (1973). *Similarity methods in engineering dynamics: Theory and practice of scale modeling.* Rochelle Park, NJ: Hayden Books.

Becker, R.A. (1954). *Introduction to theoretical mechanics.* New York: McGraw-Hill.

Beek, P.J. (1989). *Juggling dynamics.* Amsterdam: Free University Press.

Beek, P.J., & Bingham, G. P. (1991). Task-specific dynamics and the study of perception and action: A reaction to von Hofsten (1989). *Ecological Psychology, 3,* 35–54.

Bingham, G.P. (1987a). Dynamical systems and event perception: A working paper parts I-III. *Perception/Action Workshop Review, 2,* 4–14.

Bingham, G.P. (1987b). Kinematic form and scaling: Further investigations and the visual perception of lifted weight. *Journal of Experimental Psychology: Human Perception and Performance, 13,* 155–177.

Bingham, G.P. (1988). Task specific devices and the perceptual bottleneck. *Human Movement Science, 7,* 225–264.

Bingham, G.P. (1993a). Perceiving the size of trees: Form as information about scale. *Journal of Experimental Psychology: Human Perception and Performance, 19,* 1-23.

Bingham, G.P. (1993b). Perceiving the size of trees: Biological form and the horizon ratio. *Perception & Psychophysics, 54,* 485-495.

Bingham, G.P. (In Press). Dynamics and the problem of visual event recognition. In R. Port. and T. van Gelder (Eds.), *Mind as motion: Dynamics, behavior and cognition.* Cambridge, MA: MIT Press.

Bingham, G.P., Rosenblum, L.D. & Schmidt, A.C. (In Press). Dynamics and the orientation of kinetic forms in visual event recognition. *Journal of Experimental Psychology: Human Perception and Performance.*

Bingham, G.P., & Muchisky, M.M. (1992). Perceiving size in events via kinematic form. In J. Kruschke (Ed.), *Proceedings of the 14th Annual Conference of the Cognitive Science Society.* (pp. 1002–1007). Hillsdale, NJ: Lawrence Erlbaum Associates.

Bingham, G.P., & Muchisky, M.M. (1993a). Center of mass perception and inertial frames of reference. *Perception & Psychophysics, 54,* 617-632.

Bingham, G.P., & Muchisky, M.M. (1993b). Center of mass perception:

Perturbation of symmetry. *Perception & Psychophysics, 54,* 633-639.

Bingham, G.P., Schmidt, R.C., & Rosenblum, L.D. (1989). Hefting for a maximum distance throw: A smart perceptual mechanism. *Journal of Experimental Psychology: Human Perception and Performance, 15,* 507–528.

Bingham, G.P., Schmidt, R.C., Turvey, M.T., & Rosenblum, R.D. (1991). Task dynamics and resource dynamics in the assembly of a coordinated rhythmic activity. *Journal of Experimental Psychology: Human Perception and Performance, 17,* 359–381.

Fearing, R.S. (1983). *Touch processing for determining a stable grasp.* Unpublished Doctoral Dissertation, Department of Electrical Engineering and Computer Science, MIT, Cambridge, MA.

Feldman, A.G. (1986). Once more on the equilibrium-point hypothesis (model) for motor control. *Journal of Motor Behavior, 18,* 17–54.

Gibson, J. J. (1979). *The ecological approach to visual perception.* Boston: Houghton Mifflin.

Hildebrand, M., Bramble, D.M., Liem, K.F., & Wake, D.B. (1985). *Functional vertibrate morphology.* Cambridge, MA: Harvard University Press.

Hogan, N., Bizzi, E., Mussa-Ivaldi, F.A., & Flash, T. (1987). Controlling multijoint motor behavior. *Exercise and Sport Sciences Reviews, 15,* 153–190.

Hollerbach, J.M. (1982). Computers, brains and the control of movement. *Trends in Neuroscience,* pp. 189–192.

Iberall, T., Bingham, G.P., & Arbib, M.A. (1986). Opposition space as a structuring concept for the analysis of skilled hand movements. In H. Heuer and C. Fromm (Eds.), *Generation and modulation of action patterns.* (pp. 158–173). Berlin: Springer Verlag.

Kugler, P.N. (1983). *A morphological view of information for the self-assembly of rhythmic movement: A study in the similitude of natural law.* Unpublished Doctoral Dissertation, Department of Psychology, University of Connecticut, Storrs, CT.

Kugler, P.N. (1986). A morphological perspective on the origin and evolution of movment patterns. In M. G. Wade and H.T. Whiting (Eds.), *Motor development in children: Aspects of coordination and control.* (pp. 459-525). The Hague: Martin Nijhoff.

Kugler, P.N., & Turvey, M.T. (1987). *Information, natural law, and the self-assembly of rhythmic movement.* Hillsdale, NJ: Lawrence Erlbaum Associates.

Mark, L.S. (1987). Eyeheight-scaled information about affordances: A study on siting and stair climbing. *Journal of Experimental Psychology: Human Perception Performance, 10,* 683-703.

Mark, L.S., Balliet, J.A., Craven, K.D., Douglas, S.D., & Fox, T. (1990). What an actor must do in order to perceive the affordance for sitting. *Ecological Psychology, 2,* 325-366.

Muchisky, M.M., & Bingham, G.P. (1991). Center of mass perception for the visual guidance of grasping: A GSD problem. Presented at the 6th International Conference on Event Perception and Action, Amsterdam, August 29th.

Napier, J. (1956). *Hands.* New York: Pantheon Books.

Pittenger, J.B. (1985). Estimation of pendulum length from information in motion. *Perception, 14,* 247–256.

Pittenger, J.B. (1990). Detection of violations of the law of pendulum motion: Observers' sensitivity to the relation between period and length. *Ecological Psychology, 2,* 55–81.

Raibert, M.C. (1986). *Legged robots that balance.* Cambridge, MA: MIT Press.

Riccio, G.E., Martin, E.J. & Stoffregen, T.A. (1992). The role of balance dynamics in the active perception of orientation. *Journal of Experimental Psychology: Human Perception and Performance, 18,* 624-644.

Rock, I. (1973). The perception of disoriented figures. In I. Rock (Ed.), *Readings from Scientific American: The perceptual world.* (pp. 113-126). New York: W.H. Freeman.

Runeson, S. (1977). *On the visual perception of dynamic events.* Uppsala, Sweden: University of Uppsala.

Runeson, S. & Frykholm, G. (1981). Visual perception of lifted weight. *Journal of Experimental Psychology: Human Perception and Performance, 7,* 733–740.

Runeson, S. & Frykholm, G. (1983). Kinematic specification of dynamics as an informational basis for person and action perception: Expectations, gender recognition, and deceptive intention. *Journal of Experimental Psychology: General, 112,* 585–615.

Saltzman, E. & Kelso, J. A.S. (1987). Skilled actions: A task dynamic approach. *Psychological Review, 94,* 84-106.

Sears, F.W., Zemansky, M.W., & Young, H.D. (1987). *College physics (6th Ed).* Reading, MA: Addison-Wesley.

Shaw, R.E., Mark, H.S., Jenkins, H., & Mingolla, E. (1982). A dynamic geometry for predicting growth of gross craniofacial morphology. In *Factors and mechanisms influencing bone growth.*

(pp. 423–431). New York: Liss.

Solomon, H.Y., & Turvey, M.T. (1988). Haptically perceiving the distances reachable with hand-held objects. *Journal of Experimental Psychology: Human Perception and Performance, 14,* 404–427.

Solomon, H.Y., Turvey, M.T., & Burton, G. (1989a). Gravitational and muscular variables in perceiving extent by wielding. *Ecological Psychology, 1,* 265–300.

Solomon, H.Y., Turvey, M.T., & Burton, G. (1989b). Perceiving rod extents by wielding: Haptic diagonalization and decomposition of the inertia tensor. *Journal of Experimental Psychology: Human Perception and Performance, 15,* 58–68.

Stein, R.B. (1982). What muscle variable(s) do the nervous system control in limb movements? *Behavioral and Brain Sciences, 5,* 535–577.

Stoffregen, T.A., & Flynn, S.B. (1994). Visual perception of support surface deformability from human body kinematics. *Ecological Psychology, 1*(6), 33–64.

Stoffregen, T.A., & Riccio, G.E. (1988). An ecological study of orientation and the vestibular system. *Psychological Review, 96,* 3–14.

Szücs, E. (1980). *Similitude and modeling.* Amsterdam: Elsevier.

Thelen, E. (1989). Coupling perception and action in the development of skill: A dynamic approach. In H. Bertenthal and B. Bloch, (Eds.), *Sensory-motor organization and development in infancy and early childhood.* Dordrecht: Kluwer.

Todd, J.T. (1981). Visual information about moving objects. *Journal of Experimental Psychology: Human Perception and Performance, 7,* 795–810.

Turvey, M.T., Shaw R.E., Reed, E.S., & Mace, W.M. (1981). Ecological laws of preceiving and acting. *Cognition, 9,* 237–304.

Warren, W.H. (1984). Perceiving affordances: Visual guidance in stair climbing. *Journal of Experimental Psychology: Human Perception and Performance, 10,* 883–903.

Warren, W.H. (1988). Action modes and laws of control for the visual guidance of action. In O.G. Meijer and K. Roth (Eds.), *Complex movement behavior: The motor-action controversy.* (pp. 370–387). Amsterdam: North-Holland.

Warren, W.H., & Whang, S. (1987). Visual guidance of walking through apertures: Body-scaled information for affordances. *Journal of Experimental Psychology: Human Perception and Performance, 13,* 371–383.

Author Index

Subject Index

Printed and bound by CPI Group (UK) Ltd, Croydon, CR0 4YY

17/10/2024

01775683-0010